Design of Contemporary Inland Waterway Vessels

Dejan Radojčić · Aleksandar Simić ·
Nikola Momčilović · Milorad Motok ·
Benjamin Friedhoff

Design of Contemporary Inland Waterway Vessels

The Case of the Danube River

Springer

Dejan Radojčić
Department of Naval Architecture
Faculty of Mechanical Engineering
The University of Belgrade
Belgrade, Serbia

Aleksandar Simić
Department of Naval Architecture
Faculty of Mechanical Engineering
The University of Belgrade
Belgrade, Serbia

Nikola Momčilović
Department of Naval Architecture
Faculty of Mechanical Engineering
The University of Belgrade
Belgrade, Serbia

Milorad Motok
Department of Naval Architecture
Faculty of Mechanical Engineering
The University of Belgrade
Belgrade, Serbia

Benjamin Friedhoff
Development Centre for Ship Technology
and Transport Systems—DST
Duisburg, Germany

ISBN 978-3-030-77324-3 ISBN 978-3-030-77325-0 (eBook)
https://doi.org/10.1007/978-3-030-77325-0

This Springer imprint is published by the registered company Springer Nature Switzerland AG
The registered company address is: Gewerbestrasse 11, 6330 Cham, Switzerland

Foreword

Currently, about 45 million tonnes of goods is transported on the Danube waterway and its navigable tributaries. This value is far from corresponding to the economic potential of the Danube region with its almost one hundred million inhabitants. In the last years, EU-funded projects delivered improvements in waterway maintenance and infrastructure, but an intensive cooperation of all Danube states and a joint development strategy are needed to make the Danube waterway the core of a sustainable and cost-effective transport system of the region.

The Green Deal and the recently adopted Mobility Strategy of the European Union represent both a challenge and an opportunity: the volume of traffic on the Danube is to be increased significantly, and at the same time navigation is to become a zero-emission sector by 2050.

To achieve these goals, reliable and predictable fairway conditions and climate change resilient infrastructure are needed, together with zero-emission vessels on a smart waterway infrastructure that can be integrated into digitally controlled intermodal logistics chains. This requires significant investment, together with a modern regulatory framework that ensures fair competition, facilitates the transition to the green and digital future and ensures a safe and attractive workplace for people.

Against this backdrop, this book could not have been published at a better time. It offers a comprehensive view of the element ship in the complex system of Danube navigation and discusses both the initial situation and the intended course of the modernization of the Danube fleet comprehensively and clearly. The main objectives of the modernization, greening (near-zero emissions), digitalization and automation as well as the adaptation to new logistical requirements are analysed with their impact on ship design in the hydrological and technological context of Danube navigation. Solutions are proposed on the basis of already available and foreseeable technologies. This approach takes into account that in the next two decades there will probably only be solutions for specific vessel types and operating conditions and no "fits-all" technology that can generate a business case.

With the book, the authors have created an important decision-making aid for the upcoming modernization of Danube navigation. It not only provides important knowledge bases for prospective naval architects, but also gives valuable input in an understandable form for the barging sector facing investment decisions for the

Danube fleet. The book also offers valuable insights for representatives of public authorities responsible for creating the necessary regulatory framework for successful Danube navigation.

The authors of the book therefore deserve special thanks and recognition for this important contribution to the further development of Danube navigation in the interest of the economy and the people in the Danube region.

Budapest, Hungary
April 2021

Manfred Seitz
General Director,
Secretariat of the Danube Commission

Preface

This book is devoted exclusively to the topics related to design of inland waterway (IW) vessels. It is based on the work conducted for various EU projects as well as authors' topical reports and papers. At the outset, the intent was to base the book on a study titled "Environmentally Friendly Inland Waterway Ship Design for the Danube River", completed by the first author in 2009 for World Wide Fund for Nature (WWF). The objective of this study was to develop conceptual ship design targeting specific conditions on the Danube. However, a simple compilation of previous work was not possible, because amongst other reasons, the results of several related EU studies became available in the meantime. Moreover, the overall perception of transport, and therefore of the IW transport too, including ecology, sustainability, efficiency, etc., has changed dramatically during the past decade, impacting many IW transport strategies, policies and related technologies. Consequently, the volume of the book tripled relative to the initial expectation in spite of adhering to the original ideas organized around environmentally sustainable and efficient IW transport. The somewhat unusual structure of this book, and the necessity to include a variety of subjects usually presented in separate books (e.g. waterway implications on ship design, intermodal loading units, energy efficiency, hydrodynamics, new greening technologies, etc.), is a consequence. The common denominator for the diverse topics that have been included is the applicability of innovative technologies and design solutions for improving energy efficiency and environmental performance of IW vessels.

In summary, the aim of this book is to present an accumulation of all relevant work in the field of IW vessel design. Note, however, that although the book covers a wide range of engineering topics, it is not a stand-alone source; it is intended to be a companion to many existing references that are mostly focused on deep-water, seagoing vessels.

We would like to express our gratitude to our colleague, and now also a reviewer of this manuscript, Dr. Juha Schweighofer from via donau, who is one of the few individuals who has participated in almost all recent Danube projects. We are also grateful to our colleague Dr. Milan Kalajdžić from the Department of Naval Architecture, Faculty of Mechanical Engineering, University of Belgrade, for permitting us to publish his mathematical model for power evaluation of pushed convoys (Complex MM, Sect. 8.4.2). We would like to thank the colleagues from the Development

Centre for Ship Technology and Transport Systems (DST, Duisburg) for their contributions (e.g. Sect. 9.2). The authors acknowledge the help of the colleagues and their companies who have supplied us with illustrations not achievable from the Internet, project reports and journal papers, and specifically Franz Dosh (Austria); Mercurius, Concordia Damen and ThyssenKrupp Veerhaven BV (all from the Netherlands); SDG, NAVROM (Romania); and Ukrainian Danube Shipping Company (Ukraine).

Belgrade, Serbia — Dejan Radojčić
Belgrade, Serbia — Aleksandar Simić
Belgrade, Serbia — Nikola Momčilović
Belgrade, Serbia — Milorad Motok
Duisburg, Germany — Benjamin Friedhoff
July, 2021

About This Book

Inland waterway (IW) or river vessels are in every respect different from the seagoing ships. The professional literature is mostly focused on conventional seagoing fleets, leaving a gap in the documentation of design practices for IW vessels. The principal attribute that differentiates river vessels from the seagoing ships is the low, or shallow, draught due to water depth restrictions.

This book addresses key aspects for design of contemporary, shallow draught IW vessels for the transport of dry cargo (containers and bulk cargo). Most of the logic that is presented is applicable to the design of river vessels for any river, but the material that is presented is focused on vessels for the River Danube and its tributaries.

The term "contemporary river vessel" assumes that the present-day technology and current Danube river infrastructure are taken into consideration in its design. It is believed that the technologies and concepts that are proposed here are applicable for all new vessel designs for the next 10–15 years. Other innovative technologies should be considered for designs beyond that horizon. Moreover, nowadays contemporary IW vessel must be in harmony with the Environmentally Sustainable Transport (EST) policies, and hence special attention is paid to both ecology and efficiency. Note, however, that shipowners and ship operators usually tend to choose the conventional cost-effective transport technologies. Given the potential divergence of interests, the concepts and technologies treated here may be regarded as innovative.

This book is aimed at a broad readership, including naval architects and marine engineers, as well as students who would like to get acquainted with the particularities of river vessels. Non-essential issues, mathematical derivations and similar rigour are avoided here as these are presented in other references. However, it is assumed that the reader does have the basic knowledge of naval architecture which is necessary for understanding some sections. Since this book also presents design guidelines specific for IW vessels, something that is rare in general naval architecture textbooks, it may be used as a kind of a design handbook.

The book's subject is presented in the following three parts:

Part I Factors that Influence the Design of Contemporary Danube Vessels
Part II Design Guidelines with the Basics on Shallow Water Hydrodynamics

Part III Innovative Technologies and Design Solutions for Contemporary Inland Waterway Vessels.

Contents

Symbols

A_R	Rudder lateral area (m^2)
b	Waterway breadth (m)
b_c	TEU container width with clearance (m)
B	Breadth of ship (m)
B_b	Breadth of barge (m)
B_{OA}	Breadth of ship overall (m)
B_p	Breadth of pushboat (m)
C_a	Model-ship correlation allowance coefficient
C_{AA}	Air resistance coefficient
C_B	Block coefficient
C_D	Delivered power coefficient
C_F	Frictional resistance coefficient
CF	Carbon emission or conversion factor (g CO_2/g fuel)
C_M	Midship section coefficient
C_P	Prismatic coefficient
CT	Total resistance coefficient
C_{Th}	Thrust loading coefficient
CV	Viscous resistance coefficient
CW	Wave-making resistance coefficient
D	Propeller diameter (m)
EEC	Economic efficiency coefficient
$EEDI$	Energy efficiency design index (g CO_2/t·nm)
F	Ship freeboard (m)
F_n	Froude number
F_{nh}	Depth Froude number
g	Acceleration due to gravity (m/s^2)
GM	Metacentric height (m)
h	Fairway depth (m)
H	Ship depth (m)
h_a	Air draught (m)
h_c	Height of container layer (m)
h_d	Height of double bottom (m)

J	Advance coefficient
k	Form factor
k_c	Cost of transport (€/t)
k_f	Freight rate (€/t·km)
K_Q	Propeller torque coefficient
K_T	Propeller thrust coefficient
l	Nozzle length (m)
L	Length of ship (m)
$L/\nabla^{1/3}$	Slenderness ratio
l_C	Container length with the average clearance (m)
L_{OA}	Length of ship overall (m)
L_{PP}	Length of ship between perpendiculars (m)
L_{WL}	Length of ship at waterline (m)
L_{WLb}	Length of barge at waterline (m)
L_{WLp}	Length of pushboat at waterline (m)
m	Number of barges in the transverse direction of convoy
M_B	Vertical bending moment of hull girder (kNm)
m_c	Available average container mass (t)
m_{CON}	Average container mass (t)
m_{DWT}	Deadweight mass (t)
m_{DWTb}	Deadweight mass of barge (t)
n	Engine rate of revolutions (RPM)
n	Number of containers
n	Number of barges in the longitudinal direction of convoy
n_B	Number of TEU containers abreast
n_H	Number of container layers
n_L	Number of TEU containers along the ship
p	Profit coefficient
P	Power (kW)
P	Propeller pitch (m)
P_B	Engine brake power (kW)
P_D	Delivered power (kW)
P_E	Effective power (kW)
r	Ratio of shallow-water to deep-water wave-making resistance
R_e	Reynolds number
R_h	Hydraulic radius (m)
R_T	Total resistance (kN)
R_V	Viscous resistance (kN)
R_W	Wave-making resistance (kN)
S	Wetted surface area (m^2)
S_b	Bow squat (m)
S_s	Stern squat (m)
SFC	Specific fuel consumption (g fuel/kWh)
t	Thrust deduction coefficient
T	Draught of ship (m)

T	Propeller thrust (kN)
TEC	Transport efficiency coefficient (kWh/t·km)
TEI	Transport efficiency index (t·km/kWh)
V	Ship speed through water (km/h)
V_S	Ship speed over ground (km/h)
V_A	Propeller speed of advance (m/s)
V_C	River current speed (km/h)
w	Wake fraction coefficient
W	Canal bottom width (m)
$W_{lightweight}$	Lightweight mass (t)
W_{steel}	Steel weight mass (t)
β	Drift angle (°)
δ_R	Applied rudder angle (°)
Δ	Displacement (t)
∇	Volumetric displacement (m^3)
η_O	Propeller open-water efficiency
η_D	Quasi-propulsive efficiency
η_{DWT}	Deadweight ratio
η_H	Hull efficiency
η_P	Overall propulsive efficiency
η_R	Relative rotative efficiency
η_S	Transmission efficiency
η_I	Ideal efficiency of an actuator (ideal propeller)
ν	Kinematic viscosity (m^2/s)
Πt	Profit rate indicator (€/h)
ρ	Water density (t/m^3)

Subscripts

h	Finite water depth (shallow water)
∞	Infinite water depth (deep water)

Abbreviations

ABS	American Bureau of Shipping
ACES	Air Chamber Energy Saving
ADN	European Agreement Concerning the International Carriage of Dangerous Goods
ADNR	Rules for the Transport of Dangerous Goods on the Rhine
ADR	Rules for the Transport of Dangerous Goods by Road

AGN	European Agreement on Main Inland Waterways of International Importance
AIS	Automatic Identification System
ANN	Artificial Neural Network
ATM	Advising Tempomat
BTL	Biomass to Liquid
BV	Bureau Veritas
CCNR	Central Commission for Navigation on the Rhine
CEMT	European Conference of Ministers of Transport
CESNI	European Committee for Drawing up Standards in the Field of Inland Navigation
CEVNI	European Code for Inland Waterways
CFC	Convoy Formation Coefficient
CFD	Computational Fluid Dynamics
CNG	Compressed Natural Gas
CO	Carbon Monoxide
CO_2	Carbon Dioxide
CPP	Controllable Pitch Propeller
CRP	Contra Rotating Propeller
DC	Danube Commission
DNV	Det Norske Veritas
DNV-GL	Det Norske Veritas-Germanischer Lloyd
DOC	Diesel Oxidation Catalyst
DP	Ducted Propeller
DPF	Diesel Particle Filter
DST	Development Centre for Ship Technology and Transport Systems
DUT	Delft University of Technology
EC	European Commission
EC	European Council
ECA	Emission Control Area
ECDIS	Inland Electronic Chart Display and Information System
ECE	Economic Commission for Europe
ECMT	European Conference of Ministers of Transport
EEA	European Environment Agency
EEDI	Energy Efficiency Design Index
EEN	Energy-Efficient Navigation
EEOI	Energy Efficiency Operational Indicator
EGR	Exhaust Gas Recirculation
EM	Electric Motor
ERI	Electronic Reporting International
ESD	Energy-Saving Device
EST	Environmentally Sustainable Transport
ES-TRIN	European Standards for Technical Requirements for Inland Navigation
ETA	Estimated Time of Arrival
EUSDR	The EU Strategy for the Danube Region

FAME	Fatty Acid Methyl Ester
FASRB	The Framework Agreement on the Sava River Basin
FC	Fuel Cell
FEM	Finite Element Method
FIS	Fairway Information Services
FPP	Fixed-Pitch Propeller
GA	General Arrangement
GHG	Greenhouse Gas
GTL	Gas to Liquid
GVU	Gas Valve Unit
GWP	Global Warming Potential
H_2	Hydrogen
HC	Hydrocarbon
HNWL	High Navigable Water Level
HTS	High Tensile Steel
HVO	Hydrotreated Vegetable Oil
IACS	The International Association of Classification Societies
ICE	Internal Combustion Engine
ICPDR	The International Commission for the Protection of the Danube River
IENC	Inland Electronic Navigational Charts
ILU	Intermodal Loading Unit
IMO	International Maritime Organization
IPCC	Intergovernmental Panel on Climate Change
IT	Information Technology
ITTC	The International Towing Tank Conference
IW/IWW	Inland Waterway
IWA	Inland Waterway Auxiliary Engine
IWP	Inland Waterway Propulsion Engine
IWT	Inland Waterway Transport
LCO	Lithium Cobalt Oxide
LFP	Lithium Iron Phosphate
LMG	Liquefied Methane Gas
LMO	Lithium Manganese Oxide
LNG	Liquified Natural Gas
LNWL	Low Navigation Water Level
LOHC	Liquid Organic Hydrogen Carrier
Lo-Lo	Load on–Load off
LPG	Liquefied Petroleum Gas
LR	Lloyd's Register
MARIN	Maritime Research Institute Netherlands, Wageningen, The Netherlands
MARPOL	The International Convention for the Prevention of Pollution from Ships
MCR	Maximum Continuous Rating
MD	Main–Danube Canal
MEPC	Marine Environment Protection Committee

MM	Mathematical Model
MSC	Maritime Safety Committee
NCA	Lithium Nickel Cobalt Aluminium Oxide
NCM	Lithium Nickel Cobalt Manganese Oxide
NMVOC	Non-Methane Volatile Organic Compound
NO_X	Nitrogen Oxides
NRE	Non-Road Engine
NRMM	Non-Road Mobile Machinery
NtS	Notices to Skippers
PEM	Polymer Exchange Membrane
PM	Particulate Matter
PN	Particle Number
PTL	Power to Liquid
RDP	Rim-Driven Propeller
RID	Rules for the Transport of Dangerous Goods by Railway
RIS	River Information Services
Ro-Ro	Roll on–Roll off
RRR	Russian River Register
RS	Russian Maritime Register of Shipping
RVIR	Rhine Vessel Inspection Regulations
SCR	Selective Catalytic Reduction
SOLAS	The International Convention for the Safety of Life at Sea
SO_X	Sulphur Oxide
SPP	Surface-Piercing Propeller
SPS	Sandwich Plate System
SRA	Strategic Research Agenda
SSP	Side-by-Side Propeller
TDP	Tip-Driven Propeller
TEC	Transport Efficiency Coefficient
TEU	Twenty-foot Equivalent Unit
TRL	Technological Readiness Level
TSRA	Towards a Strategic Agenda for IWT and Ports
TTP	Tank to Propeller
TTW	Tank to Wheel
UNECE	United Nations Economic Commission for Europe
UNFCCC	United Nations Framework Convention on Climate Change
VBD	Versuchsanstalt für Binnenschiffbau (see DST)
via donau	Österreichische Wasserstraßen-Gesellschaft mbH.
WHSC	World Harmonized Stationary Cycle
WP	Work Package
WTP	Well to Propeller
WTT	Well to Tank
WTW	Well to Wheel
XTL	X to Liquid
ZES	Zero Emission Services

Projects' Acronyms

COVADEM	Collaborative Navigation Depth Measurement (Dutch acronym)
COVEDA	COntainer VEssels for the DAnube Waterway
CREATING	Concepts to Reduce Environmental impact and Attain optimal Transport performance by Inland NaviGation
DE-LIGHT	Developing Lightweight Modules for Transport Systems Featuring Efficient Production and Lifecycle Benefits at Structural and Functional Integrity using Risk Based Design
ECCONET	Effects of Climate Change on the Inland Waterway Networks
EUDET	Evaluation of the Danube Waterway as a Key European Transport Resource
GRENDEL	Green and Efficient Danube Fleet
IDV	Innovative Danube Vessel
INBAT	Innovative Barge Train
INBISHIP	New Opportunities for Inland Waterway Transport
LASS	Lightweight Construction Applications at Sea
MoVeIt!	Modernization of Vessels for Inland Waterway Freight Transport
MUNIN	Maritime Unmanned Navigation through Intelligence in Networks
MUTAND	Multimodal Ro-Ro Transport on the Danube River
NAIADES	An Integrated European Action Programme for Inland Waterway Transport
NEWS	Development of a Next generation European Inland Waterway Ship and logistics system
NOVIMAR	NOVel Iwt and MARitime transport concepts
PELS	Project Energy-saving air-Lubricated Ships
PLATINA	Platform for the implementation of NAIADES
PROMINENT	Promoting Innovation in the Inland Waterways Transport
SMOOTH	Sustainable Methods for Optimal design and Operation of Ships with air-lubricated Hulls
SPIN	European Strategies to Promote Inland Navigation
SYNCHRO-NET	Synchro-modal Supply Chain Eco-Net
TOPOFAHRT	Reduction of fuel consumption and pollutant emissions through "topography-oriented" navigation, training concept for boat masters (German acronym)
VEBIS	Improvement of the Efficiency of Inland Water Transportation (German acronym)

Chapter 1
Introduction

1.1 Background: Strategies and Statistics Relevant for European IWT

The global challenge, so European too, is to reduce poisonous emissions, and in the first place the GHG emissions. This means that serious measures for achieving energy efficient but also environmentally sustainable transport have to be undertaken. This process should be executed systematically and it requires a transport policy framework at the European level. Amongst the strategies is to strengthen IWT compared to other transport modes, as IWT enables significant reduction of GHG emissions (see Statistics, Sect. 1.1.3). A short list of the most relevant documents of the kind follows.

1.1.1 Transport Policies with Impact on Design of IW Vessels

White Paper on Transport and Mannheim Declaration

The European transport strategy, defined in 2011, in *White Paper* named "Roadmap to Single European Transport Area—Towards a competitive and resource efficient transport system" (see EC 2011), gives short-term and long-term targets for GHG emission reduction of at least 60% by 2050 with respect to 1990 (corresponding to emissions cuts of around 70% below 2008 levels). Accordingly, future development must rely on several measures, as for instance: efficiency performance of vehicles has to be improved, multimodal logistic chains optimized, transport and infrastructure have to be used more efficiently etc. Amongst others, it suggests shifting 30% of road freight travelling over 300 km to other transport modes, e.g. waterborne, by 2030 and more than 50% by 2050! Solutions that will facilitate modal shift of cargo from road to waterborne transport are suggested in SRIA (2020).

D. Radojčić et al., *Design of Contemporary Inland Waterway Vessels*,
https://doi.org/10.1007/978-3-030-77325-0_1

Concerning inland navigation, the *Mannheim Declaration* (see the Mannheim Declaration 2018) goes further by tasking the CCNR to develop a roadmap in order to reduce both GHG and pollutant emissions by 35% compared with 2015 by 2035, and almost eliminate them by 2050. Amongst others, in the Mannheim Declaration the development of modern technologies is suggested with the purpose to increase competitiveness, safety and sustainability of inland navigation.

Inland Waterway Strategies—NAIADES and PLATINA

Inland navigation is promoted by the NAIADES programme (2006–2016), which defines the policy of the EU through various legislative and supportive measures. Follows the NAIADES II Action Program "Towards quality inland waterway transport" for period 2014–2020, which includes the development of infrastructure and fleet. PLATINA (2008–2012) is a platform for the implementation of NAIADES and the follow-up project is PLATINA II (2013–2016) for implementation of NAIADES II. Recommendations by the NAIADES II Implementation Expert Group that concern the future, are given in "An Inland Waterway Transport Agenda for Europe 2021–2027". PLATINA II represented an essential framework for the Danube Region Strategies. Recently PLATINA 3 was initiated (time frame Jan. 2021–June 2023) see PLATINA 3 (2020); it makes a ridge towards future research, innovation and implementation needs within IWT in Europe.

Danube Region Strategies—EUSDR, ICPDR and FASRB

IWT in the Danube Area is promoted by a macro-regional Strategy of the EU for the Danube Region (EUSDR) which defines actions that should be undertaken in particular Danube region countries (see EC Action Plan 2020). Danube River Protection Convention (ICPDR 2015) is relevant for the hydro-morphological situation, i.e. potential engineering measures which influence navigation. Management priorities until 2021 are published in the Danube River Basin Management Plan. Concerning the largest navigable Danube tributary, the Sava river, see Framework Agreement on the Sava River Basin (FASRB 2002). Besides the abovementioned, IWT is also promoted by various national action programs or national transport master plans.

1.1.2 Strategic Research Agenda for IWT

A Strategic Research Agenda (SRA 2010), for the IWT mode, was compiled in the PLATINA project. This SRA defined research and development priorities for future research with the purpose to strengthen IWT. It provides an overview of general and sector-specific challenges and strategic goals with corresponding research topics. Most of the research topics are based on conventional technologies and solutions aiming to increase the efficiency of the IWT sector. Topics and expected research outcome are systematically sorted in three thematic pillars aiming to:

- Strengthen competitiveness;
- Ensure environmental sustainability;
- Match growth and change trade patterns.

Although the SRA should be considered as an overall concept, for this book, only the first two pillars (research topics) are of interest, as they match the book's subject. Therefore, a somewhat reduced overview of challenges and strategic goals in IWT, with corresponding research topics clustered in two thematic pillars, are depicted in Fig. 1.1. An overview of research topics and expected research outcome is shown in Fig. 1.2.

The activity that followed the SRA is the work in progress named "Towards a strategic agenda for IWT and ports" (TSRA). It presents a vision up to 2050 and expectations of research in the IWT sector. Note that subjects treated in the book concern short-term and medium-term objectives based on present-day technologies. The SRA was followed by the SRIA (Strategic Research and Innovation Agenda, see SRIA 2020) whose objective is to provide and demonstrate zero-emission solutions for *all main ship types and services* before 2030, and zero-emission waterborne transport by 2050. The SRIA is divided into six activities, out of which these four are of interest for the book: use of sustainable alternative fuels, electrification, energy efficiency, and design and retrofitting. Note however that PLATINA's SRA is focused

	Research Pillars	Strengthening competitiveness						Environmental sustainability					
Challenges	**Strategic Goals** / **Research topics**	Refitting vessels	Fleet adaption to climate change	Innovative vessel design	Shipbuilding technology	Structural strength of the hull	Handling, maintenance, repair	Means to reduce emissions of existing technologies	New technologies for power supply on board	Energy efficient navigation	Hydrodynamics	Equipment on board	Advanced safety standards
Infrastructure conditions and limits	Better exploitation of available infrastructure		X	X		X				X	X		
Over-ageing of large parts of the fleets	Modernisation of the fleet	X		X	X		X	X	X		X	X	X
Lack of nautic. person. and increasing requirements to nautical personnel	Coping with increasing requirements on professionals			X			X						X
Predicted increases in transport demand	Releasing other modes of transport	X		X								X	
	Intelligent IWT-approaches for general cargo	X		X	X	X						X	X
	Improvement of data management and information systems												X
Climate change impact	Reduction of greenhouse gas emissions							X	X	X	X		
	Coping with expected infrastructure conditions		X	X			X			X	X		X
	Maintain manoeuvrability and safety standards		X	X									X
Finiteness of the resource "fossil fuel"	Transition to the "after-fossil-fuel-era"			X				X	X	X	X		
Limitation of hazardous emissions	Further reduction of fuel consumption and emission	X		X				X	X	X	X	X	
	More pronounced role of IWT within intermodal transport chains			X									

Fig. 1.1 Overview of challenges and strategic goals, with corresponding research topics. *Source* SRA, 2010

on IWT, thus is more appropriate for the present book, while the SRIA is directed to all ship types.

Summarizing, research topics and expected research outcome shown in Fig. 1.2, systematically sorted in SRA's first two pillars, are also the subject of this book and a part of its contents, though here they are presented in a different order.

Strengthening competitiveness

Refitting vessels of the existing fleet
Strategies to refit existing vessels
Decision support for owners

Adaptation to climate changes
Matching vessel particulars to infrastructure
Service strategies during the lifespan of ship
Decision support for owners

Innovative vessel design
Matching vessel particulars to market demands
Matching vessel design to new prime energy sources
Matching design to propulsion and steering technologies
Ship design with "flexible particulars"
Ship design to match extreme nautical conditions

Shipbuilding technology
Strengthening position of the EU shipbuilding industry
New approaches for the modernisation of the IWT fleet
Modular shipbuilding concepts and solutions
Building tech. suitable for new materials such as composites

Structural strength of the hull
Increasing payload efficiency
Applicability of light structure concepts

Handling, maintenance and repair
New materials
New engineering methodologies and standards

Environmental sustainability

Improving existing technologies to reduce emissions
Further improvement of diesel powered engines
Implementation of LNG as fuel
Increase of propulsion efficiency

New technologies for power supply on board
Gas engine solutions
High energy storage density electric cells
Fuel cell solutions
Waste heat recovery
Solar energy for onboard electricity

Energy efficient navigation
Onboard tool for support of energy efficient navigation
Voyage planning based on real-time fairway information
Voyage planning with pre-booked timeslots

Hydrodynamics
Reducing power requirements
Reducing wash

Equipment on board
Increase reliability and safety in ship's handling
Preventing air and water pollution

Advanced safety standards
Active and passive safety of vessels
Safety of crew and cargo

Fig. 1.2 Overview of research topics (green boxes) and expected research outcome (white boxes). *Source* SRA, 2010

1.1.3 *Selected Statistics on IWT*

Energy Consumption, Freight Volume and Emissions in the Transport Sector of EU Member States, 2018 (Source: EEA, 2020)

(a) The transport sector accounts for 31% of total energy consumption.
(b) The energy consumption in the transport sector is 32% higher than in 1990.
(c) The largest absolute increase in energy consumption occurred in road transport and is 34% higher than in 1990. Road transport accounts for 73% of the total energy demand.
(d) Rail and IWT were the only modes of transport for which an absolute decrease in energy consumption was recorded between 1990 and 2017. Energy consumption in the IWT sector declined and was 17% lower than in 1990.
(e) The energy consumption by transport mode in million terajoules (TJ), rounded to the first decimal place follows: Rail transport 0.3; IWT 0.2; Maritime 1.9; Road transport 12.8; Aviation 2.4 ($1 TJ = 10^{12}$ J).
(f) The freight transport volume in t · km and modal split within the EU is shown in Fig. 1.3. The modal share of freight transported over land within the EU (Table 1.1) is almost constant over the last 20 years. The modal share varies to a great extent from one country to another, e.g. modal share of IWT in the

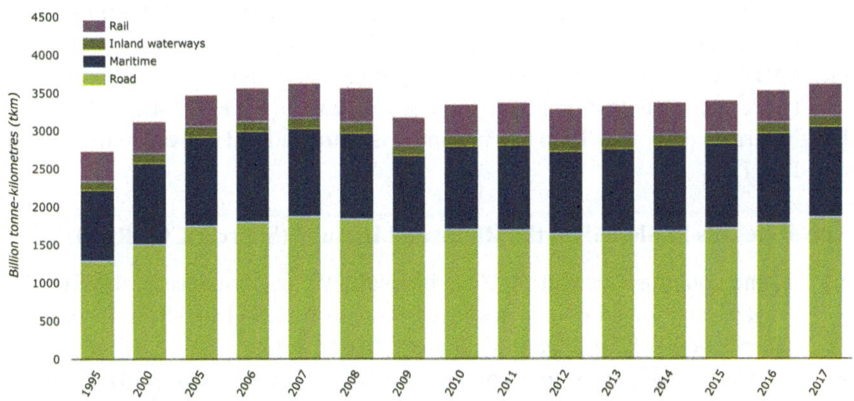

Fig. 1.3 Freight transport volume and modal split within the EU. *Source* EEA, 2020

Table 1.1 Transport over land statistics for 2017 (Data from: EEA, 2020)

Trans. mode	Energy Consumption		Transport Volume	
	10^6 TJ	%	10^9 t·km	%
Rail	0.3	2.3	421	17
IWT	0.2	1.5	147	6
Road	12.8	96.2	1870	77

Netherlands is almost 40%, Romania 20%, Germany 12%, while for more than half EU countries it is less than 1%.

(g) Air pollutant emissions from all transport types have significantly declined since 1990, despite the general increase in activity within the sector. The following emissions were decreased: nitrogen oxides (NOx) by 40%, sulphur oxides (SOx) by 66%, carbon monoxide (CO) and non-methane volatile organic compounds (NMVOCs) by 87% and particulate matter (since 2000 44% and 35% for PM2.5 and PM10 respectively). However, GHG emissions since 1990 increased by 29%.

Fleet Expansion, Efficiency and External Costs—West European IW Fleet, 2017 (Source: The Power of Inland Navigation, BVB, 2017)

(h) The West European IW fleet in 2015 had a capacity of around 15 mil. t. More than half of that sails under the Dutch flag, i.e. the Netherlands 62%, Germany 18%, Belgium 12%, France 6%, all other countries 2%.

(i) Between 2005 and 2014 the fleet expanded by some 1700 vessels, 65% being Dutch. By far the largest number of all newbuildings were large vessels of over 2500 t; they were followed by medium-sized vessels of 1000–2000 t, while those of 2000–2500 t were third; small vessels of less than 1000 t took the fourth place.

(j) Increase in scale, due to their better efficiency, is an evident trend. During the last fifty years, the average growth in tonnage increased from 1000 to 3000 t per vessel.

(k) IWT is energy efficient mode and is approximately three times more efficient (per $t \cdot km$) than that of a large truck with a trailer (see Fig. 1.4).

(l) External costs of IWT are low in comparison to road and railway transport (see Fig. 1.5).

Type of Vessels Exploited on the Rhine and Danube (Source: CCNR, 2019)

(m) Rhine countries account for 84% of total IWT performance in the EU plus Switzerland. Danube countries have a share of 16%, and all other EU countries taken together have a share of almost zero percent.

(n) Share of the countries' in $t \cdot km$ in total transport performance in EU, 2018: the Netherlands 34.8%, Germany 34.6%, Romania 9.1%, Belgium 8.5%, France 5.4%, Bulgaria 3.6%.

(o) Around 75% of total transport performance on the Danube takes place on the lower Danube and only one-quarter on the Middle and Upper Danube.

(p) On the Danube 79% of goods are transported by pushed convoys, and 21% by selfpropelled cargo vessels. Transit traffic going upstream represented 57.4% of all transported volumes.

(q) In 2018, more than 15,000 cargo vessels were registered in Europe; 65% of the fleet was found in Rhine countries (dry cargo 7000, liquid cargo 1462, push and tug boats 1240), 23% in Danube countries (2652, 204, 657 respectively), and the remaining 12% in other European countries with inland waterways. The fleet structure on the Danube is dry cargo 75%, liquid cargo 6% and push and tug 19%.

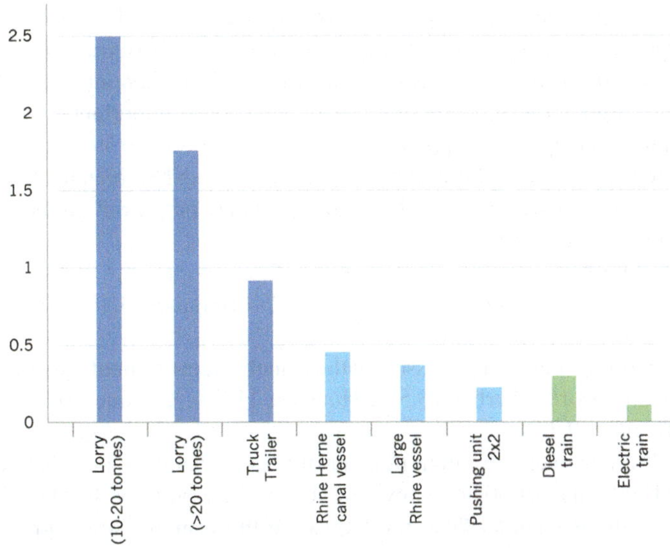

Fig. 1.4 Energy consumption per t · km of various transport modes. *Source* BVB, 2017

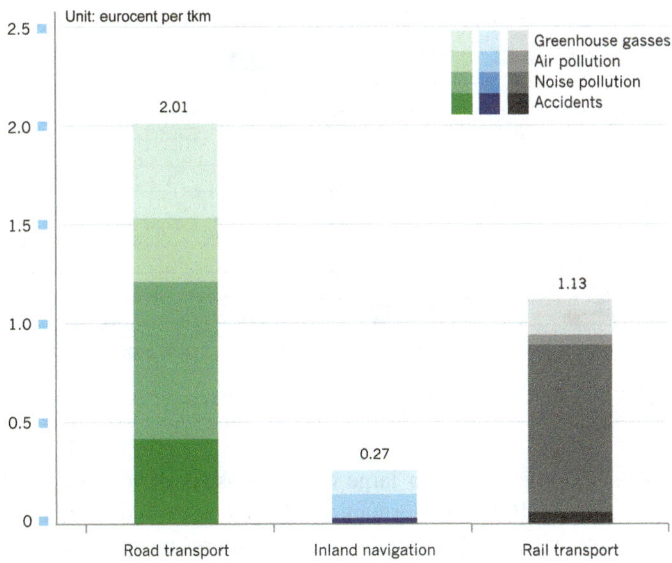

Fig. 1.5 Average external costs of modalities (Data from: BVB 2014)

(r) On the Danube more than 80% of dry cargo vessels are barges (non-selfpropelled vessels). The share of the selfpropelled vessels on the Danube is much lower than on the Rhine. Dry cargo tonnage amounts to 93%, while liquid cargo is 7% only. Though convoy transport is important on the Danube, the push and tug fleet dropped from around 730 in 2008 to 590 in 2016 (mainly due to the reduction of the Ukrainian, Hungarian and Serbian fleet; the Romanian fleet, however, grew). The average tonnage per vessel on the Danube is stable and is about 1000 t.

(s) Distribution of dry cargo vessels in the Rhine countries in 2018: the Netherlands 49.8%, Germany 22.3%, France 14.2%, Belgium 13.4%, others less than 1%.

(t) Distribution of dry cargo vessels in the Danube fleet (without Austria) in 2017: Romania 48.1%, Serbia 13.5%, Hungary 11.7%, Ukraine 10.1%, Bulgaria 7.3%, other 9.3.

(u) The Danube fleet has a high number of vessels that were built between 1960 and 1990. Since 2000 not many new vessels were built for the Danube. About half of the Romanian fleet, the largest on the Danube, was built in the last three decades (see Fig. 1.6).

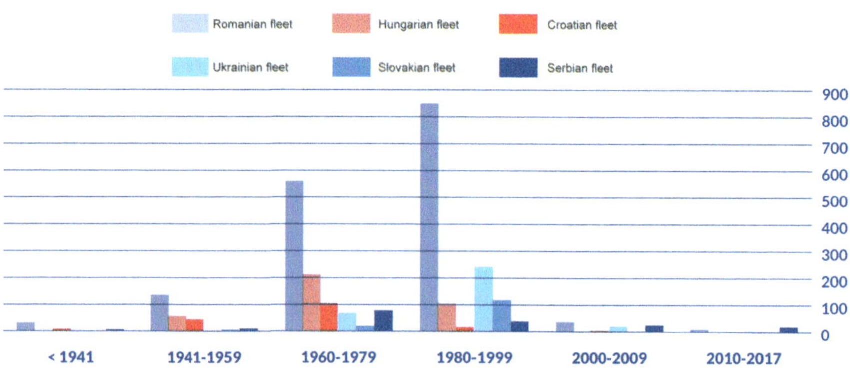

Fig. 1.6 Number of vessels per year of construction in Danube countries. *Source* CCNR, 2019

(v) The West European fleet in large extent consists of relatively old vessels. Around 50% of Belgian, German and Dutch vessels were built more than 50 years ago, while in France, these vessels represent approximately 80% of the total fleet. Moreover, even 15% of the European fleet was built more than 75 years ago. The percentage of old vessels on the Danube is much lower than on the Rhine.

(w) For the Rhine countries in 2018 only 3 tug/push boats and 17 selfpropelled newbuildings were launched (in 2009 more than 300), half of them registered under the Dutch flag, closely followed by Belgian.

(x) More than 100 new vessels built after 2014 had some greening measures, see Fig. 1.7. About half of the newly built vessels had at least one greening measure.

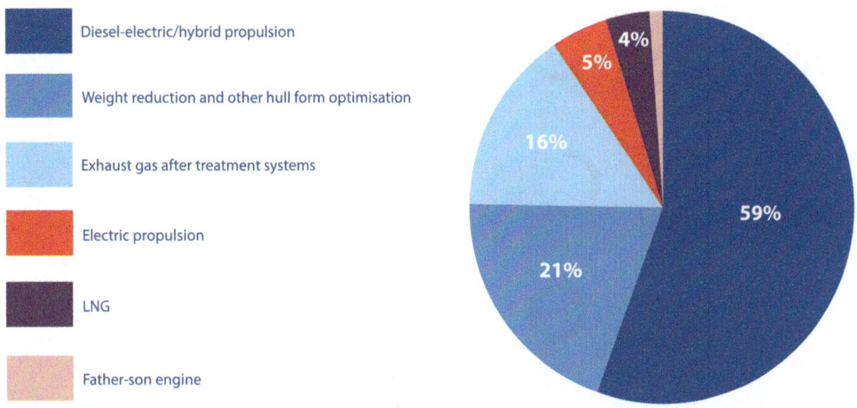

Fig. 1.7 Greening measures by type and frequency observed for the newly built vessels 2014–2016. *Source* CCNR, 2017

Who is Who in Freight Transport on the Danube (Source: via donau 2019)

(y) Freight transport on the Danube 2017 was 39.3 million tons. On the Rhine it is approximately 4 times more, and on the Mississippi it is 4 times more than on the Rhine. On the Mississippi, as on the Danube, most of the cargo is transported by pushed convoys. Romania recorded by far the largest transport volume on the Danube in 2017 with just over 19 million tons, followed by Serbia with 12.5 million tons and Austria with 9.5 million tons (Fig. 1.8).

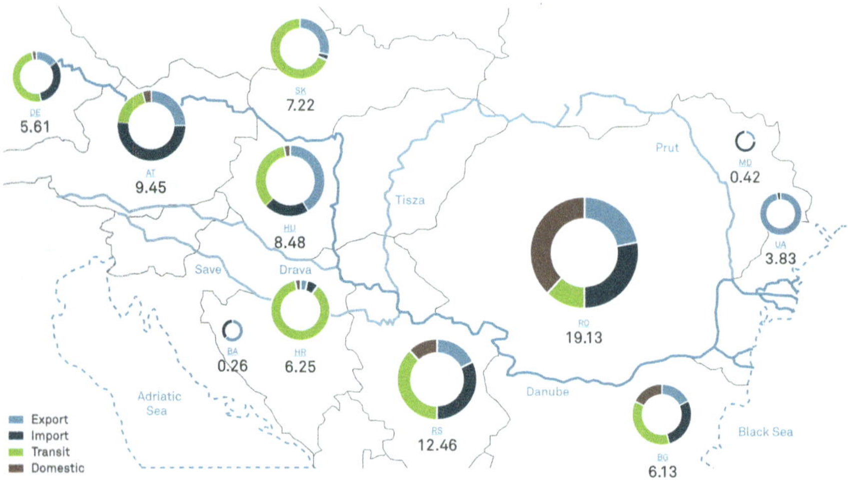

Million Tons	DE	AT	SK	HU	HR	BA	RS	RO	BG	MD	UA
Export	0.84	2.40	2.09	3.50	0.19	0.17	2.30	4.21	1.11	0.10	3.67
Import	1.81	4.82	0.10	1.81	0.33	0.09	3.96	5.40	1.73	0.32	0.15
Transit	2.78	1.84	5.01	2.92	5.67	0.00	4.76	2.20	2.20	0.00	0.00
Domestic	0.18	0.39	0.02	0.25	0.06	0.00	1.44	7.32	1.09	0.00	0.01
Total	5.61	9.45	7.22	8.48	6.25	0.26	12.46	19.13	6.13	0.42	3.83

Fig. 1.8 Freight transport on the entire Danube 2017. *Source* via donau, 2019

Summarized Statistics on IWT

- Road transport is liable for enormous energy consumption—see (a), (b), (c), (e), (f).
- Rail and IWT are energy efficient transport modes with large capacities—see (d), (e), (f), (k), (l).
- Rail and IWT are environmentally sustainable transport modes—see (g) and Table 1.1.
- External costs of IWT are low—see (l).
- IWT on the Rhine is a paradigm for the Danube—see (m), (q), (r).
- IWT frontrunners are Netherland (Rhine) and Romania (Danube)—see (f), (h), (i), (n), (r), (s), (t), (u), (w), (y).
- The majority of vessels on the Rhine: selfpropelled, on the Danube: pushed convoys—see (p), (q), (r).
- The ongoing trend to larger, greener, energy efficient vessels—see (i), (j), (x).

1.1.4 Final Notes

IWT is a sustainable, energy-efficient and safe transport mode with plenty of spare capacity. External effects of IWT, as are accidents, poisonous emissions including GHG emissions, noise etc. are low compared to other transport modes. On many corridors, IWT is the only answer to problems related to congested roads. IWT has no restrictions on freight growth (according to some sources an allowable increase on the Rhine and Danube are 4 and 10 times, respectively). Last but not least, IWT is a cost-efficient transport mode.

Concerning IWT versus other transport modes (according to Summarized Statistics on IWT):

- Road transport is responsible for enormous energy consumption.
- Rail and IWT are energy efficient transport modes with large capacities.
- Rail and IWT are environmentally sustainable transport modes.
- External costs of IWT are low.
- IWT on the Rhine is a paradigm for the Danube.
- IWT frontrunners are the Netherlands on the Rhine and Romania on the Danube.
- The majority of vessels on the Rhine is selfpropelled, while on the Danube pushed convoys are dominating.
- The ongoing trend is to larger, greener and energy efficient vessels.

It should be emphasized, although it sounds obvious, that significant differences between sea transport and IWT exist (Radojčić 2008):

- Sea vessels have no competition whatsoever—international trade cannot be done without them.
- IWT has very strong competition from alternatives—railway and road transport. Moreover, road transport inherently fulfils the door-to-door requirement.
- IWT is more constrained by ecological considerations—rivers flow through mainland population centers.

Though not mentioned earlier, it should probably be underlined now (see EEA 2015), that "policies that only focus on the incremental efficiency of vehicles are not sufficient for a long-term low-carbon transport sector. Further development must also optimize the performance of multi-modal logistic chains, and use transport and infrastructure more efficiently". These measures, however, are not subject of this book. Nevertheless, although GHG emissions from IWT are very small compared to those of other transport modes (see Figs. 1.3 and 1.4 and Table 1.1), it does not yet mean that the IWT's decarbonization is not an issue, as nowadays, according to Schweighofer (2021), transport modes that do not contribute to the reduction of GHG emissions risk losing social and political acceptance.

Research topics and expected research outcome (shown in Fig. 1.2) are subjects of this book and a kind of its contents, though here are presented in a different order.

1.2 Book Arrangement

The book's first part, titled *"Factors that Influence the Design of Contemporary Danube Vessels"* treats many standalone issues that must be considered in the design of IW vessels, such as:

- Restrictions in water depth, lock sizes and bridge heights for the Danube river that determine the main ship dimensions.
- Transhipment possibilities, intermodal loading units and related problems that influence IW vessel design.
- Waterborne transport considerations in general, and the state-of-the-art of selfpropelled vessels and barge trains on the Danube.
- An overview of the regulatory framework (classification and statutory regulations) for the design of vessels in Europe and the Danube in particular.
- Transport efficiencies, performance indicators (i.e. energy efficiency design indexes and economic indicators) for IW vessels.

The book's second part, titled *"Design Guidelines with the Basics on Shallow Water Hydrodynamics"* provides further technical details such as:

- Shallow water hydrodynamics (resistance, propulsion efficiency, power evaluation, squat, manoeuvrability) which is essential for the design and operation of every IW vessel.
- Essential expressions and guidelines needed for the preliminary design of IW vessels (ship form, power evaluation, lightship weight estimations etc.) ending with typical worked examples for selfpropelled ships and pushed convoys.

The third part—"*Innovative Technologies and Design Solutions for Contemporary Inland Waterway Vessels*" should give key answers to the book's goals, i.e. it treats subjects concerning:

- Technical measures that make inland ships environmentally sustainable and more efficient; these are a recent improvements that concern prime movers and fuels, hull resistance, propulsion, structure and better overall ship utilization.
- Three groups of concepts of contemporary and innovative IW vessels: (a) those which are researched in various projects, (b) advanced vessels that are already built, and (c) a shallow draught pushboat and two different selfpropelled concepts proposed for the Danube River.
- Conversion and retrofitting options for existing ships (with a numerical example).

References

BVB (2014) The power of inland navigation 2013–2014. https://www.edinna.eu/wp-content/uploads/2017/08/140410-the_power_of_inland_navigation_2013.pdf. Accessed 07 Feb 2021

BVB (2017) The power of inland navigation 2016–2017. https://issuu.com/bvbinnenvaart/docs/waardetransport_spreads-uk. Accessed 18 Dec 2020

CCNR (2017) Market report 2014–2017. Main features and trends of the European inland waterway transport sector. https://ccr-zkr.org/files/documents/ompublicationssp/Market-report-2014-2017_Web.pdf. Accessed 25 Dec 2020

CCNR (2019) Inland navigation in Europe—market observation. Annual report 2019. https://inland-navigation-market.org/wp-content/uploads/2019/11/ccnr_2019_Q2_en-min2.pdf.pdf. Accessed 20 Dec 2020

EC (2011) White paper. Roadmap to single european transport area—towards a competitive and resource efficient transport system. https://eurlex.europa.eu/LexUriServ/LexUriServ.do?uri=COM:2011:0144:FIN:EN:PDF. Accessed 18 Dec 2020

EC Action Plan (2020) European Union strategy for the Danube region. https://danube-region.eu/wp-content/uploads/2020/04/EUSDR-ACTION-PLAN-SWD202059-final.pdf. Accessed 18 Dec 2020

EEA (2015) Final energy consumption in Europe by mode of transport. https://www.eea.europa.eu/data-and-maps/indicators/transport-final-energy-consumption-by-mode/assessment. Accessed 18 Dec 2020

EEA (2020) Report on energy consumption. Transport: increasing oil consumption and greenhouse gas emissions hamper EU progress towards environment and climate objectives. https://www.eea.europa.eu/themes/transport/term/increasing-oil-consumption-and-ghg. Accessed 18 Dec 2020

FASRB (2002) Framework agreement on the sava river Basin. http://www.savacommission.org/dms/docs/dokumenti/documents_publications/basic_documents/fasrb.pdf. Accessed 18 Dec 2020

ICPDR (2015) Danube river protection convention. https://www.icpdr.org/main/icpdr/danube-river-protection-convention. Accessed 25 Dec 2020

Mannheim Declaration (2018) 150 years of the Mannheim Act—the driving force behind dynamic Rhine and inland navigation. Mannheimer_Erklaerung_en.pdf (ccr-zkr.org). Accessed 9 March 2021

PLATINA 3 (2020) Basic project data. https://navigation.danube-region.eu/platina-3. Accessed 18 Dec 2020

Radojčić D (2008) On engineering greener logistics on inland waterways. PP Presentation, Ro-Ro 2008, Gothenburg

SRA (2010) Strategic research agenda I for inland waterway transport. http://ria.org.ro/ria/images/strategic%20reserch%20agenda.pdf. Accessed 18 Dec 2020

SRIA (2020) Strategic research and innovation agenda for the partnership on zero-emission waterborne transport (Draft version) 201021_SRIA_Zero_Emission_Waterborne_Transport_spread.pdf. Accessed 9 March 2020

Schweighofer J (2021) Möglichkeiten zur CO_2-Reduktion in der Donauschifffahrt durch den Einsatz von alternativen Energieträgern. In: El-Motion Conference 2021, Vienna

via donau (2019) Manual on Danube navigation. http://www.viadonau.org/fileadmin/user_upload/Manual_on_Danube_Navigation.pdf. Accessed 18 Dec 2020

Part I
Factors That Influence the Design of Contemporary Danube Vessels

Chapter 2
The Danube Waterway with Its Tributaries

2.1 Impacts of Climate Change

The climate on the earth changes, it is a natural process. Yet the term *climate change* (or global warming) refers to the rise in average surface temperature on Earth mainly due to burning of fossil fuels. The share of human influence on climate change over the last decades has been dramatically increased, above all by the accumulation of GHG emissions (mostly CO_2). Thus, climate change effects are continuously amplified.

Climate change insights which are of interest for IW navigation, generally, address the assessments of the future river conditions. Expected overall changes are relative to the examined timeframe. Apart from climate change, the water level is related to the discharges and riverbed morphology which varies over time; hence, the water levels are difficult to compare historically. Therefore, in the various studies, discharges are commonly compared, rather than the water level values. Consequently, the assessment of possible impacts of climate change on IW navigation is not a trivial task.

ECCONET, KLIWAS and Other Studies

Some of the unavoidable studies on climate change effects on inland navigation are ECCONET (2012a, b) and KLIWAS (2016) projects. ECCONET investigated the past (1950–2005), the near future (mid of the twenty-first century) and the distant future (end of the twenty-first century) in the catchments of the Rhine and the Danube. KLIWAS had a purpose to develop a sound statement about the span of possible climate changes in order to adjust infrastructure and river related operations. KLIWAS was the most important step in the German strategy for adaptation to climate change (DAS), dating back to 2008, see GFC (2008). KLIWAS was recently followed by The ProWaS project (Nilson et al. 2019), founded by BMVI as an additional step in DAS, with the purpose to assess the climate change effects on inland and coastal areas.

Besides, the International Commission for the Protection of the Danube River (ICPDR) started developing an adaptation strategy for the Danube in 2012. The

D. Radojčić et al., *Design of Contemporary Inland Waterway Vessels*,
https://doi.org/10.1007/978-3-030-77325-0_2

strategy was updated in 2018, based on the study performed by LMU for the Danube river, see LMU (2018). The study explores the climate impact on a river basin and provides the recommendations for the adaptation measures, referring to the KLIWAS and ECCONET results on navigation issues. Research on adaptations strategies for the Danube, with respect to climate change, could be also found in Stolz et al. (2018).

Climate effects are influencing inland navigation throughout a variety of issues: low-water level conditions, high-water level conditions, and ice conditions. For IWT the low-water level conditions are more disruptive than high-water level conditions (see Christodouloua et al. 2020 for instance) from the economic and navigation point of view. Consequently, by far the most important topic is the low-water level condition as it significantly affects the IW vessel design primarily by limiting its draught.

Low-Water Conditions

Negative effects of low-water levels issues on the Upper Danube are expressed more than on the Rhine and are increasing from the mid towards the end of the twentieth century. Comparison to the current period is indicated in the ECCONET study. KLIWAS predicted prolonged dry summer periods with an increase of the air temperature for the Danube and the Rhine area. Amongst the conclusions is that the maintenance strategies due to river bed elevation and sediment balance could have more influence on the water depth than the lower precipitation.

A diagram in Fig. 2.1 is presented with a purpose to illustrate that low waters will occur in the future, even without a climate change, if they have already occurred in the past. It shows the low-water level recordings for the Rhine in the past 200 years.[1] It should be noted that low water practically did not appear on the Rhine for last 40 or so years, which was probably one of the reasons for the increase-in-scale-trend and optimization of the new-buildings according to high water level conditions (which increased their efficiency, see Sect. 1.1.3 ad j). As a consequence, larger vessels, even empty, were unable to reach their destination in November 2018.[2] That led to the conclusion to reconsider the vessel design and to optimize them for low water conditions too, see CCNR (2019a, b). Consequently, taking into account emerging conditions, German transport secretary delivered an action plan for the low water conditions and its effects on the transport chain. The plan includes vessel optimization for new conditions, especially considering shallow draught designs, see de Wilt (2019) and BMVI (2019).

The abovementioned year 2018 was bad for IW shipping. It initiated several studies, see van Dorsser et al. (2020) for instance, which presents an empirical model for estimation of the effect of changing water depths on the capacity of inland ships

[1] A similar diagram was not found for the Danube, so this one is given and discussed as a vivid example.

[2] Financial reports (Gordon 2019) estimated a GDP drop in Germany of around 0.7% in 2018, due to Rhine prolonged low-level water periods in which the river became unnavigable for the first time in recorded history. CCNR (2019a) reported Rhine and Danube transport performance decrease of around 27% and 10% respectively, when comparing the third quarter of 2018 to the same period in 2017. The Danube performance drop was mainly addressed to the upper Danube region.

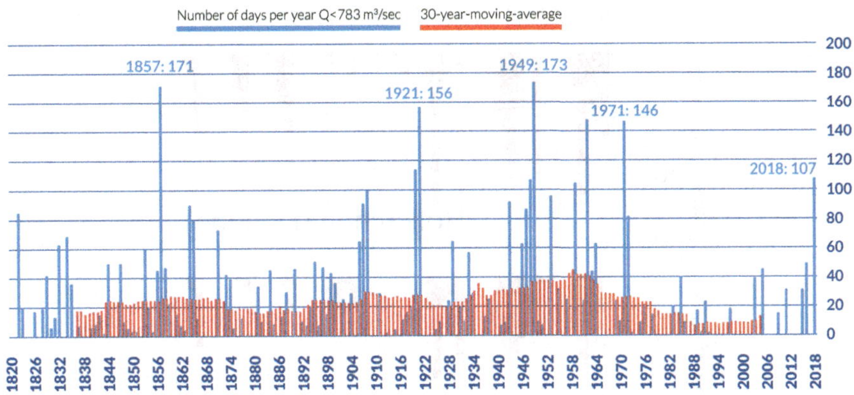

Fig. 2.1 Number of low water days per year[3] at Middle Rhine. *Source* CCNR, 2019a

and barges. This is a simple model as input variables are only vessel's hull type, length and beam; for better results data on vessel's capacity, design and empty draughts are needed too.

Fluctuations of navigational conditions for the Upper Danube are given in Fig. 2.2. Shown is the annual number of days below the low-water level threshold for the period 1950–2005, ECCONET (2012a). The frequency of low-water level peaks clearly demonstrates that these conditions are not a new issue. Concerning the future, according to the investigation performed in Beuthe et al. (2013, 2014) which was based on the ECCONET project, it was concluded that significant transport changes which would need overall transport mode changes are not expected until 2050. Transport costs on the Danube could potentially increase in the future (mid of the century) up to 15%, only due to climate change issues during the dry periods (which is in the same range as for the Rhine).

High-Water Conditions

Frequent periods of more intense rainfall (danger of high water) are expected according to via donau (2006). KLIWAS was unable to clearly predict high-water trends for the distant future due to a large number of uncertainties, although stated that increased winter precipitation periods are expected. High water, at its extreme (floods), could lead to higher river velocities and potential navigation disruption. This might also endanger the manoeuvrability of the vessels. Also, bridge heights on the Danube may become a barrier for container transport. However, it is not expected that transport costs will change significantly due to high-water conditions.

Ice Conditions

Ice formation depends on the river characteristics, potentially damages river infrastructure and prevents navigation for the majority of vessels. According to ECCONET,

[3]With a discharge smaller than 783 m^3/s, see CCNR (2019a).

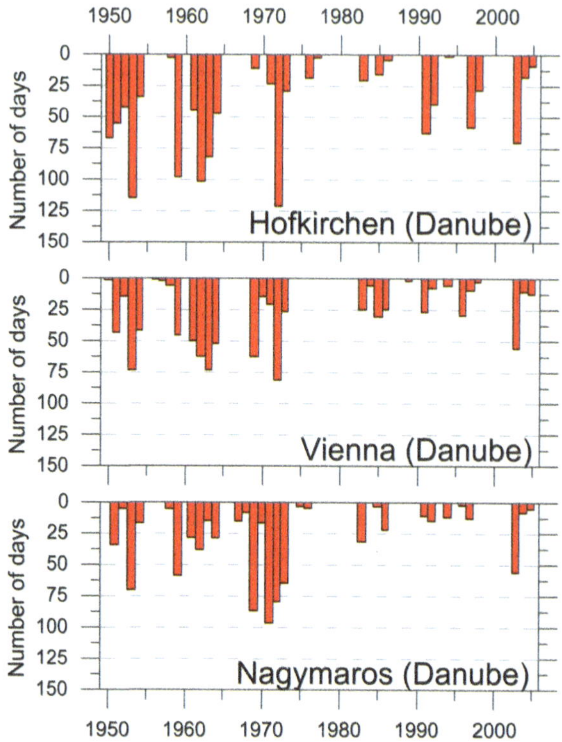

Fig. 2.2 Annual number of days below the low-water level threshold FDC_Q95$_{1961\text{-}1990}$[4] for the period 1950–2005 in the German (top), Austrian (centre) and Hungarian Danube (bottom). *Source* ECCONET, 2012a

future indices are looking towards fewer restrictions on the Danube due to ice formations (similar trends are predicted for the Rhine). Moreover, since the 1960s there were no serious issues related to navigation during the ice periods.

Navigation and Vessel Design

Due to climate change, the Danube navigation usability is not expected to change much until 2050. However, negative effects might be expected on a long-term horizon at the end of this century. Due to a large number of uncertainties, the abovementioned projects could not predict any specific climate-trend-effects on the Danube, but provided some guidance and expected possibilities—all related to more or less prolonged summer periods that could affect navigation in future, but not significantly.

According to PIANC (2008) (a study which preceded KLIWAS and ECCONET) shipping companies try to respond to the phenomena of low-water levels and floods in a way that the reliability of inland navigation is assured through "an adaptation

[4]Flow value exceeded in 95% of time steps in a reference period. Here, it is calculated based on mean monthly flow values in the period 1961–1990, and it serves as a low-water threshold.

of the fleet and new vessels of different design", as well as "light loading of current vessels and use of vessels with decreased draught". Increased and decreased water levels, and therefore water velocities too, change in timing of seasonal high and low water; ice periods also demand better manoeuvring capabilities.

Summarizing, to remain a competitive, cost-effective and reliable mode, IWT must be less vulnerable to low-water periods. Since more frequent low water periods are expected, the vessel design trend should be reconsidered. Moreover, the lifetime of IW vessels is relatively long (see for instance Sect. 1.1.3, ad u and v), hence the climate change predictions have to be accounted for. Particular attention should be paid to the design of new large vessels, as they are inherently more sensitive to low-water periods than the smaller vessels which are, on the other hand, less efficient. Concerning the existing fleet, negative effects of climate change are beyond their lifetime, as gradually it will be replaced with new vessel types.

2.2 What is an Optimal Contemporary IW Ship?

There are several definitions of an optimal ship, one of them, according to Žigić (2008), defines an optimal ship as:

- Modern and environmentally friendly.
- With low exhaust emissions.
- With low fuel consumption.
- Economical in operation.
- Highly compatible with the waterway.
- Have minimal impacts on bank vegetation and fish fauna.

Adequate Draught is a Sine Qua Non for Efficient IW Vessel Design

The available water depth, that changes along the river and during the season, should be specially analysed as it influences the transport performance through the ship carrying capacity and speed (see Fig. 2.3). Note that the diagrams depend on the vessel's characteristics; each displays a singularity:

1. The point till which the full loading capacity, i.e. full draught is reached (satisfying the keel clearance of at least 0.3 m); deeper water is irrelevant for payload.
2. The point till which achievable ship speed depends on the water depth (if sufficient power is available); deeper water is irrelevant for ship speed, i.e. maximal sailing speed is reached.

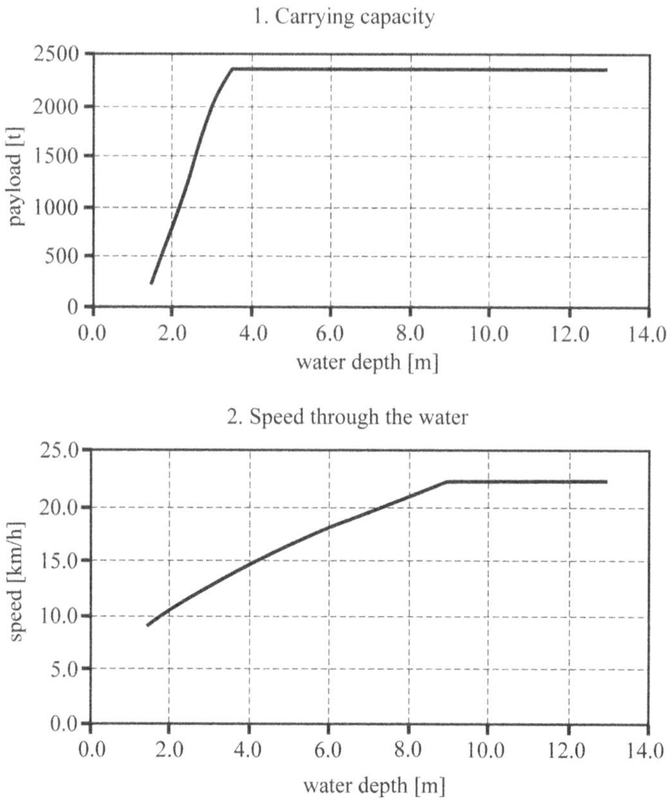

Fig. 2.3 Influence of water depth on ship payload and speed. *Source* Žigić, 2008

The transport effectiveness is increased with the increase of water depth, i.e. the larger vessel draught.[5] So, during low water levels, the ship should be able to operate with restricted economic effects. The same ship should be able to operate in deeper water too, but will then be less efficient than the ship initially designed for deep-water operation only. In other words, larger vessels with larger draught, although more effective in deep water than the smaller ones, are much more sensitive to shallow water and variation of a fairway depth. Extremely shallow water, however, is often regarded as a severe operational condition. Therefore, transport costs, however calculated, are very much influenced by water depth.

Follows that, through the loading capacity and vessel speed, water level influences the freight rate or transport price, see Fig. 2.4. The freight rate mirrors the low water; e.g. at the end of 2015 and 2016 price levels were 4 to 5 times higher than in normal times (see CCNR 2019a). The loading degree (i.e. the percentage of maximum vessel

[5]Note: To prevent grounding, fairway water depth must be larger than dynamic draught + underkeel clearance (more than 20–30 cm, depending on the riverbed). The dynamic draught = static draught + squat (see Sects. 7.5 and 8.5).

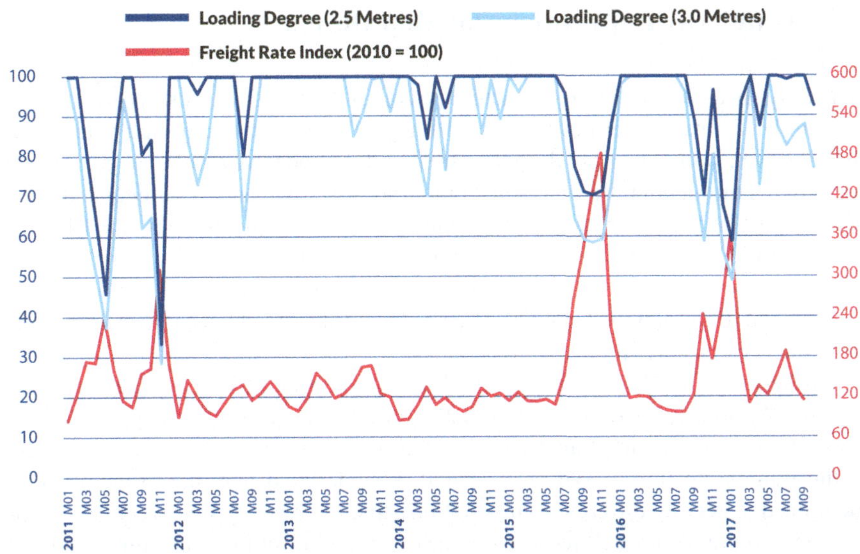

Fig. 2.4 Evolution of freight rate at Middle Rhine between 2011 and 2017, for vessels with a draught of 2.5 and 3 m compared with freight rate index. *Source* CCNR, 2019a

loading) of vessels with smaller draught was less impacted by low water periods. Note however that for the Danube region, a connection between water level and prices is less relevant because freight rates are mainly determined by bunker fuel costs (CCNR 2019b).

IWT is a Sustainable and Energy-Efficient Transport Mode

All transport modes, to a greater or smaller extent, have some negative impacts on water, air, soil biotic balance, climatic conditions, health, and economy, to name a few. Amongst them, IWT seems to have the least effects that can be quantified, for instance through direct and indirect costs (see Figs. 1.4 and 1.5). Direct costs are more or less obvious, but indirect costs are somewhat hidden and difficult to quantify. There are other impacts that are even more difficult to quantify, for instance, accidents, congestion, impacts on humans, flora and fauna etc. An optimal ship, therefore, should cause the least impacts of those mentioned above. This can be achieved by:

(a) Applying contemporary design measures, and
(b) Through making design compromises (sometimes it is necessary to sacrifice/reduce cost-effectiveness to obtain an overall good and environmentally acceptable vessel).

2.3 Waterway Implications on Ship Design

Generally speaking, the main characteristics of all inland (river) vessels are more or less similar. Primarily they have restricted draught (T) due to the restricted water depth (h). However, some rivers are deeper or are regulated and have a minimal guaranteed water depth throughout the year, while others are shallower and/or unregulated. Two most important European rivers differ mainly in the abovementioned—the Rhine is deeper while the Danube, although much longer and wider, is relatively shallow and unregulated with large variations in water depths. Consequently, the main difference between the Rhine and the Danube vessels is their draught, which has very important consequences on several other ship parameters.

Furthermore, the Rhine passes through the most developed part of Europe, probably the World, so it is quite normal that several technical solutions applied on the Rhine vessels are copied/transferred to other river vessels, in this particular case to the Danube vessels. However, it should be underlined that due to the already-mentioned waterway differences it is not possible to copy/transfer every service or technical solution. Other differences are also important, for instance, hinterland and infrastructure development along the Rhine and the Danube corridor (see Sect. 3.3), which actually dictate volume and type of cargo, transhipment facilities, intermodal loading units etc.

Characteristics of inland vessels for the Danube waterway depend very much on the waterway itself, i.e. its depth, the height of the bridges and size of locks. Therefore, the main characteristics of the Danube waterway and its tributaries (see Fig. 2.5) should be analysed next.

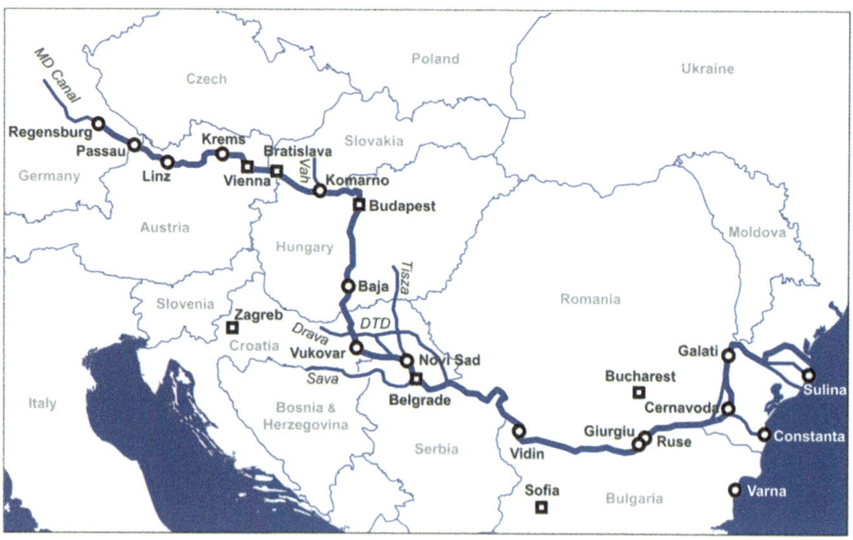

Fig. 2.5 Danube waterway

2.4 Restrictions of the Danube Waterway[6]

2.4.1 The Danube

Danube's total length is 2850 km, hence after the Volga, it is the largest European river; it flows through no less than 10 riparian countries. The Danube (often denoted as a South-East European IWW Corridor) is an important segment of the Rhine-Main-Danube waterway with a total length of 3504 km that connects the North Sea and the Black Sea as well as 15 European countries. Its navigable length from Kelheim in Germany to Sulina in Romania is even 2411 km.

According to its physical and geographical characteristics, the river Danube is officially divided by the Danube Commission into three main sectors: Upper Danube (Sector I), Middle Danube (Sector II) and Lower Danube (Sector III). Each of these sectors is subdivided into sections according to different navigational conditions (Table 2.1). The EUDET Project (EUDET 1998) showed, however, that such division is partly out of date, and proposed a new division of the Danube, which differentiates the canalised (articulated) sections from the free-flowing parts of the waterway. Although the EUDET division relates better to the present state of Danube waterway, there is still not enough statistical data (especially concerning water depth) to cover it properly. Even in the EUDET study, waterway statistics are mostly given according to the Danube Commission classical subdivision. A descriptive diagram showing very large variations of water depths along the Danube is given in Fig. 2.6. Accordingly, the current speed along the Danube is also very uneven, varying between 3.0 and 10.0 km/h (IDV 2013).

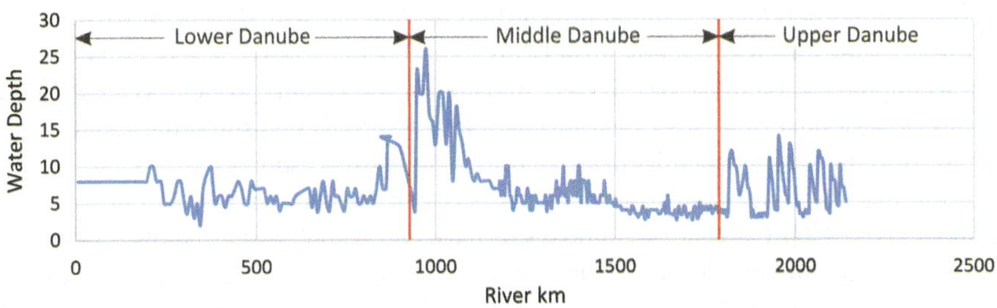

Fig. 2.6 Danube water depths at LNWL, the distance between Sulina and Linz. *Source* IDV, 2013

The largest convoys which can operate on particular sections of the Danube and MD canal, according to the Danube Commission, are shown in Fig. 2.7. UNECE waterway classes, i.e. size of vessels and convoys, are discussed later in Sect. 4.1.1.

[6]The main sources of information in this section are SPIN (see Radojčić 2004), COVEDA (COVEDA 2003), Manual on Danube Navigation (via donau 2019).

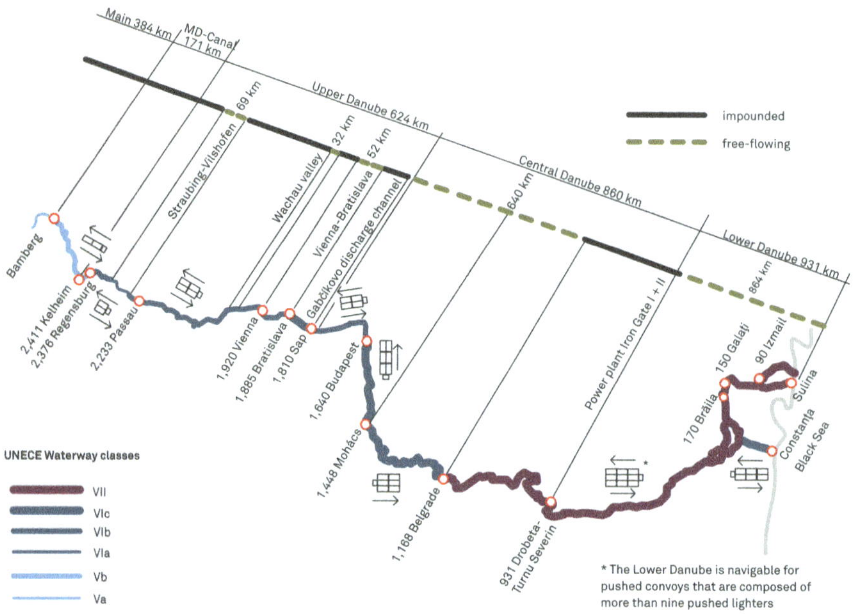

Fig. 2.7 Maximal possible dimensions of convoys on the Danube waterway according to waterway classes. *Source* via donau, 2019

The most important statistical information, from the point of view of vessel design, is the waterway depth (already discussed in Sect. 2.2) and the air clearance under the bridges. Both are delicate as they vary from one source to another.

So, an attempt is made to re-examine different sources[7] with the purpose to deduce the appropriate data for water depth at LNWL[8] and critical bridge heights at HNWL,[9] and to implement them to the EUDET division of the Danube waterway (see Table 2.2). Given depths by LNWL and air clearance over HNWL show the range from the most to the least critical values found in the examined sources.

Note that there are few disagreements between the data shown in Fig. 2.7 and Table 2.2 (e.g. UNECE waterway classes on the Central and Lower Danube). Detailed information about navigational characteristics on the Danube and its tributaries are listed in the Blue Book (where the Danube as main European waterway is designated as E 80 and, for instance, the Sava as E 80–12).

---

[7]EUDET, WESKA (2002), Blue Book (2017) and via donau (2019). Note that data vary in different editions of the same source.

[8]LNWL: *Low Navigation Water Level* is the water level that corresponds to the flow available for 94% of duration of the navigable season, i.e. excluding the winter periods of break of navigation affected by ice.

[9]HNWL: *High Navigable Water Level* is the water level that corresponds to the flow occurring at 1% of duration of the navigable season.

Table 2.1 Division of the waterway by the Danube commission

Section	From	Danube km	To	Danube km
Upper Danube – Sector I Gradient/km (average) 37 cm; 16 locks (MD canal 16 locks, Main 34 locks) Travel speed: upstream 9-13 km/h; downstream 16-18 km/h				
I-1	Kelheim	2415	Passau	2227
I-2	Passau	2227	Linz	2135
I-3	Linz	2135	Vienna	1929
I-4	Vienna	1929	Gonyu	1791
Middle Danube – Sector II Gradient/km (average) 8 cm; 2 locks Travel speed: upstream 9-13 km/h; downstream 18-20 km/h				
II-1	Gonyu	1791	Budapest	1646
II-2	Budapest	1646	Moldava Veche	1048
II-3	Moldava Veche	1048	Drobeta	931
Lower Danube – Sector III Gradient/km (average) 4 cm; No locks (Danube Black Sea canal 2 locks) Travel speed: upstream 11-15 km/h; downstream 18-20 km/h				
III-1	Drobeta	931	Braila	170
III-2	Braila	170	Sulina	0

The Danube Bottlenecks

In order to improve navigation conditions, the loaded draught[10] of the vessels, minimum height under the bridges, fairway width and minimal curve radius according to Danube Commission recommendations from 2013 should be respectively at least: 2.5 m below LNWL on free-flowing sections, 5.25 m in relation to HNWL, 150–180 m and 1000 m.[11] Namely, waterways of class IV and higher, i.e. with vessel/convoy minimum draught of 2.5 m, called also *E waterways* are of importance for international freight transport; lower classes are of regional or national significance. "Fairway Rehabilitation and Maintenance Master Plan for the Danube and its Navigable Tributaries" (see Fairway 2019) gives the guidelines for achieving these targets.

However, due to various reasons, the abovementioned recommendations could not be met. Several infrastructural strategic bottlenecks in the waterway network of

[10]Note that previous versions of the Danube Commission required minimum water depth of 2.5 m, i.e. approximately 30 cm less.

[11]Amongst the targets are minimum requirements set out by the RIS directive, but that does not influence vessels' dimensions.

Table 2.2 The EUDET division of the Danube with main restrictions of the waterway

Section	Danube km	ECE Class	Remark	Depth by LNWL (m)	Air clearance over HNWL (m), if lower than 7.5m	Minimal lock dimensions (m) Beam × Length
MD Canal	171 – 2414	Vb	canalised	2.7	6.00	1 x 12 x 190
Kelheim – Straubing	2414 – 2324	Vb VIb	canalised	2.9	5.25 – 5.75	1 × 12 × 190
Straubing – Vilshofen	2324 – 2249	VIa	free-flowing (shallow)	1.6 – 2.0	4.74 - 5.00	
Vilshofen – Melk	2249 – 2038	VIb	canalised	2.8	4.61 - 5.15	2 × 24 × 230
Melk – Durnstein	2038 – 2008	VIb	free-flowing (shallow)	2.3 - 2.5		
Durnstein – Vienna	2008 – 1921	VIb	canalised	2.8		2 × 24 × 230
Vienna – Cunovo	1921 – 1853	VIc	free-flowing (shallow)	1.7 - 2.5		2 × 24 × 230
Cunovo – Palkovicovo	1853 – 1811	VII	canalised	2.5		2 × 34 × 275
Palkovicovo – Budapest	1811 –1646	VII	free-flowing (shallow)	1.7 – 2.5	6.70 - 7.29	
Budapest – Slankamen	1646 – 1215	VII	free-flowing (good)	2.0 - 2.5		
Slankamen – Iron Gates II	1215 – 863	VII	canalised	well over 2.5		2 × 34 × 310
Iron Gates II – Bala Arm	863 – 346	VII	free-flowing	2.0 - 2.3		
Bala/Borcea Arm – Giurgeni	346 – 240	VIc	free-flowing (good)	2.7		
Giurgeni – Braila	240 – 170	VII	free-flowing	2.0 - 2.4		
Braila – Sulina	170 – 0	VII VIc VIa	maritime section	7.3		
Bala Arm – Cernavoda	346 – 299	VIc	free-flowing (shallow)	could be bypassed		
Cernavoda – Giurgeni	299 – 240	VII	free-flowing (good)	Over 2.5		
Cernavoda – Constanta	64 – 0	VIc	navigable canal	well over 2.5		2 × 25 × 310
Chilia Arm – Black Sea	116 – 0	VII	free-flowing (good)	Over 2.5		

Note: Navigability of the fairway is also influenced by the natural profile of a watercourse - thalweg (river path with maximum depth and/or velocity).

the Danube region are still recognized by UNECE, mainly in the Upper and Lower Danube. It should be noted that the sectors which are regarded to be the bottlenecks are fairly long; e.g. the entire Lower Danube (the stretch of some 700 km) is considered as a strategic bottleneck. Consequently, fairway maintenance primarily consists of dredging of shallow areas (fords).

The Lower Danube is usually more affected by dry seasons than the Upper Danube due to natural deposition of sediments at slower river sections (average gradient/km is, respectively, 4 and 37 cm on the Lower and Upper Danube; see Table 2.1). Concerning the economic impacts of low-water periods on the Danube, see Scholten and Rothstein (2016).

Of particular interest is the water depth which is on several sectors less than recommended. Just one example of critical sectors is Danube 375–175 km (Calarasi–Braila). According to the Blue Book, strategic bottlenecks, amongst several others, are:

- From 375 to 300 km, with fairway depth limited to 1.4–2.5 m for 61–126 days a year,
- From 300 to 175 km, with fairway depth limited to 2.0–2.5 m for 5–32 days a year.

The Danube between 346 and 300 km (Bala Branch outlet to Black Sea Canal at Cernavoda, Fig. 2.8) is particularly critical for navigation during the dry seasons, with water depth on some sectors of around 1.5 m only. Consequently, ships often have to use a detour—via the Bala-Borcea Branch—which increases the navigation length to the Black Sea Canal for around 110 km. Moreover, one-way navigation and convoy dismantling are often necessary in the Borcea Branch.

Fig. 2.8 Critical points on Calarasi-Braila sector

Note that navigation may be suspended not only due to low-water, high-water or icy conditions, but also due to other extreme weather conditions which are considered to be unsafe for navigation (e.g. fog and wind). Extreme wind condition is a wind speed of 17–18 m/s which, depending on a type, size and loading conditions of a vessel, may be considered as unsafe for navigation.

2.4.2 *The Danube Tributaries and Canals*

The Danube has more than 30 navigable tributaries, but only those having the ECE class III and above are given in Table 2.3. Concerning the canals, of primary interest are obviously the 171 km long Main-Danube (MD) canal and 64 km long Danube-Black sea canal; their data of interest are given in Table 2.2. Table 2.3 thus shows the data for only one canal—the Canal system Danube-Tisa-Danube (DTD) which essentially consists of three main canals that connect the Danube with the River Tisa (Tisza).

Table 2.3 Navigable tributaries of the Danube with UNECE class III and above

Name	Danube (km)	Navigable length (km)	UNECE classification	No. of locks
Vah	1770 Left bank	79	VIa	2
Canal System Danube-Tisa-Danube	1363 1253 1076 all Left bank	90 39 147	III - IV	7
Drava	1382 Right bank	68	III - IV	0
Tisa/Tisza	1214 Left bank	519	IV	3
Sava	1170 Right bank	586	III - Va	0

Data from: Manual on Danube Navigation, via donau (2019)

2.5 Concluding Remarks

Water Depth

According to UNECE waterway classes, along the entire Danube, a minimal allowed vessel/convoy draught of 2.5 m is assumed. Shipping companies typically assume 2.2–2.3 m, depending on the season. Nevertheless, according to Table 2.2 the water depth on some critical Danube sectors is as follows:

(a) On the Upper Danube, the most critical 69 km stretch is between Straubing and Vilshofen with $h_{W\text{-}LNRL} < 2$ m (according to some statistics even 1.6 m).
(b) Several sectors on the Upper Danube (upstream of Budapest) have $h_{W\text{-}LNRL} <$ 2.5 m.
(c) A few sectors on the Lower Danube (downstream of Iron Gates II) have $h_{W\text{-}LNRL}$ = 2.0–2.3 m. According to some statistics, water depth on critical sectors is as low as 1.5 m, so detour via other (longer) routes is necessary (Fig. 2.8).
(d) Elsewhere, $h_{W\text{-}LNRL} > 2.5$ m. On the Middle Danube, the depth is often over 5 m.

It should be underlined, however, that the *fairway depth fluctuates* through the year and from year to year, hence the number of days of low water in a year on the particular sector should also be identified.

Bridge Height or Air Clearance

The most critical bridge heights are also on the Upper Danube, i.e. the bridges in Deggendorf and Passau. Depending on the source of data, $h_{A\text{-}HNWL}$ = 4.74–5.00 m and 4.61–5.15 m, respectively. The heights of MD canal bridges are around 6 m. Few other bridges upstream of Budapest are below 7.5 m. Downstream from Budapest, all bridges are above $h_{A\text{-}HWL} > 7.5$ m. In general, the bridge height is important for the container vessels as it may limit the number of container layers.

Size of Locks

Most of the Danube locks have standard European dimensions. The most critical one is upstream of Straubing with breadth and length of 12 m × 190 m; the same are all locks of MD canal, while the rest on the Upper Danube are 2 m × 24 m × 230 m. The locks built by ex-East-European countries are even 2 m × 34 m × 275 (310) m and on the Danube-Black Sea canal 2 m × 25 m × 310 m.

Implications on Ship Design

Taking into account that (a) an IW vessel should be designed according to the particular waterway, and (b) that all-around clearance between the vessel (or her cargo) and bridge/river-bottom/lock-side should be at least 0.3 m, the maximal allowed vessel dimensions (minimal operational for draught), with possible minor restrictions in sailing during the dry seasons, are:

- For the whole Danube including the stretch upstream of Straubing-Vilshofen, as well as through the DM Canal: $T \leq 1.7$ m, $B \leq 11.45$ m.
- Downstream of Vilshofen: $T \leq 2.0$ m, $B \leq 23.4$ m.

The length of selfpropelled vessels is practically unrestricted, while coupling train formation will be discussed later in Sect. 4.2. The air draught depends on the bridges (see Tables 2.2 and 4.1) and, as mentioned earlier, primarily concerns container ships.

Note that the abovementioned recommendations for operational minimal draughts are for all through the year navigation. Shallow water periods are rare, though they happen, see for instance footnote no. 2, Table 2.2, Figs. 2.1, 2.2 and especially

2.4. Consequently, the vessels should be designed to operate at higher draughts, but special attention should be paid to low draught operation (see Sects. 2.2 and 10.3).

References

Beuthe M, Jourquin B, Urbain N et al (2013) Impacts of climate change on transport by inland navigation: a multimodal approach for the Rhine and the Danube. In: NECTAR 2013 international conference, Azores

Beuthe M, Jourquin B, Urbain N et al (2014) Climate change impacts on transport on the Rhine and Danube: a multimodal approach. Transp Res Part D 27:6–11. https://doi.org/10.1016/j.trd.2013.11.002

Blue Book (2017) Inventory of main standards and parameters of the E waterway network, UN. https://doi.org/10.18356/233cb552-en

BMVI (2019) Unterzeichner der gemeinsamen Erklärung des BMVI zur Sicherstellung zuverlässig kalkulierbarer Transportbedingungen am Rhein (8-Punkte-Plan). https://www.bmvi.de/SharedDocs/DE/Anlage/WS/8-punkte-plan.html. Accessed 11 Nov 2020

CCNR (2019a) Inland navigation in Europe—market observation. Annual Report 2019. https://inland-navigation-market.org/wp-content/uploads/2019/11/ccnr_2019_Q2_en-min2.pdf.pdf. Accessed 20 Dec 2020

CCNR (2019b) Low water effects on Rhine navigation. CCNR Press Release, Strasbourg. https://ccnr.eu/files/documents/cpresse/cp20191218en.pdf. Accessed 20 Dec 2020

Christodoulou A, Christidis P, Bisselink B (2020) Forecasting the impacts of climate change on inland waterways. Transp Res Part D 82. https://doi.org/10.1016/j.trd.2019.10.012

COVEDA (2003) Container vessels for the Danube waterway. DPC Project

de Wilt H (2019) Germany's Rhine low water contingency planning. World Cargo News. https://www.worldcargonews.com/news/news/germanys-rhine-low-water-contingency-planning-62583. Accessed 10 Nov 2020

ECCONET (2012a) Deliverable 1.5: impact of climate change on hydrological conditions of navigations. Report for European Commission, Brussels

ECCONET (2012b) Deliverable 3.3: consequences of climate change for inland waterway transport. Report for European Commission, Brussels

EUDET (1998) Evaluation of the Danube waterway as the key European transport resource

Fairway (2019) National action plans. https://navigation.danube-region.eu/wp-content/uploads/sites/10/sites/10/2020/01/FRMMP_national_action_plans_Oct2019.pdf. Accessed 18 Dec 2020

GFC (2008) German strategy for adaptation to climate change. The Federal Government of Germany

Gordon P (2019) Germany's dried-up rivers cut growth but the rebound is coming. Bloomberg. https://www.bloomberg.com/news/articles/2019-01-23/germany-s-dried-up-rivers-cut-growth-but-the-rebound-is-coming. Accessed 10 Nov 2020

IDV (2013) WP2, analysis of technical solutions. Vienna Model Basin

KLIWAS (2016) Impacts of climate change on waterways and navigation in Germany. Concluding report of the BMVI, Technical conclusions from the results of the KLIWAS research programme, Federal Ministry of Transport and Digital Infrastructure

LMU (2018) Integrating and editing new scientific results in climate change research and the resulting impacts on water availability to revise the existing adaptation strategies in the Danube River basin. Final Report

Nilson E, Ehlers BM, Abalichin J et al (2019) ProWaS climate projection service for waterways and navigation in Germany. Pilot study, EGU General Assembly 2019

PIANC (2008) Waterborne transport, ports and waterways: a review of climate change drivers, impacts, responses and mitigation. Climate Change and Navigation. Report

Radojčić D (2004) Innovative transport vehicles on the Danube and smaller waterways. In: Proceedings of the SPIN-TN working group 3—intermodality and interoperability. GROWTH Programme, Vienna

Scholten A, Rothstein B (2016) Navigation on the Danube—limitations by low water levels and their impacts. JRC Technical Report. EUR 28374 EN; https://doi.org/10.2788/236234

Stolz R, Prasch M, Weber M et al (2018) Climate change impacts on the water resources in the Danube river basin and possibilities to adapt—the way to an adaptation strategy and its update. J Environ Geograph 2(3–4):13–24. https://doi.org/10.2478/jengeo-2018-0009

van Dorsser C, Vinke F, Hekkenberg R, van Koningsveld M (2020) The effect of low water on loading capacity of inland ships. Eur J Transp Infrastruct Res (EJTIR) 20(3):47–70. https://doi.org/10.18757/ejtir.2020.20.3.3981

via donau (2006) Prospects on the development of infrastructure and navigation on the Danube

via donau (2019) Manual on Danube navigation. http://www.viadonau.org/fileadmin/user_upload/Manual_on_Danube_Navigation.pdf. Accessed 18 Dec 2020

WESKA (2002) Westeuropaeischer schiffahrts-und hafenkalender. binnenschifffahrts-verlag GmbH, Duisburg-Ruhrort

Žigić B (2008) Optimal ship design for shallow water operation. 4th Danube summit, Constanza

Chapter 3
Intermodality and IWT

3.1 Lo-Lo and Ro-Ro Technologies[1]

By definition, intermodal[2] freight transport means the transport of goods in one loading unit, using two or more modes of transport successively, without handling the goods themselves. As such, intermodal transport has its main potential in long-distance transport. Transport chain links depicted in Fig. 3.1 are generally necessary for the intermodal freight transport that integrates IWT. Intermodal possibilities that involve waterborne transport are shown in Fig. 3.2 (where ISO containers and swap-bodies[3] should be regarded as stackable and unstackable Intermodal Loading Unit—ILU, respectively).

For both Lo-Lo (container) technology and Ro-Ro technology, the first and last links (land transport) are more or less the same and are unavoidable. They are called *pre-haulage* and *end-haulage*, respectively. Therefore, further consideration is limited to the rest of the transport chain, i.e. *two transhipments and waterborne transport* (the bottom line of Fig. 3.1). If transportation is executed directly from departure to destination sites without changing the transport mode, that is a *direct-transport* (or door-to-door, i.e. top line in Fig. 3.1). Note that bulk cargo (ore, construction material etc.) is usually transported by a single mode, while general cargo uses multimodal transport options which require additional implementation of logistics.

The main source of information in this chapter is Radojčić (2005)

[1]The transhipment of containers and other stackable Intermodal Loading Units (ILUs) to/from container vessels has to be done vertically (hence the abbreviation Lo-Lo for Load on—Load off), contrary to the horizontal transhipments (on wheels) of various kinds of cargo (Roll on—Roll off).

[2]Strictly speaking, *Intermodal* and *Multimodal* are different terms. The main difference is in the responsibility for the movement of cargo. As that is not important for the book's subject, these differences will be ignored.

[3]A swap body is a type of a standardized container used in road and rail transport.

D. Radojčić et al., *Design of Contemporary Inland Waterway Vessels*,
https://doi.org/10.1007/978-3-030-77325-0_3

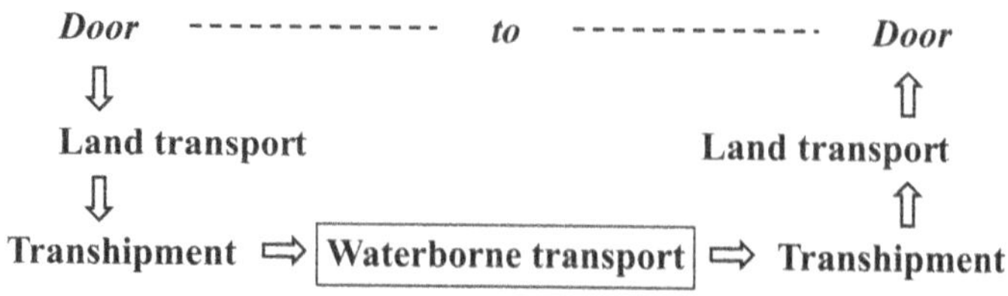

Fig. 3.1 Door-to-Door transport–necessary transport chain links

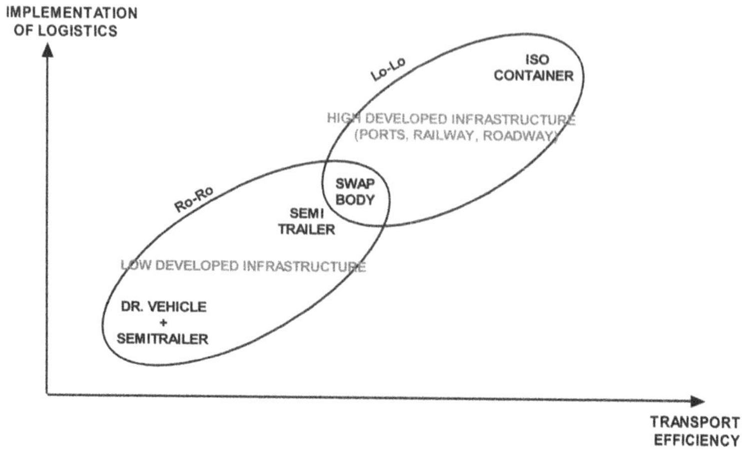

Fig. 3.2 IWW multimodal possibilities

The Intermodal Loading Units in IWT are containers, semi-trailers or swap-bodies. Consequently, the payload is the gross mass of the loaded container, semi-trailer or a swap-body. For Ro-Ro vessels, however, it is the deck area, not the mass of payload, that is critical: in fact, the key metrics is the length of the standard lane, i.e. it is the lane-meters that are sold. Thus, for Ro-Ro vessels the *payload* should be considered as *pay-area*, whereas the key parameter for the container vessel is the number of TEUs, i.e. the payload should be considered as *pay-volume*. Therefore, it would be appropriate to use *TEU·km* and *lane-meter·km* for container and Ro-Ro vessels respectively, instead of usually accepted, transport performance metrics *t·km*. In any case, proper relation between cargo space and cargo weight is of utmost importance for ship design.

Logistics

The efficiency of intermodal transport, transhipment, distribution of goods and other related subjects, very much depends on the organization, i.e. on the *logistics* which is practically omitted here. Essentially, the objective of logistics is to reduce transport costs. Unfortunately, in accomplishing this, environmental standards are often overlooked (see Radojčić 2008). In that context, some of the topics related to *greener logistics* are in the focus of this book, e.g.:

- Environmentally acceptable transport mode, i.e. IWT.
- Techniques to increase the capacity ratio of IW vessels.
- Shipment of cargo at appropriate speed.
- Suitable type of loading units, transhipment technologies, etc.

It should be noted that certain topics, some even being in the root of environmentally unsustainable logistics, are beyond the book's scope, e.g.:

- Transfer of dirty industries to undeveloped countries (resulting in massive transport growth)
- Transport and packaging techniques which result in additional waste, etc.

Furthermore, subjects that concern the amount and type of transported products on the Danube, logistic chains etc. are also ignored here, although they are important logistics issues that influence the design of IW vessels.

Transhipments and Cargo Handling Equipment

Worldwide container transport requires massive investments in all chain components to provide fast, on schedule and consequently economical door-to-door service. It does not tolerate bottlenecks such as prolonged port stays. Thus Lo-Lo transhipment of containers is efficient only if dedicated equipment is used, for instance, *spreaders*, expensive *gantry cranes* (for massive transhipment of containers), *reach stackers* (for smaller terminals), etc. Otherwise, transhipment will be relatively slow and therefore inefficient.

Furthermore, for successful application of container technology, it is often necessary to have transhipment equipment both at the final destination and origin point in the hinterland, rather than just in the ports. Necessary expertise and experience are required at all points. Non-stackable swap bodies are sometimes also transhipped vertically, by *grappler arms* (side-lift instead of top-lift). In that case, some additional space between swap-bodies is needed. Further discussion about the transhipment possibilities of various ILUs is beyond the scope of this book, although waterborne transport cannot be implemented without two transhipments.

Note, however, that the transhipment is a relatively expensive link of the transport chain. Namely, in average, *roughly estimated*, 50% of total intermodal costs are due to waterborne transport itself, $15 \div 20\%$ transhipment costs and rest are related to pre- and end-hauling. As transhipment costs are fixed, regardless of the transportation distance, waterborne share within the intermodal transport chain will be more profitable for longer than for the shorter segments.

3.2 State-Of-The-Art of Intermodal Loading Units (ILU)

Stackable ISO containers—often called maritime containers—are assumed here, and their 20 feet equivalent units (TEU) are 6.06 × 2.44 × 2.44 m (accordingly, 40 feet container (FEU) is 12.19 m long with identical cross-section), see Table 3.1. Other ILUs (*domestic* or *continental containers*, *swap-bodies*, *semi-trailers*) are slightly wider—having 2.50 to 2.60 m outer width—and allow more Euro-pallets to be packed into them (see Tables 3.2 and 3.3). That is: maritime containers have lower pallet capacity, ranging from 76 to 82% utilization, while the capacity of the domestic containers ranges from 93 ÷ 98% (see SPIN 2003).

This may amplify a deficiency caused by incompatible standards, brought by different authorities in a long period, see Fig. 3.3. Summarizing, these are dimensions of:

(a) Wooden pallets (Euro pallets 1200 × 800 mm and ISO pallets 1200 × 1000 mm)
(b) Stackable ISO (maritime) containers (outer width 2438 mm, inner width 2330 mm)
(c) European locks (min. chamber width 12 m)
(d) Ship-to-lock clearance (min. prescribed 0.3 + 0.3 m)
(e) Gangway width (safety rules require 2 × 650 mm).

Therefore, (c) + (d) outlines maximal (standard) IW container vessel breadth to 11.4 m, but an additional 5 cm (rarely more) are often added, resulting in 11.45 m.

Table 3.1 20‘ and 40‘ ISO-1 containers (outer dimensions and allowed gross weight)

Container type	Length (mm)	Width (mm)	Height (mm)	Height (feet)	Allowed gross-weight (t)
20' ISO-1 container	6.058	2.438	2.438	8	24.00
			2.591	8.5	
			2.743	9	
			2.896	9.5	
40' ISO-1 container	12.192	2.438	2.438	8	30.48
			2.591	8.5	
			2.743	9	
			2.896	9.5	

Table 3.2 Domestic containers (outer dimensions and allowed gross weight)

Type	Outer length (mm)	Outer width (mm)	Outer height (mm)	Maximum gross weight (t)
20' Domestic	6.058	2.500	2.591	24.00
7.15 m Domestic	7.150	2.500	2.600	16.00
30' Domestic	9.125	2.500	2.591	25.40
40' Domestic	12.192	2.500	2.591÷2.896	30.48

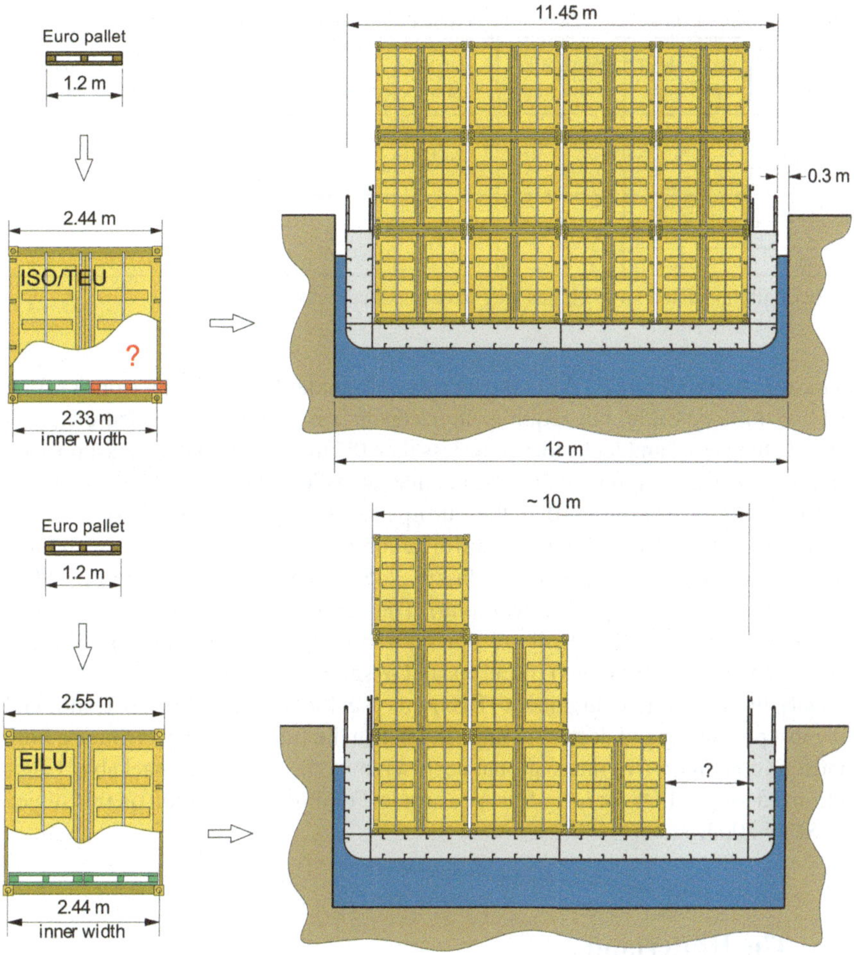

Fig. 3.3 Problems with incompatible standards

Subtracting e) from the vessel's breadth results in internal hold width of 10 m (max. 10.10 m).[4] It is a pure coincidence that this space can accommodate four ISO containers abreast,[5] probably one of them being a 2.5 m ILU too. However, the majority of pallet-wide ILUs are 2.55 m wide because packing of pallets into 2.50 m outer width ILUs is somewhat difficult. Pallet-wide ILUs, however, require a hold width of some 10.25 m.

A feasible solution to this ship-hold-size vs. ILU-size problem could be mixed stowing of maritime containers (2.44 m wide) and ILUs (2.55 m wide), the second

[4]Vessels with the maximal breath of 9.5 m (class IV) have a hold width of about 7.5 m, thus have room for three ISO containers abreast.

[5]Some 50 mm tolerance between the containers is required for container stowing.

Table 3.3 Swap bodies, Type C and Type A (outer dimensions and gross weights)

Length (mm)	Size-code of swap body	Maximim width (mm)	Height (mm)	Maximum gross weight (t)	Number of standard pallets 1x1.2 / 0.8x1.2 (m)
7150	C 715	2550/2600 (*)	2670	16.0	14/17
7450	C 745				14/18
7820	C 782				14/19
12192	A 1219	2550/2600 (*)	2670	34.0	24/30
12500	A 1250				24/30
13600	A 1360				26/33

* Thermally insulated swap bodies may have maximal width of 2.60 m

on the top above all others. Also, the increase of standard vessel beam from 11.45 m to 11.65 m is possible, but requires an acceptance of the recommended clearance between the vessel and lock sides to be less than 0.3 m. The last but the least possible is an introduction of a bit smaller "sea container-pallets" that would be just 30 mm narrower and 35 mm shorter than the Euro-pallet (Žigić and Bison 1999).

European ILUs (EILU),[6] which fulfil all requirements for compatibility with Euro-pallets, are stackable swap-bodies Class C 745 which are 7.45 × 2.55 × 2.90 m (longer type would be up to 45 feet long). The dominant European ILUs are swap-bodies, however, most of them are not stackable. It is expected that the share of stackable ILUs (swap-bodies) in European IW shipping will increase.

Only the maritime containers (TEUs) are usually considered in IWT, although the most promising and dominant units in IWT in the future are expected to be both, maritime containers (for international deep-sea/overseas trade) and stackable, pallet-optimized long EILUs for intra-European trade. For more information on this subject see SPIN 2003.

3.3 The Hinterland

The industry and infrastructural development of the Danube hinterland are often neglected in the review of new transport projects, such as, for example, modal shift projects and intermodal transport. Furthermore, intermodal transport possibilities are often compared to those on the Rhine—although there is enormous difference between the Rhine and the Danube hinterland, see Table 3.4. These differences are at least as important for IWT, as are the waterway differences which are more often discussed.

The Rhine passes through the most developed parts of the World, while the development of the Danube hinterland varies, but is generally below that of the Rhine. The hinterland of the Upper Danube (Germany, Austria and partly Slovakia and Hungary) is generally well developed, while that of the Middle and Lower Danube is not. The

[6] EILUs are mostly implemented in inter-European trucking, but also in road-rail combined transport.

Table 3.4 An overview of the characteristics of the Rhine and the Danube corridor

The Rhine corridor	The Danube corridor
• The Rhine is maintained river, guaranteed depths often 3.5 m • Developed hinterland and transport infrastructure. • 850 km navigable • Many inland ports (on average every 20 km) • Developed logistics • Industrial production in the vicinity of the river • General knowledge about IWT potentials does exist • Awareness about EST exist • Seaport Rotterdam – 450 mil. t, 14+ mil TEU Inland port Duisburg – 50-60 mil. t • 84% of the total EU IWT performance	• The Danube is partially regulated river, shallows on many sectors, occasionally 2.5 m • Undeveloped hinterland and transport infrastructure • Long river, 2400 km navigable • Not enough ports (on average every 60 km) • Undeveloped logistics concepts • Industrial production not in the river's vicinity • Sufficient knowledge about IWT potentials does not exist • Awareness about EST is not sufficient • Seaport Constanta – 50-60 mil. t, 0.7+ mil TEU Inland port Linz around 4 mil. t • 16% of the total EU IWT performance

Rhine's Port of Rotterdam and the Danube's Constanta are important seaports for both rivers. However, the transhipped cargo in Constanta is mostly for the area in the vicinity of Bucharest, hence the Danube as a whole is not as well utilized IWT resource as it could be expected. Upgrades in the infrastructure are required for initialization of modal shift projects. This important fact is often forgotten, and proven intermodal solutions which "work on the pattern river" Rhine, are suggested for the Danube. Table 3.4 provides an overview of the Rhine corridor and so-called South-east corridor (the Danube) from the hinterland viewpoint.

3.4 Comparison of Ro-Ro and Lo-Lo Technologies for the Danube[7]

An analysis of the cost/benefit tradeoffs for Ro-Ro vessels is somewhat different from that for container vessels. The reason for this is that Ro-Ro vessels have to compete with road transport, which nowadays absolutely dominates in Europe and particularly in the Danube corridor. Nevertheless, since Ro-Ro vessels should transport semitrailers and trucks, they should not be considered simply as an alternative to the road transport, but as an "intermodal partner" to the road mode, which enables the long-distance haulage to be more environmentally and economically sustainable. A brief comparison of Ro-Ro and Lo-Lo technologies is given in Table 3.5.

Summarizing, the comparison of container and Ro-Ro services is a bit complex, mostly because these are two different intermodal services with different good and bad points. Concerning the Danube, it may be expected that both of them can find the

[7]The main source of information in this section is Radojčić (2006).

Table 3.5 Pros and cons of container and Ro-Ro vessels in the Danube context

Pros & Cons for **Container** vessels	Pros & Cons for **Ro-Ro** vessels
Service Well established service and technical solutions already exist on the Rhine – that experience can be applied to the Danube. Nevertheless, establishing successful service on the Danube demands that containerization is widely accepted, especially in the hinterland. IWT of containers along the Danube corridor presently has to compete with railway transport more than with road transport.	**Service** Public Ro-Ro service, open to all road users, could be organized/designed for the Danube corridor. There are no similar solutions which could be copied from other rivers but the first generation of very successful vessels already exist on the Danube. Ro-Ro service + pre- and end-haling (short distance trucking to/from origin/destination) is an alternative to the long-distance trucking.
Hinterland Developed hinterland, transport infrastructure and an established tradition of using containers are required. Application of logistic concepts is essential. Neither of these is the case with the Danube yet (with exception of the Upper Danube regions).	**Hinterland** No need for the developed hinterland, logistics etc. since the only necessary Ro-Ro "partner" – the truck – already enables door-to-door service.
Cargo Stackability of containers (and other stackable ILUs) is a great advantage – this increases the efficiency of waterborne transport. Besides the maritime ISO containers (for international/intercontinental trade), use of long, pallet-wide ILUs (stackable swap-bodies) is expected to increase in inter European trade.	**Cargo** Cargo does not need to be of uniform size or weight, nor is its deck location essential. However, wheeled cargo is unstackable, which is a disadvantage. Transportation of non-standard cargo (in size and weight) is also possible. Semitrailers optimized for sideloading/unloading of pallets is easier and cheaper to use than usual front-loading of other ILUs. Possibility to carry a variety of cargo types and ILUs is also an advantage.
Transhipment and Port facilities Vertical Lo-Lo transhipments are expensive. For best results dedicated transhipment equipment is necessary. Container terminals should have access to railway and road infrastructure (trimodal container terminals). Nevertheless, the main problem on the Danube is an insufficient number of containers required to amortize the investment needed for the dedicated equipment of any type.	**Transhipment and Port facilities** Horizontal Ro-Ro transhipment is simple and cheap. Transhipment equipment (the ramp) usually belongs to the vessel. Ro-Ro terminals are large parking places and are, therefore, minor infrastructural investment.

(continued)

Table 3.5 (continued)

Pros & Cons for **Container** vessels	Pros & Cons for **Ro-Ro** vessels
Vessels Danube container vessels can transport two and occasionally three container layers only, due to low water depth and insufficient bridge heights (on the Upper Danube). Only ILUs with a max width of 2.5 m can be stowed in the standard 11.4 m beam vessels - this is an increasing hindrance to use of present vessels.	**Vessels** Ro-Ro vessels are inherently low draught vessels ($T < 1.7$ m, so navigation during dry seasons is enabled) with large flush deck-area. Larger vessels with unaccompanied, point-to-point service (driver does not accompany transport), equipped with own transhipment equipment (the ramp), are more efficient than alternative types of Ro-Ro vessels (accompanied, bus-stop service).
Transport costs Waterborne part of transport is cheap (due to the stackability of ILUs) but transhipment costs might be higher than expected, presently depending on the ILU type used. Roughly in IWT, 2 container layers are necessary to achieve the breakeven and at least 3 layers are needed for sustainable profit.	**Transport costs** Cost of Ro-Ro transport, compared to the long-distance trucking (the main competitor) might be very attractive along the Danube corridor since origin/destination distances are often more than 1000 km. Ship travel time is not much longer than road-time. There are other benefits (compared to the road transport mode) which are not expressed in monetary terms. Due to low transhipment costs, the overall transport costs might be lower than expected.

Source Radojčić (2006)

place—the question is when. The Ro-Ro technology may be deployed immediately, until the containerization and intermodality become the norm, not only in the Danube corridor itself but also in its hinterland. Furthermore, taking into account that Ro-Ro vessels are also universal vessels, their usability and efficiency look very promising throughout their lifetime.

The Ro-Ro technology, however, is not in the book's focus and hence will not be discussed further; for more information see MUTAND (2002) and CREATING (2006) studies.

3.5 Concluding Remarks

The dramatic difference between sea and IWT, already mentioned in Chap. 1, should be stated again:

(a) Sea vessels have no competition (without them international trade is impossible).
(b) IWT has very strong competition from alternatives (railway and road transport).
(c) IWT is more constrained by ecological considerations as rivers flow through the mainland.

The land transport modes "dictate" the cost of transport, but also the type of intermodal loading units that should be used. So, if standard ISO containers (TEU, FEU) work well on the sea, that does not yet mean they will be so competitive in IWT. IWT should adapt itself to other modes and hence standards that are broadly used in Europe; in particular to the pallet-wide containers (EILUs) which are just 6 ÷ 16 cm wider than the usual sea containers. This small difference of only a few centimetres, poses problems in IWT, resulting often in un-competitiveness compared to land transport modes.

IWT of containers is inherently a more efficient intermodal solution (than Ro-Ro) due to their stackability and stowability. Nevertheless, given that the Danube is not fully regulated and that it has shallows on several sectors, container vessels with lower carrying capacity (than on the Rhine, for instance) should be considered as a feasible solution. Barge trains with partly loaded barges can give good results in restricted water depths too. In any case, a sufficient number of containers is necessary for successful IWT, and that depends very much on the development of the regional economy along the Danube corridor.

Efficient cargo handling (transhipment), not only in the ports/hubs but in the hinterland also, is essential for successful intermodal transport. In other words, development of the hinterland, transport infrastructure, knowledge about IWT possibilities, EST, logistics etc. are decisive factors for the application of intermodality. These explain why containerization is fully accepted on the Rhine, while the Ro-Ro technology seems to be more appropriate for the Danube.

References

Creating (2006) WP2, Concepts to Reduce Environmental Impact and Attain Optimal Transport Performance by Inland Navigation

MUTAND (2002) Multimodal Ro-Ro transport on the Danube river. DPC & DST project

Radojčić D (2005) Danube intermodal ships—container vs. Ro-Ro. The ship in intermodal traffic, 26th Duisburg Colloquium, Duisburg

Radojčić D (2006) New opportunities on the Danube corridor. Ro-Ro 2006, Ghent

Radojčić D (2008) On engineering greener logistics on inland waterways. PP Presentation, Ro-Ro 2008, Gothenburg

SPIN (2003) Intermodal loading units. Working paper for WG3

Žigić B (1999) Bison G (1999) Barriers to higher integration of inland navigation into inter-modal transport chains and some ideas how to overcome them. RINA Int Conf on Coastal Ships and Inland Waterways, London

Chapter 4
Waterborne Transport—Representative Ship Types

4.1 Background

According to statistical data for IWT on the Danube (Sect. 1.1.3) follows that around 80% of goods are transported by pushed convoys and 20% by selfpropelled cargo vessels. Concerning the fleet structure, 75% are vessels for dry cargo, 6% tankers and 19% are pushboats and tugs. The majority of these vessels were built between 1960 and 1990. The share of selfpropelled vessels, however, is slowly increasing due to (a) decommissioning of old barges and pushboats, and (b) acquisition of pre-owned selfpropelled vessels from the Rhine Corridor.

Few typical vessel types and their important technical characteristics will be reviewed in this section. In particular, attention will be paid to (a) *pushed convoys* (i.e. *pushboats* and *barges*) and (b) *selfpropelled* vessels, both for dry cargo, as the liquid cargo transport is out of the scope of this book. Concerning the design of new dry cargo vessels, transport of containers should also be taken into account, although containerization is not (yet) sufficiently developed along the whole Danube corridor.

Development of Pushing Technology

It should be noted that although the Rhine and Danube fleets are relatively old, the pushed convoys (i.e. barges and pushboats) are relatively new. Namely, pushing technology was copied from the Mississippi River at the end of the fifties on the Rhine and beginning of the sixties on the Danube. It gradually replaced traditional towing technology (towboats and towed vessels/barges) as it is more efficient. The first pushboat on the Rhine Wasserbüffel was launched in 1957 (2×630 HP, $L =$ 36 m, $B = 8.2$ m, $H = 2.5$ m; barges of 1300 t, $L = 64$ m, $B = 9.2$ m, see Heuser and Renner 2007). On the Danube in Yugoslavia, now Serbia, in 1961 a towboat Kosmaj was converted to a pushboat[1] (see Fig. 4.1); the first newly-built pushboat

[1] Similarly as on the Rhine, i.e. before the first pushboat Wasserbüffel was built, a towboat President Herrenschmidt was converted to a pushboat.

D. Radojčić et al., *Design of Contemporary Inland Waterway Vessels*,
https://doi.org/10.1007/978-3-030-77325-0_4

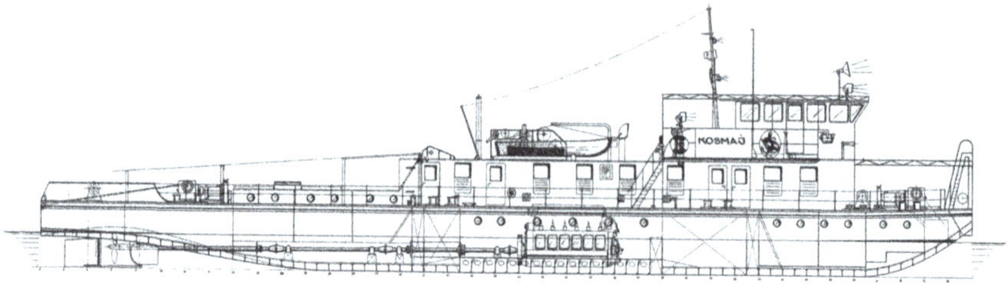

Fig. 4.1 The first converted pushboat on the Danube Kosmaj (2 × 370 kW, $L = 39$ m, $B = 7.6$ m, $T_{max} = 1.4$ m)

Fig. 4.2 The first newly-built pushboat on the Danube Kablar (2 × 220 kW, $L = 22.1$ m, $B = 6.3$ m, $T_{max} = 1.4$ m, built 1961, decommissioned 2012)

was Kablar, Fig. 4.2 (MF 2016). Soon after, the pushing technology was introduced in other Danube corridor countries. The pushing technology nowadays dominates the Danube waterway.

The main advantage of pushing versus towing is that less power is needed for pushing, which may be explained through the use of so-called *convoy formation coefficient—CFC*. Namely, $CFC = R_T/\Sigma R_i$, where R_T is the total resistance of barge convoy and R_i is individual resistance of each barge in a pushed formation. So, *CFC* is always less than one and is between 0.65 and 0.85 (lower values are for slender barge convoy configurations) and is around 0.75 for a typical coupling train. *CFC* for towing configurations is a bit higher, or in other words, with the same power pushing speed is around 10% higher than the towing speed. Moreover, steering of towed barges was often necessary, requiring extra manpower onboard, so overall savings maybe even up to 50%. Nevertheless, the towing technology, although obsolete, was never quite abandoned on the Danube as it has some advantages, particularly during dry seasons when the water level is low (towing tugs have smaller draught than contemporary pushboats).

EU Project PROMINENT

Within the project PROMINENT (see PROMINENT 2016) the Danube fleet, as an important part of the European fleet, was examined. A macro model of this fleet was created and this was considered to be a starting point for the introduction of new technologies. Several databases were combined and good insights of number and share of vessels in various fleet families were obtained.

4.1.1 ECE Classification of European Inland Waterways

The Danube is European inland waterway class IV and higher (classes below IV are waterways of regional or national importance), see Sect. 2.4.1. This determines the maximal dimensions of vessels which can operate on the Danube, i.e. restricts the length and width of vessels and convoys, as well as their minimal water and air draughts. The maximal dimensions of IW vessels and convoys which can navigate the waterway of the respective class according to ECE classification of European inland waterways are given in Tables 4.1a (motor cargo vessels) and 4.1b (pushed convoys).

4.2 Pushed Convoys

There are two pushed convoy (or pushed train) types: (a) *Push convoy* (pushboat + barge/barges), Fig. 4.3, and (b) *Couple convoy* or coupling train (selfpropelled ship + barge/barges that may be pushed or coupled on the ship's side), Fig. 4.4. Coupled convoys are in general more efficient than pushed convoys, but in the Danube region, the selfpropelled vessels (or motor cargo vessels) rarely push more than one barge. In both cases the barges are firmly connected by a wire/steel rope to the pushing/coupling unit and if more than one, between themselves too. Manually driven coupling winches are used for tensioning the wire ropes, see Fig. 4.5. The mooring equipment used on the Danube is standardised.

Coupling between the selfpropelled ship and a barge may be flexible, i.e. articulated with hydraulic cylinders. This device enables curving of the coupling convoy during navigation through sharp and narrow river bends, see Fig. 4.6 (Žigić 2006). A disadvantage of flexible hydraulic connection is an additional cost and a lack of universality concerning the usage of any barge, as both, the ship and a barge have to be equipped with this device. Nevertheless, encouraging results were obtained on the river Main.

Table 4.1 ECE classification of European inland waterways
(a) Motor cargo vessels

Motor cargo vessels					
Type of vessel: general characteristics					
Waterway class	Max. length L (m)	Max. width B (m)	Draught d (m)	Deadweight T (t)	Min. height under bridges H (m)
IV	80-85	9.5	2.5	1,000-1,500	5.25 / 7.00
Va	95-110	11.4	2.5-2.8	1,500-3,000	5.25 / 7.00 / 9.10
Vb	95-110	11.4	2.5-2.8	1,500-3,000	5.25 / 7.00 / 9.10
VIa	95-110	11.4	2.5-2.8	1,500-3,000	7.00 / 9.10
VIb	140	15.0	3.9	1,500-3,000	7.00 / 9.10
VIc	140	15.0	3.9	1,500-3,000	9.10
VII	140	15.0	3.9	1,500-3,000	9.10

(b) Pushed convoys

Pushed convoys						
Type of convoy: general characteristics						
Waterway class	Formation	Length L (m)	Width B (m)	Draught d (m)	Deadweight T (t)	Min. height under bridges H (m)
IV		85	9.5	2.5-2.8	1,250-1,450	5.25 / 7.00
Va		95-110	11.4	2.5-4.5	1,600-3,000	5.25 / 7.00 / 9.10
Vb		172-185	11.4	2.5-4.5	3,200-6,000	5.25 / 7.00 / 9.10
VIa		95-110	22.8	2.5-4.5	3,200-6,000	7.00 / 9.10
VIb		185-195	22.8	2.5-4.5	6,400-12,000	7.00 / 9.10
VIc		270-280	22.8	2.5-4.5	9,600-18,000	9.10
	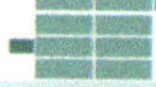	195-200	33.0–34.2	2.5-4.5	9,600-18,000	9.10
VII	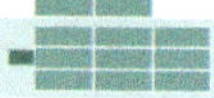	275-285	33.0-34.2	2.5-4.5	14,500-27,000	9.10

Note: Min. height under the bridges of 5.25 m, 7.00 m and 9.10 m is required for 2, 3 and 4 container layers respectively (assuming that approximately 50% of containers are empty, otherwise, ballasting is necessary).

Source via donau (2019)

Possible vessel formations with up to three barges are depicted in Fig. 4.7; for larger pushed convoys see Table 4.1. Note that for the upstream navigation, convoys are longer and narrower to reduce resistance; downstream navigation requires shorter convoy formations that enable better manoeuvrability and stopping ability. The difference in total resistance (R_T) between narrow and wide barge formations are significant. For instance, for Europe IIb barge formations[2] $2+2+2$ versus $3+3$ (corresponding to ECE classification VIc, see Table 4.1) with a payload of 10,500 tdw, the difference in R_T is around 45%. For the analogous smaller formations of 2 and 3

[2]Draught 2.5 m, water depth 5 m and convoy speed 12 km/h—may be considered as typical for the Danube.

Fig. 4.3 Push convoy consisting of 8 tank barges and a pushboat—about 265 × 33 m, 12,000–15,000 tdw, depending on the draught. *Source* JRB

Fig. 4.4 Typical Danube coupled convoy, selfpropelled vessel Pocerina with Danube barge

barges (with 3500 and 5250 tdw) differences in R_T are even 120% and 95% respectively, in favour of narrower formations (see Bilen and Bilen-Katić 1997). Optimization of convoy formations, push speed etc. will be discussed later in Sects. 6.3.3 and 8.7.2.

Large push convoys, particularly on the Middle and Lower Danube, consist often of 9, 12 or even 16 barges. When oil was transported, before the pipelines were built along the Danube corridor, the convoys were often huge,[3] transporting more than 20,000 t, see Fig. 4.8.

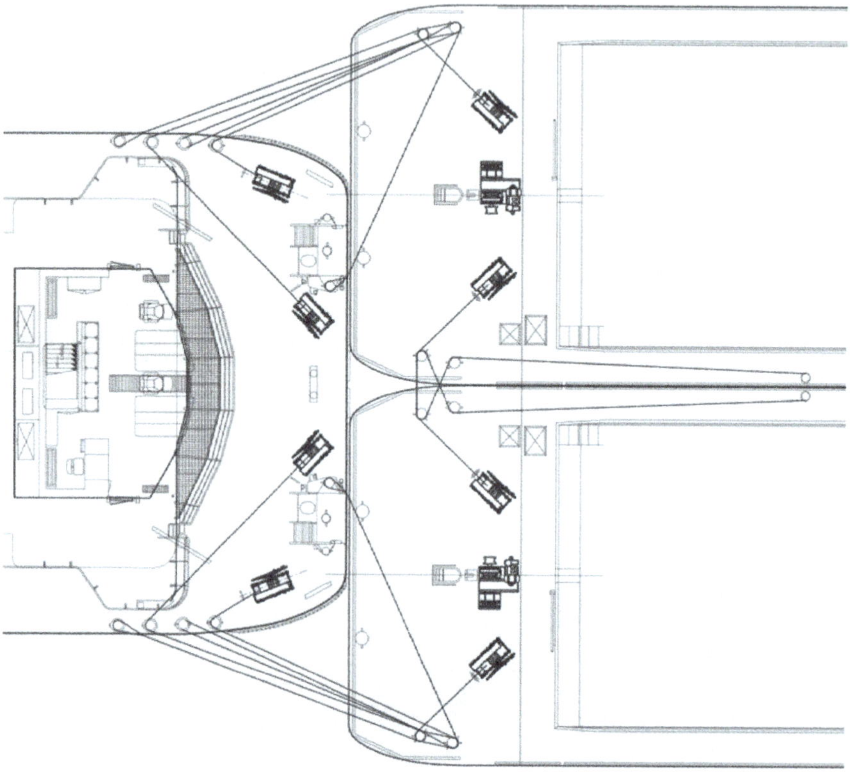

Fig. 4.5 Mooring equipment (pushboat pushing two barges). *Source* Kooiman (2020)

a) b)

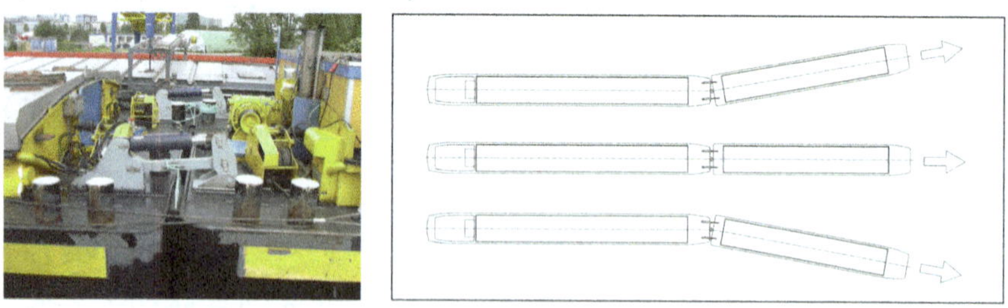

Fig. 4.6 **a** Hydraulically operated flexible coupling, **b** schematic presentation of hydraulically operated flexible coupling. *Source* Žigić (2006)

[3]Mississippi record is 60 coal barges, totaling around 67,000 t! On the Mississippi, which is wider and deeper, large barge convoys (tows as called in the USA) consisting of even 40 barges (around

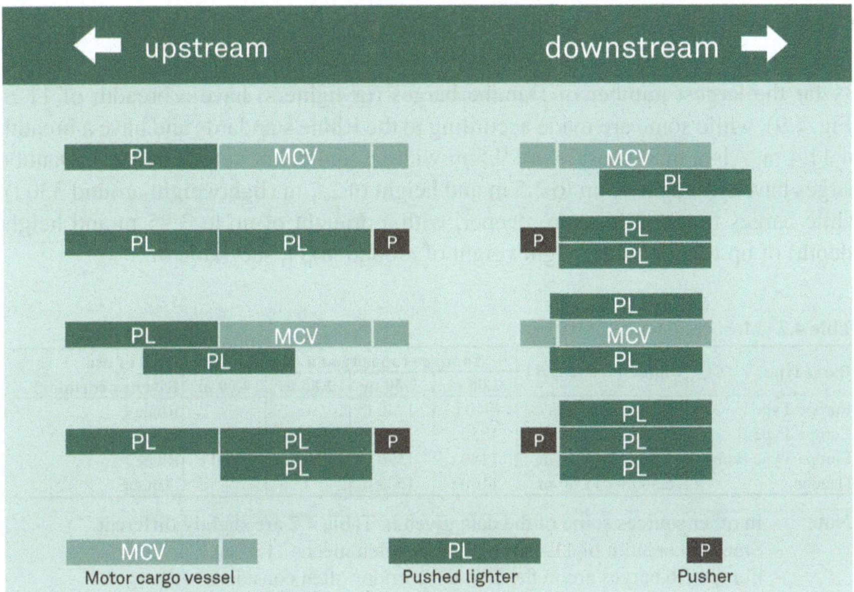

Fig. 4.7 Possible vessel formations. *Source* via donau (2019)

Fig. 4.8 One of the larger convoys on the Danube—upstream sailing, load 20,500 t, tank barges, pushboat Kumanovo 3300 HP, Danube km 970, 1974. *Source* MNT (2018)

370 × 60 m excluding the push boat) are not unusual. Tows are sometimes expressed in acres (unit of land measurement).

4.2.1 Barges

By far the largest number of Danube barges (or lighters) have a breadth of 11 m (Fig. 4.9), while some are made according to the Rhine standards and have a breadth of 11.4 m. Also, many barges are 9.5 m wide. Usually, but not necessarily, Danube barges have a draught of up to 2.5 m and height of 2.7 m (lightweight around 330 t), while barges for the Rhine are deeper, with a draught of up to 3.95 m and height (depth) of up to 4 m, with a lightweight of around 400 t, see Table 4.2.

Table 4.2 Most common river barges

Barge type	Dimensions (L x B)	Tonnage capacity at a draught of				Area of use
		2.00 m	2.50 m	2.80 m	4.00 m	River or cortidor
Europe Type I	70.00 m x 9.50 m	940 t	1240 t	-	-	Rhine
Europe Type II	76.50 m x 11.40 m	1250 t	1660 t	1850 t	-	Rhine, Danube
Europe Type IIa/b	76.50 m x 11.40 m	1140 t	1530 t	1800 t	2800 t	Rhine
Danube	76.50 m x 11.00 m	1100 t	1500 t	-	-	Danube

Note:
- In other sources some of the data given in Table 4.2 are slightly different.
- Standard breadth of 11.4 m in practice often means 11.45 m
- Europe IIb barges are in the Danube corridor often considered as barges with a side height of 2.7 m, i.e. the Danube barges but with a breadth of 11.4 m.

Source CREATING (2007)

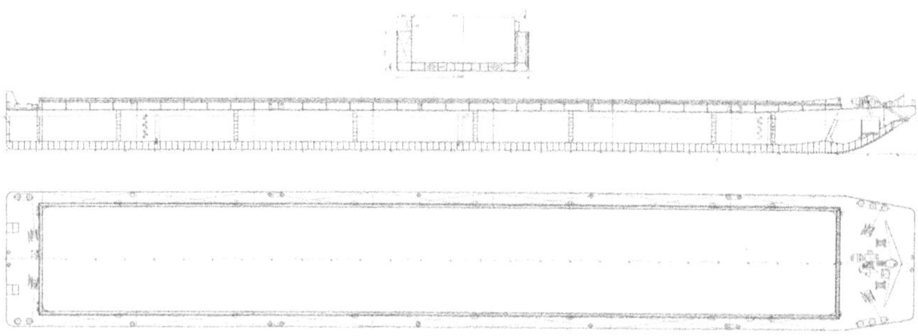

Fig. 4.9 Danube barge for a general cargo of 1600 tdw, $L = 76.5$ m, $B = 11.0$ m (hold width 8.54 m), $H = 3.2$ m, $T = 2.7$ m, lightweight 342 t

Danube-sea barges[4] are 38.25 m long so that two coupled barges correspond to one standard Danube barge of 76.5 m (other characteristics are $B = 11$ m, $H = 3.9$ m, $T = 3.3$ m, corresponding to $m_{DWT} = 1070$ t with a lightship weight of 240 t). Nevertheless, there are several other barge types along the Danube corridor, not to mention the old towed barges still in use. Note however that the Danube barges ($B = 11$ m) cannot accommodate 4 containers abreast. For 4 containers abreast internal hold width of at least 10 m is necessary, see Sect. 3.2.

[4]The Interlighter concept—oceangoing barge-carriers were of Yulius Futcik class.

The difference between Europe II and IIa/IIb is in the side height (3.5 vs. 4 m respectively), and between IIa and IIb in the bow shape, the last one having bit better resistance characteristics (see Heuser 1985). Further development are barges with a side height of 2.7 m (relevant for the Danube) which are designated as Europe IIc barges (Müller 2003). The draught of empty barges is 50÷60 cm and immersion is about 8 t/cm, while the block coefficient for all mentioned barges is around 0.93÷0.95. Moreover, there are double-ended barges and those with the stump-end stern (vertical, flat transom, see Fig. 4.9), which are better for packing in the convoys. Stern-to-stern or "stump" connection results in smaller resistance of a convoy.

The abovementioned dimensions of all types of barges were not changed for last 50 or so years, as they follow from the standard European lock chamber size (see Sects. 2.4 and 3.2). Examining possibilities for improvements, two types of Large Barges (LB—named Europe III barges) were developed from Europe IIb barges (Heuser and Renner 2007). Their dimensions are 100 × 14 m and 100 × 15 m, both with a side height of 4 m. The second one was developed for container transport and on draught of 2.5 m (interesting for the Danube), it can accommodate 159 TEUs of 17 t (on full draught of 3.5 m even 242 TEUs). Two convoy formations consisting of 2 + 2 and 2 + 2 + 2 Europe IIb barges were compared with, respectively, 1 + 1 and 2 + 2 LBs carrying a comparable load. In both cases, LBs showed to be superior from the viewpoint of transport efficiency (t km/kWh) and specific power requirement (kW/tdw). Large barges for the Danube (DLB) for a payload of 2490 tdw (lightweight is around 450 t) are mentioned in the IDV (2014) project. DLB's dimensions are 80 × 15 m and have a design draught of 2.7 m.

Summarizing, three main barge breadths exist, 9.5, 11 and 11.4/11.45 m. On the Danube 11 m breadth is dominant. Barge lengths are less important, there is a variety of them in service, but 70 and 76.5 m dominate. Employment of larger barges would be profitable but it seems that there is not (yet) need for that.

4.2.2 *Pushboats*

Long-range and *harbour* pushboats (or pushers)[5] were built on the Danube and most of them had two propellers, but large pushboats with three propellers were not rare. Besides the draught restriction, Danube pushboats generally differ from those on the Rhine as they have more accommodation space due to larger crews that work in shifts (the Danube is a much longer river than the Rhine, i.e. more time is spent sailing).

[5] In the USA the pushboats are called towboats as they push the tow of barges. They may be up to 60 × 17 m having over 8000 kW. Typical barges are a bit shorter than in Europe, i.e. 61 × 11 m.

Typical Danube Pushboats

During the 1970s,[6] after some experience was gained, a kind of standard or recommendation emerged in (then) Eastern Bloc shipping companies concerning the long-range Danube pushboats. Accordingly, the pushboat should have the following characteristics: up to 2 × 1000 kW, a length of around 35 m, a breadth of 11 m (like Danube barges) and a draught of less than 1.9 m with a displacement around 400–500 m^3. These pushboats were built in series, in all Danube countries downstream of Austria. A typical example is a pushboat Brest built in the mid-1980s in Serbian shipyards, now Ukrainian vessel (class Valentin Solovljev) shown in Fig. 4.10.

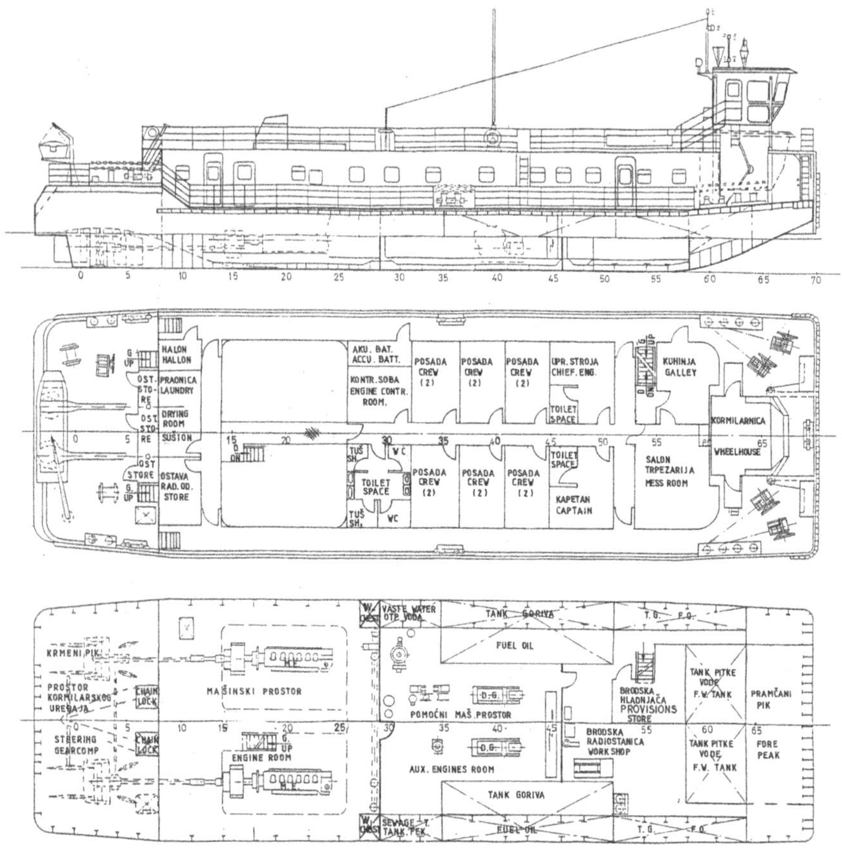

Fig. 4.10 Pushboat Brest—2 × 835 kW, $L_{OA} = 37.25$ m, $B_{max} = 11.30$ m, $H = 2.7$ m, $T = 1.8$ m, $V = 13$ km/h with 6 barges of 1750 tdw, autonomy 360 h, crew 16

[6]By far the most of the pushboats now in use on the Danube were built in the 1970s and 1980s (see DC 2016), thus this period merits to be mentioned.

Note that the main engines are placed behind the midship, close to the propellers, which is relatively unusual as it complicates even-keel balancing of a pushboat. Namely, in one hand, the buoyancy at the stern is reduced due to propeller tunnels (thus the tunnels should be short) and in another, a good water inflow to propellers is essential (thus long tunnels are needed).[7] On the other side, considerable weights, that should be counterbalanced by the buoyancy, are positioned at the stern: anchoring equipment with the windlasses, set of common and flanking rudders (2 + 2 per shaftline), steering gears and propellers in nozzles. Therefore, most pushboats have the main engines around the midship or even in the forepart (thus with long shaftlines which may also be problematic). Another peculiarity for a long-range pushboat is a single-deck superstructure that enabled a low air-draught.

Worth mentioning are probably the strongest pushboats on the Danube—Serbian pushboats Karadjordje and Karlovac, see Fig. 4.11. They were designed to push large convoys of 12–16 barges on the Lower Danube. One of the pushboats was equipped with a special system (device) for rudder unloading. The reason for this innovation was that (floating) logs and other debris were often wedged in the nozzles and/or main or flanking rudders, which blocked or damaged other rudders also. So, the purpose of the rudder unloading device was to permit the rest of the rudders (those which were not directly blocked) to remain in function. Although the purpose of this invention sounds logical, due to the poorly-developed mechanism (prototype) and the need for frequent interventions, the unloading device was soon replaced by the conventional connecting rod system. The second reconstruction consisted of removal of the middle powertrain (shaftline) thus only two propellers/engines were left operational.

Fig. 4.11 Pushboat Karadjordje—3 × 1280 kW, $L_{OA} = 40.45$ m, $B = 13.0$ m, $H = 2.8$ m, $T = 1.95/2.15$ m, Speed with 12 barges of 1700 tdw 14 km/h, autonomy 300 h, crew 16

Small pushboats for packing the barges into convoys, moving barges on short distances, for harbour operations etc., were produced in series, e.g. see Fig. 4.12.

[7] See discussion about the tunnels in Sect. 8.2.2.

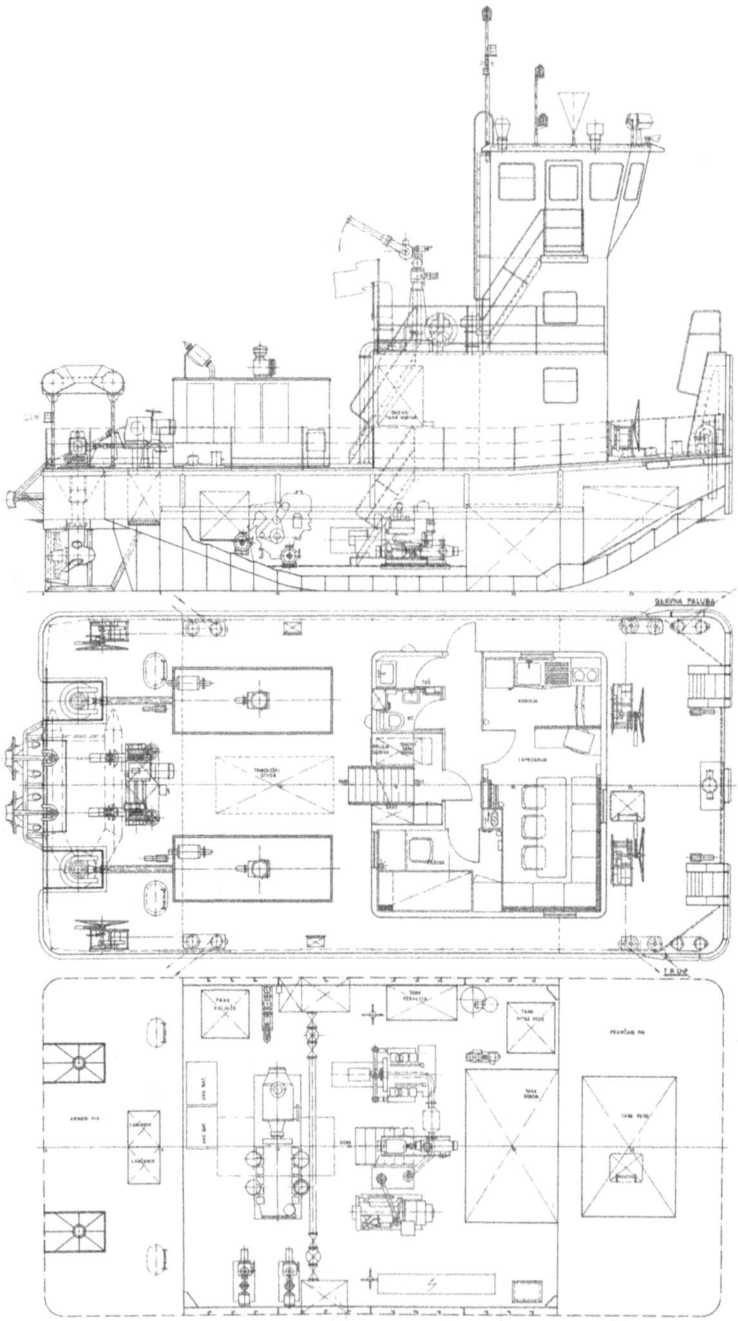

Fig. 4.12 Small (harbour) pushboat—2 × 150 kW, L_{OA} = 15.5 m, B_{OA}= 7.2 m, H = 2.5 m, T = 1.5 m

In a particular case, good manoeuvring capabilities were accomplished with two deck-mounted rudder-propellers.

Several pushboats built in the last century were retrofitted in the meantime. In the first place, the old engines were replaced with the new ones, which significantly reduced their fuel consumption (probably by some 20/30%) and thus prolonged their life.

Romanian pushboats of Mercur class are probably the most modern ones on the Danube. Several vessels were retrofitted, and notable are Series 200 (Fig. 4.13) and Series 300. The second is bigger: two decks, 2 × 1200 kW, $L = 34.6$ m, $B = 11$ m, $H = 2.8$ m, lightweight around 500 t. Note that within the same series different engine brands were installed so data vary.

Fig. 4.13 Pushboat Mercur Series 200, 2 × 900 kW, $L_{OA} = 34.6$ m, $B = 10.1$ m, $H = 2.65$ m, $T_{min/max} = 1.65/1.77$ m, lightweight around 400 t. *Source* NAVROM (2020)

State-of-the-Art Design from the Rhine

Pushboat Veerhaven IV (see GA plan, Fig. 4.14) is delivered in 2013 and is the last in the series of several successful Veerhaven pushboats built for ThyssenKrupp Veerhaven B.V. It is intended to push 6 loaded barges on the Rhine (route Rotterdam-Duisburg). The builder is a Dutch shipyard Kooiman Marine Group and main particulars that follow are taken from Kooiman (2020).

- Main dimensions: $L_{OA} = 40.0$ m, $B_{OA} = 15.0$ m, $H = 2.75$ m, $T = 1.74$ m with 50% consumables.

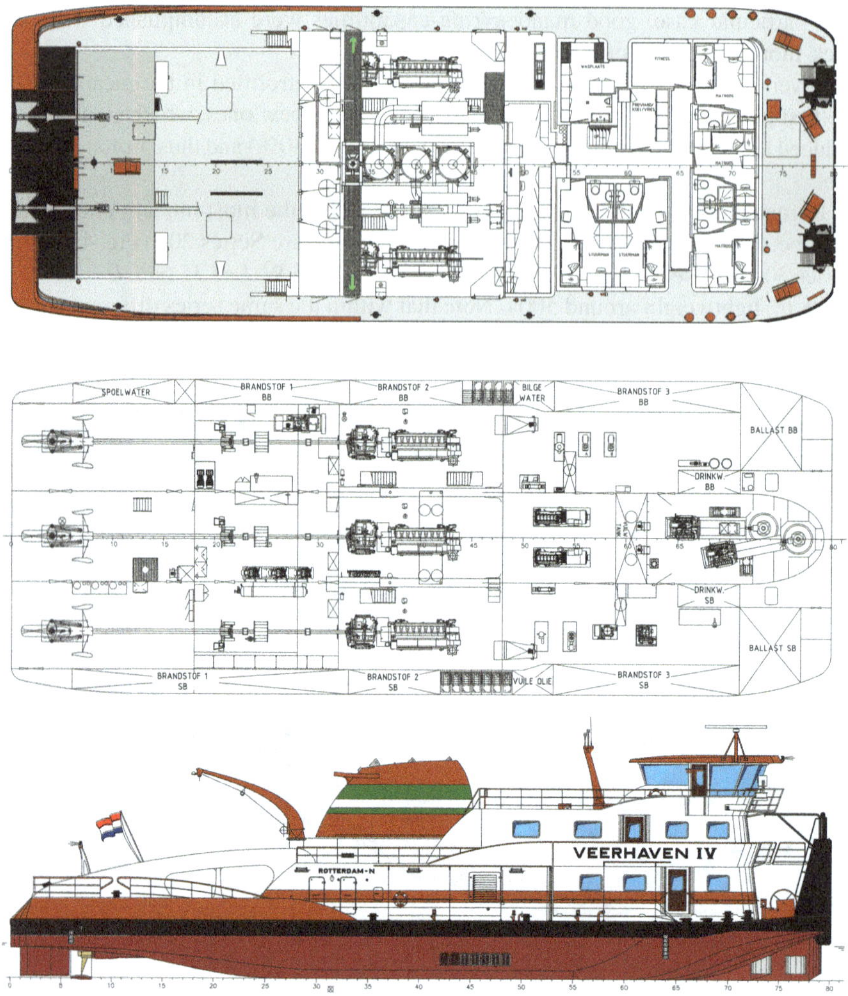

Fig. 4.14 Pushboat Veerhaven IV—3 × 1380 kW, $L_{OA} = 40.0$ m, $B_{OA} = 15.0$ m $H = 2.75$ m, $T = 1.74$ m with 50% consumables. *Source* ThyssenKrupp Veerhaven B.V.

- Triple screw propulsion system (in total 4140 kW) with three nozzles and three fishtail rudders. Propellers 5-bladed, $D = 2.05$ m in HR type nozzle.
- Two three-channel bow thruster units, each driven by a Diesel engine of 441 kW.
- Auxiliary engines 2 × 211 kW (gen. 240 kVA) and 1 × 95 kW (gen. 100 kVA).
- Accommodation is located in the two deck superstructure and consists of 9 cabins, fitness room, wet room, mess, galley, wheelhouse and control room. The accommodation is spring mounted on pneumatic shock absorbers.
- Other notable features: one-man watch operation, economic cruise control (enable fuel saving 2/5%), anchors can be dropped from the wheelhouse etc.

4.3 Selfpropelled Vessels

There are several ways to classify dry cargo, selfpropelled vessels, as for instance:

- According to the type of cargo they are designed to transport. Those for the *bulk cargo* (gravel, sand, coal etc.) usually have hold shape with the hopper sides (Fig. 4.4). *General cargo* ships (which are of main interest in this book) have vertical hold sides; those which are container-wide should also have a hold width for compact abreast stowing of standard containers (see Sect. 3.2). Tankers are out of the book's scope.
- Classification based on the *main dimensions* is usually presented by common European ship types and correspond to ECE classification of IWW, see Table 4.3. Note that the large cargo vessels of GMS 135 and Jowi type do not exist nowadays on the Danube; these are the Rhine vessels, but except the draught restrictions, there are no other technical reasons for them not to be present on the Danube too.
- Classification based on the range of operation (similarly to the pushboat classification), i.e. *long-range* and *short-range*, e.g. see Figs. 4.15 and 4.4 respectively.

In this chapter only the following two types will be further examined:

- Selfpropelled Lo-Lo vessels (general cargo, long-range, ECE classification IV and above) and
- Selfpropelled Ro-Ro vessels (long-range).

Table 4.3 Typical European vessel types (Data from: ECCONET (2012), VBW 2016)

Vessel type	max L x B (m) approx. P_B (kW)	Draught T (m)	Payload approx. (t)	ECE classif.
Gustav Koening (extended version)	80 x 8.2 580	2.5 1.1	1100 250	III
Johan Welker (Europe type)	85 x 9.5 750	2.6 1.2	1400 300	IV
Large cargo vessel (GMS 110)	110 x 11.45 2000	3.5 1.35	2900 400	VIa
Large cargo vessel (GMS 135)	135 x 11.45 2000	3.5 1.35	3800 670	VIb
Large cargo vessel (Jowi type)	135 x 16.8 3000	3.5 1.6	5200 1300	VIb

$L_{OA} = 95$ m
$B_{OA} = 11.4$ m (11.0 m)
$H = 3.2$ m
$T = 2.7$ m

Highest fixed point 6.5 m above basis line
Cargo capacity 1960 t (1634 t)
$P_B = 2 \times 600$ kW (2×770 kW, V=18 km/h)
Bow thruster 130 kW.

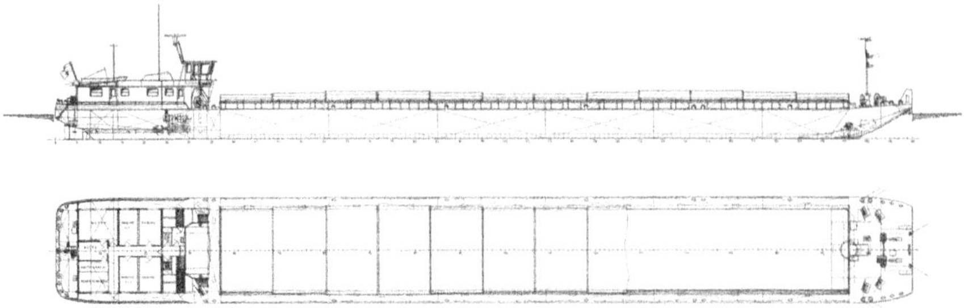

Fig. 4.15 General arrangement of MGSS Jochenstein

4.3.1 *Selfpropelled Lo-Lo Vessels*

Danube Vessels

A very successful selfpropelled *general cargo and container* vessel on the Danube is of the class MGSS Jochenstein (Fig. 4.15). These ships were built in the Osterreichische Schiffswerften AG in Linz-Korneuberg more than 30 years ago for German (10 vessels) and Soviet—now Ukrainian—shipping companies (15 + 4 vessels). The Ukrainian ships' class name is Kapitan Antipov and Kapitan Širkov; their particulars, if different from Jochenstein, are given below in the brackets. Note that Jochenstein class has a cargo hold width of 8.8 m, thus like the Ukrainian ships do not have enough space for stowing 4 containers abreast. The Ukrainian ships have a crew of 12 as are intended for long voyages along the whole Danube. These vessels belong to the large cargo vessels type (GMS 110), see Table 4.3.

Rhine Corridor Vessels

On the river Rhine, and the Rhine Corridor in general, a large variety of general cargo ships exist. Most of them are customized for container transport too, which is not the case on the Danube. The largest container ships are of Jowi type and probably the smallest of Hopper Class, for 6 and 2 containers abreast respectively (Figs. 4.16 and 4.17). Note however that a large variety of standard length and/or breadth combinations exist beyond the typical vessel types given in Table 4.3. For instance, vessels with a breadth of 9.5 m can have a length of 70, 80, 86, 105 or even 110 m. Or with $B = 11.45$ m can be of $L = 83, 86, 105, 110$ or 135 m (in this respect see Sect. 8.7). Common for all of them, and particularly for the newer ones, is container space optimization, i.e. increase of container capacity, see Figs. 4.18 and 4.19. The second figure is indicative as it gives a vivid comparison of conventional GMS 110 type cargo vessel, with the recently built ship which has several technological

Fig. 4.16 Jowi—$L = 135.0$ m, $B = 17.4$ m, 500 TEU, 3200 kW. *Source* Müller (2003)

Fig. 4.17 Amer Hopper—$L = 86$ m, $B = 7.03$ m, $H = 2.86$ m. *Source* Mercurius

innovations (both have the same dimensions $L = 110$ m, $B = 11.45$ m). The innovative ship has better stowing capacity thanks to the electric propulsion (see Sendo Liner, Sect. 10.2.2) as she can accommodate longitudinally 14 instead of usual 13 TEUs.

Note, however, that recent trend is towards scale enlargement (see Sect. 1.1.3 ad 'i' and 'j'). This raises a question about the upper limits of IW vessels. Present limits

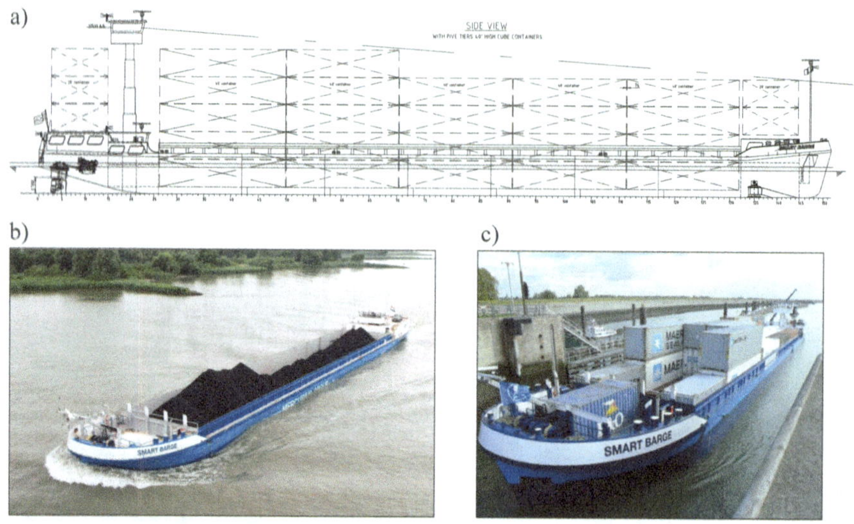

Fig. 4.18 Ship smart barge $L = 86$ m, $B = 11.45$ m, $H = 4.25$ m, $T_{max} = 3.2$ m, $P_B = 2 \times 500$ kW. *Source* Mercurius; **a** GA plan (*Note* 3 × 3 TEUs in the fore end and on the top of the accommodation superstructure); **b** Smart Barge carrying bulk cargo (coal); **c** Smart Barge carrying containers

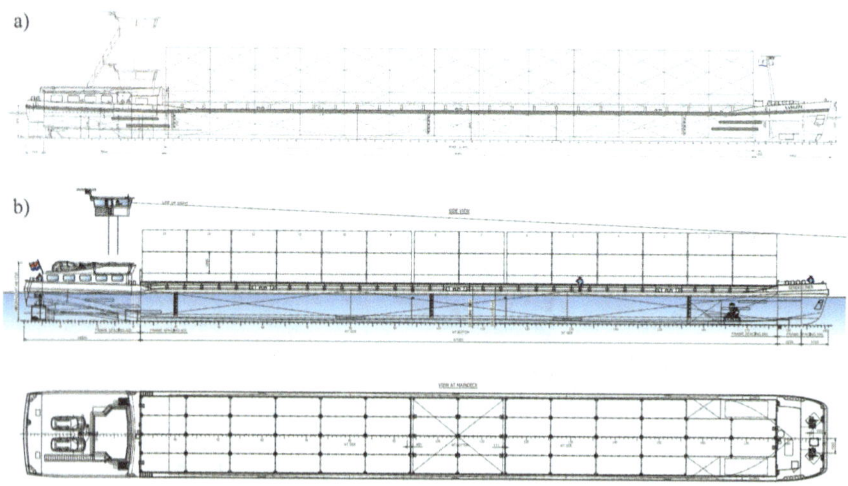

Fig. 4.19 **a** Container ship Project 821, $L = 110$ m, $B = 11.45$ m, $T = 3.55$ m (diesel propulsion). *Source* Concordia DAMEN; **b** Container ship Sendo liner, $L = 110$ m, $B = 11.45$ m, $T = 3.7$ m (electric propulsion). *Source* Concordia DAMEN

are determined by CEMT classification (Table 4.1a; classes below IV are ignored). The length is limited to 85, 110 and 140 m, and the breadth to 9.5, 11.4 and 15 m, for CEMT classes IV, Va and VIb, respectively. Other regulations, which also have to be obeyed, are those issued by CCNR, which already allowed length extension for the 'indivisible vessels' from 110 to 135 m (that, in a way, overruled CEMT's class VIb of 140 m). Amongst other issues, they define minimum crew member based on ship length. Summarizing, by far the largest percentage of IW ships nowadays, already pushed their dimensions to the maximal limit, concerning the length and breadth. This parallels to size classification of the seagoing ships (e.g. Panamax, Suezmax, Aframax etc.) or to "paragraph ships" that take advantage of a 'paragraph' of some regulation.

The selfpropelled ship Mercurius Amsterdam (and the larger Transferium, data given in parenthesis) should be mentioned as it is the container ship equipped with its own transhipment equipment at the expense of a bit smaller container capacity, see Fig. 4.20. This vessel is suitable for short-haul container transport and transhipment directly onto the quay (no need for port or hub). Crane's lift capacity is 30 t/30 m (40 t/30 m), transhipment ~20 (22) TEU/h.

Besides 4 containers abreast (Figs. 4.18, 4.19 and 4.20), smaller ships with 3 containers abreast are also very common, see Fig. 4.21. As the recent trend is to purchase the second-hand vessels from the Rhine Corridor and employ them on the Danube, rather than to build the new ones, probably the first container optimized vessels for the Danube will be of this type.

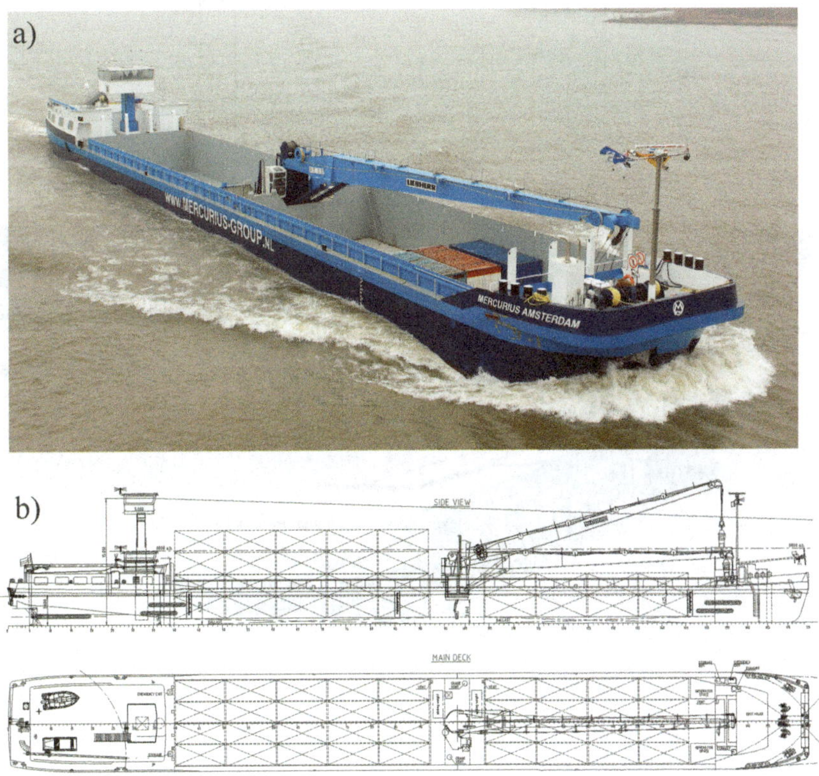

Fig. 4.20 **a** Crane barge Mercurius Amsterdam, $L = 86$ m, $B = 11.5$ m, 144 TEU, $P_B = 2 \times 860$ HP. *Source* Mercurius; **b** GA Transferium, $L = 96.5$ m, $B = 11.5$ m, 164 TEU, $P_B = 2 \times 1500$ HP. *Source* Mercurius

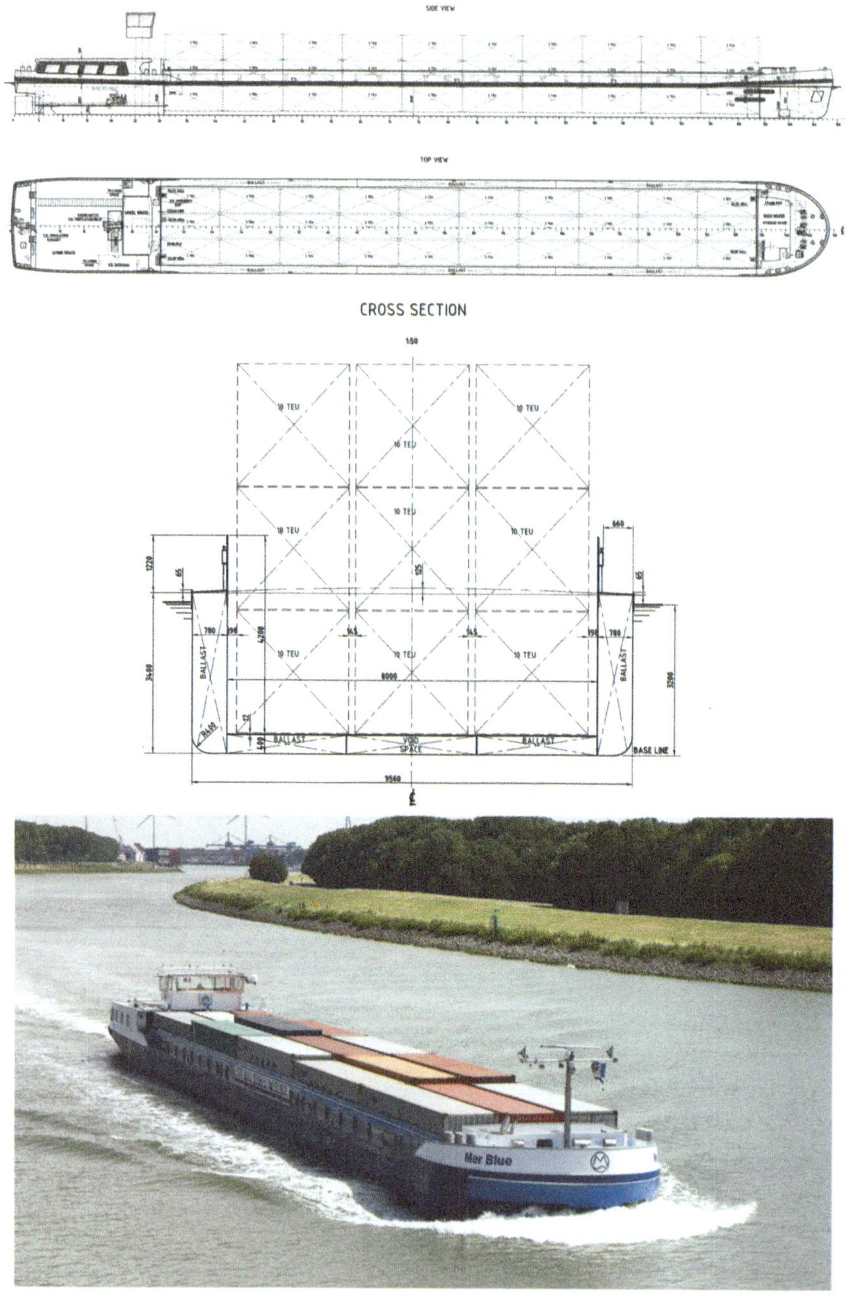

Fig. 4.21 Container ship Mer Blue, $L_{OA} = 84.8$ m, $B_{OA} = 9.6$ m, $H = 3.4$ m, $T_{\max} = 3.2$ m, $P_B = 552$ kW, 90 TEUs in 3 layers. *Source* Mercurius

4.3.2 *Selfpropelled Ro-Ro Vessels*[8]

The unaccompanied Ro-Ro service with point-to-point operation has been functioning very successfully on the Danube for decades. This service transported semitrailers, and occasionally trucks, on the route Vidin-Passau-Vidin (one direction distance 1440 km) with optional stops in Vienna and Linz. Approximately 4500 trailers/year were transported in 90 round trips per year, with a fleet of four dedicated Ro-Ro vessels with a capacity of 50 trailers each. The round trip lasts two weeks but could be shortened to 11–12 days. Two of the abovementioned four dedicated Ro-Ro vessels on the Danube were built in Deggendorf (Fig. 4.22) and were of semi-catamaran hull type, and two (full catamarans) were built in Serbia (Fig. 4.23). Their main characteristics are very similar and are summarized in Table 4.4.

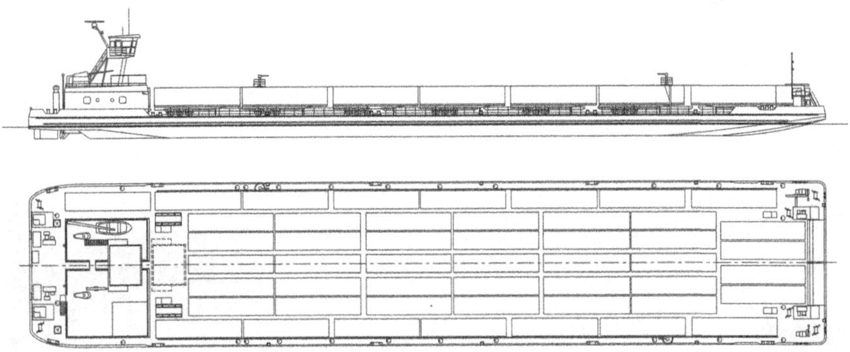

Fig. 4.22 Danube semi-catamaran built in Deggendorf, Germany

The main cargo of the abovementioned Ro-Ro vessels are loaded trucks and semitrailers of a standard weight allowed on the European roads. From a shipping point of view, this is light cargo. Thus, river Ro-Ro vessels with a single flush deck have a very low draught of less than 1.7 m, permitting operation throughout the year (including dry season), which is of utmost importance for successful Ro-Ro service. In other words, properly parked loaded semitrailers onboard a single deck vessel, inherently result in well-balanced ships with a draught between 1.6 and 1.7 m. Larger vessel will have a larger deck area, allowing a larger number of vehicles, but the draught will still be more-or-less the same.

Note however that these vessels were also used for transport of heavy cargo, e.g. electric power transformers of more than 600 t. In this case, their advantage was easy loading and unloading (horizontal, on the wheels) as ports were not equipped with adequate heavy cranes for Lo-Lo transhipment.

There are other Ro-Ro services on the Danube, mainly for transportation of smaller vehicles (newly built cars and vans) onboard double-deck Ro-Ro barges (barge-convoy usually consists of two barges and a pushboat).

[8]The main source of information in this section is Radojčić (2006).

Table 4.4 The main particulars of Ro-Ro vessels shown in Fig. 4.22

Length overall	114.0 m
Breadth maximal	22.8 m
Depth	3.3 m
Draught fully loaded	1.65 m
Deadweight	1530 tons
Cargo capacity	1372 tons (50 road trailers, 28 tons each)
Engine output	2 x 910 kW
Service speed	18 km/h (in calm and deep water)
Crew accommodation for	12 - 16 persons

Note: There are small differences between catamaran and semi-catamaran Ro-Ro vessels

Fig. 4.23 Ro-Ro terminal in Vidin, Danube km 793. *Source* Popov (2005)

The Ro-Ro terminals are equally important (see Fig. 4.23) and are located in all Danube riparian countries. Since such terminals require relatively modest financing (essentially they are just large parking places), they are a minor infrastructural investment and thus can be easily built when and where needed.

As the Ro-Ro service seemed to be a very promising solution for the Danube (see Sect. 3.4), in the MUTAND and CREATING Projects (see MUTAND 2002 and CREATING 2006) the Ro-Ro vessels were further analysed.

4.4 Concluding Remarks

Pushed convoys are by far the best choice for the transport of large quantities of relatively cheap cargo, like bulk cargo (coal, ore, gravel, sand, grain etc.). Due to convoy's capability to:

- transport a different kind of cargo in each barge of the convoy,
- push different kinds of barges/vessels in the same convoy and
- leave out or add one or more barges on the way,

a pushed convoy is a very flexible transport option, in a way comparable to the railway. Note that only the first feature of those above is common for other vessel types.

The selfpropelled vessels are faster and therefore more suitable for container transport (which has to compete with land transport modes, i.e. railway and truck). Design of the selfpropelled vessels for the Rhine and the Danube, particularly the large ones, differs very much. The first ones have a deep draught, thus transportation at reduced draught is a bad option and cannot be profitable. Consequently, the future Danube selfpropelled vessels can't be cloned from the Rhine, they have to be custom designed for the Danube conditions—in the first place for shallow water and long-range navigation.

Partly loaded barges can be the simplest and cheapest answer to dry seasons and restricted draught navigation, taking into account that power needed to push an additional barge (or few of them) rises slightly, while cargo volume can increase rapidly. Therefore, the problem usually poses the draught of a pushboat which cannot be reduced. So, a shallow draught pushboat would be advantageous in these situations.

Furthermore, knowing that the considered Danube vessels are already few decades old, and having in mind innovations on the Rhine, the following vessel types require further attention (excluding tankers which are out of the scope of this book):

- Low-draught, long-range pushboats, and
- Low-draught, long-range and relatively fast selfpropelled dry cargo ships.

Design concepts of both vessel types are analysed in Chap. 10, i.e. of a low draught pushboat in Sect. 10.2.2 and of low draught narrow- and wide-body selfpropelled ships in Sects. 10.3.1 and 10.3.2, respectively.

References

Bilen B, Bilen-Katić B (1997) Modified pusher tug design for the Danube river. Int Shipbuilding Prog 44(438)

CREATING (2006) WP2, Concepts to reduce environmental impact and attain optimal transport performance by Inland navigation

CREATING (2007) WP5, Concepts to reduce environmental impact and attain optimal transport performance by Inland navigation

DC (2016) Danube navigation statistics in 2015–2016. Danube Commission. ttp://www.danubecommission.org/uploads/doc/STATISTIC/en_stat_2015_2016.pdf. Accessed 19 Dec 2020

ECCONET (2012) Deliverable 2.1.1: IWT fleet and operation. Report for European Commission Directorate-General for Energy and Transport, Brussels

Heuser H (1985) Auswirkungen der Leichter-Formgebung auf Widerstand und Leistungsbedarf von Schubverbänden, Wahl von Zwei- oder Dreischraubenschub-booten für den Vortrieb. Binnenschiffahrt und Wasserstraßen No. 11/85

Heuser H, Renner V (2007) 50 Jahre Schubsschiffahrt in Europa. Binnenschiffahrts-Verlag GmbH, Duisburg

IDV (2014) Innovative Danube vessels—main project results. Danube region strategy

Kooiman (2020) Kooiman coupling winches. Speciffication sheet. https://kooimanmarinegroup.com/assets/flyer-a4-coupling-winches-en.pdf. Accessed 28 Dec 2020

MF (2016) Short history of Serbian shipbuilding (in Serbian). University of Belgrade, Faculty of Mechanical Engineering

MNT (2018) Danube in Serbia—voyage through technical museums. Museum of Science and Technology, Belgrade

Müller E (2003) Innovative types of inland ships and their use on the river Rhine, its tributaries and the adjacent canals. VBD Duisburg

MUTAND (2002) Multimodal Ro-Ro transport on the Danube river. DPC & DST project

NAVROM (2020) Mercus series 200. Technical data. https://navromshipyard.ro/portofoliu/mercur-seria-200/. Accessed 28 Dec 2020

Popov I (2005) Das Ro-Ro-Schiff im intermodalen verkehr auf der Donau. 26th Dusburg Colloquium 2005

PROMINENT (2016) List of operational profiles and fleet families: Identification of the fleet, typical fleet families and operational profiles on European inland waterways, Deliverable 1.1

Radojčić D (2006) New opportunities on the Danube corridor. Ro-Ro 2006, Ghent, 2006

VBW (2016) Driving dynamics of inland vessels—vessel behaviour on European inland waterways and waterway infrastructure with special respect to German waterways. ISBN 978-3-939230-49-6 (Online)

via donau (2019) Manual on Danube navigation. http://www.viadonau.org/fileadmin/user_upload/Manual_on_Danube_Navigation.pdf. Accessed 18 Dec 2020

Žigić B (2006) Modernization of the Danube fleet—matching the future requirements. 3rd Danube summit, Budapest

Chapter 5
Overview of Regulatory Framework from a Viewpoint of the Design of Danube Vessels

5.1 Introduction

Naval architects are usually more familiar with rules and regulations for the design of the sea-going ships than with those aimed for inland waterway vessels. In the big picture, there is an analogy between the two—but, actually, significant differences are present. Maritime legislation is much more unified. For decades, rules of various classification societies have been harmonized and in 2006 first Common Structural Rules were published by IACS, i.e. on the UN level. Maritime statutory regulations are also unified on the UN level through the documents of IMO, MARPOL, SOLAS, etc. Due to various reasons, such a level of unification and standardization in the IW vessels domain is still not obtained. Instead, there is a range of national, regional, and international rules in use that are based on different principles and use different tools, methods and criteria (see Bačkalov 2017). In the last decade or so, considerable effort was devoted to the improvement of that on the European level, especially in the field of statutory regulations. This is accomplished through a number of UN and EU resolutions, directives, and standards—an extensive collection of documents that are not always easy to grasp. It wouldn't be wrong to say that naval architects are even a bit startled when facing a (long) list of their acronyms. In Sect. 5.3 most relevant of those documents are listed and their correlation explained. Some deficiencies observed in them are also stated.

The main sources of documents and explanations presented in this chapter are official web sites of UNECE, CESNI, CCNR, DC, BV, DNV-GL, and LR.[1] Discussed issues are subject to continuous enhancement and upgrade. The reader is strongly advised to refer to updated versions of relevant rules and regulations before applying them in any practical task.

[1] See UNECE (2020), CESNI (2020), CCNR (2021), DC (2020), BV (2020), DNV-GL (2020) and LR (2020).

D. Radojčić et al., *Design of Contemporary Inland Waterway Vessels*,
https://doi.org/10.1007/978-3-030-77325-0_5

5.2 Classification

5.2.1 *Classification Societies' Rules for IW Vessels—Short History and State of the Art*

A more comprehensive presentation of this subject can be found in Grgić (2017).

For inland navigation vessels the classification begins to apply much later than for maritime vessels. The inspection of ships and issuing of navigation licenses are carried out mainly by the national authorities. In the international navigable rivers, those are the so-called Basin administrations, the Rhine Commission on the Rhine and the Danube Commission on the Danube (see Sect. 5.3), which mainly deal with navigation regulations. The rules were first made by classification societies in the countries that had a developed inland waterway network, GL, and BV. LR first issued rules for the construction of inland navigation vessels in the early 1980s. Rules have been drawn up at the Rotterdam office, and plans for inland navigation vessels are still being verified in Rotterdam. DNV has never made rules for inland navigation vessels, but that fact became obsolete after the merge with GL in 2013. ABS has rules adapted to US conditions. The specific situation is in Russia where the classification of ships is divided between two institutions: Russian Maritime Register of Shipping (RS) and Russian River Register (RRR). A list of classification societies recognized by the EU is given in Directive 2006/87/EC (see Sect. 5.3).

Classification Societies' Rules focus on the structural strength of the vessel hull and the reliability of its machinery and equipment. It must be emphasized that the class of ship is not yet an obligation prescribed by international regulation and is not mandatory in all European countries. The ship must have the so-called statutory testimony, issued by the administration of the flag state. However, classification societies are, most commonly, involved in that also—as explained in the following section. Furthermore, the naval architects in design bureaus are very much keen to comply with classification societies' rules for better safety and "their own peace of mind"—whether the class for the particular ship is officially required or not.

5.2.2 *Classification Societies Acting on Behalf of Flag States*

Familiarity and deep understanding of various aspects of ships, a wide all-around-the World network of offices, skilled inspectors and surveyors—make most of the classification societies perfect bodies for issuing all kinds of attestations, inspection reports, and assessment of compliance with statutory regulations presented in the following Sect. 5.3. That is why it is most common that flag state administrations empower them to act on their behalf in those affairs. As an example just let us note that, at present, BV acts in this way regularly on behalf of 13 countries, including Belgium, the Netherlands, Germany, France, Poland, Romania (and as a peculiarity, even Brazil).

5.3 Statutory Regulations

As already noted in the Introduction of this chapter, on the European waterways in rule is an extensive collection of national, regional, and international statutory documents. Although some difficulties do exist in their interpretation and fusion in any practical task, more careful examination reveals that they actually are converging, though that process still did not meet its end. Hopefully, the list of most relevant statutory documents given in this chapter will help in understanding that. As a foreword, some general notes on statutory legislation are outlined first.

All countries with significant waterways lengths are to incorporate ES-TRIN standards in their national statutory regulations.

Regional rules (CCNR, DC) are harmonized (and even issued jointly) with UNECE and EC. It should be noted that the significant length of European inland waterways is not entirely EU bounded and the Danube is a prominent example of that. Thus, harmonization has to be done on the UN level also (UNECE taking the role that IMO has in the maritime sector). However, river Rhine is by far the most important and the most intensive Europe‘s inland waterway, with the longest history of respective legislation. (Actually, CCNR founded by Switzerland, Belgium, Netherland, Germany, and France in 1815 stands as the oldest international organization in modern history, see CCNR 2015). Thus a majority of CCNR documents are not only taken as the basis for regulations issued by EC, but even replicated without modification in those of UNECE and DC and, via that, practically implemented in national rules of the non-EU countries also. Since other waterways (Danube) may significantly differ from the Rhine in many aspects, in (limited) number of cases this practice led to some inadequacies.

As noted in Sect. 5.2, the prime focus of all classification societies' rules is on ship strength and definition of minimal hull structure scantlings. They prescribe and provide very detailed procedures for that. That leaves the opportunity for the statutory field to treat those aspects in a reduced manner (in some cases limited to demand that "hull has to be built according to good shipbuilding practice"). On the other hand, statutory regulations may consider some other aspects of ship safety (stability, flooding, fire protection, etc.) in a more comprehensive way. However, it is important to underline that, at present, class is not mandatory for IW vessels in all European countries—but is expected to be in the future.

Follows the list of the most important statutory documents sorted by their issuers and with brief explanation of their contents.

UNECE—United Nations Economic Commission for Europe

The most relevant issues (see UNECE 2020) are:

UNECE 2006Resolution No 61—Recommendations on Harmonized Europe-Wide Technical Requirements for Inland Navigation Vessels. The Resolution classifies European waterways into 3 ZONES (Canals are classified as Zone 4). Design wave height (a significant parameter for ship hull design) is assigned to each zone. However, some confusion is still left about the definition of this parameter (in various classification societies' rules it may be taken as diverse as $h_{1/3}$, $h_{1/10}$, $h_{1\%}$, $h_{5\%}$).[2] The whole length of Danube is proclaimed as Zone 3 ($h_? = 0.6$ m). The Resolution also defines procedures and criteria for recognition of Classification Societies; these items are replicated in 2006/87/EC and CCNR ADN (see in following).

UNECE 2015 Resolution No 24—European Code for Inland Waterways (CEVNI). The Resolution defines the core rules applicable to the traffic on inland waterways in the UNECE region such as marks and draught scales on vessels, visual signals on vessels, sound signals and radiotelephony, waterway signs and markings, rules of the road, berthing rules, signaling and reporting requirements as well as prevention of pollution of water and disposal of waste. These harmonized rules constitute the legal and technical basis for national inland waterway codes in UNECE member States.

CEMT—European Conference of Ministers of Transport (in English ECMT)

The most relevant issue is *CEMT 1992 Resolution No 92/2*—on New Classification of Inland Waterways (see CEMT 2020). Among other things, the Resolution introduced the classification of motor vessels and pushed convoys presented in Table 4.1.

EC—European Council

The most relevant issues are Directives of the European Parliament and of the Council (see EC 2020):

Directive 2003/25/EC—On specific stability requirements for Ro-Ro passenger ships.

Directive 2005/44/EC on harmonised river information services on the EU's inland waterways (referred to as the RIS Directive) requires Member States to implement RIS according to certain standards (for more about RIS see Sect. 9.6.1).

Directive 2006/87/EC—Laying down technical requirements for IW vessel; requirements are replicated in CESNI ES-TRIN and CCNR RVIR documents (see in following); among other things the Directive gives the list of classification societies recognized by the EU.

CCNR—Central Commission for Navigation on the Rhine

The most relevant issues (see CCNR 2021) are:

RVIR—Rhine Vessels Inspection Regulations. The first Rhine Vessel Inspection Regulations dates back to 1905. This regulation has been continually updated to reflect technological progress. The inspection regulations have undergone substantial revisions over the years. The most recent revisions were carried out in 1976, 1995, 2006 and 2021.

ADN 2021—European Agreement Concerning the International Carriage of Dangerous Goods (issued jointly with UNECE). Since the nineteenth century (1838) the Central Commission has drawn up specific rules for the transport of dangerous goods on the Rhine (cannon powder, explosives, poisonous and corrosive substances, etc.).The ADNR regulation dates from 1971. A new version came into force on 1 January 1995. This new version underwent in-depth restructuring to ensure improved

[2] h_x is the averege height of x portion of the highest observed waves.

harmonisation with the ADR (road) regulations and RID (railway) regulations. Restructuring was completed in 2002, and the new restructured version came into force in 2003. In parallel to these developments, the European Community adopted a Directive on the inland transport of dangerous goods (Directive 2008/68/EC of the European Parliament and of the Council of 24 September 2008). The new directive applies the ADN regulation to transport of dangerous goods by inland waterway within or between Member States by 30 June 2011 at the latest.

It should be noted that official Commission documents are published in French, German and Dutch. English translation is available—but it does not have the status of original. Some additional caution must be taken then to avoid misinterpretation, as in some topics it is possible (see Bačkalov 2017).

DC—Danube Commission

The Danube Commission (see DC 2020) is an international intergovernmental organization established by the Convention regarding the regime of navigation on the Danube signed in Belgrade in 1948.

The main objectives of the Danube Commission's activity are to provide and develop free navigation on the Danube for the commercial vessels flying the flag of all states in accordance with interests and sovereign rights of the Member States of the Belgrade Convention, as well as to strengthen and develop economic and cultural relations of the said states among themselves and with the other countries.

The Member States of the Danube Commission are: the Republic of Austria, the Republic of Bulgaria, Hungary, the Federal Republic of Germany, the Republic of Moldova, the Russian Federation, Romania, the Republic of Serbia, the Slovak Republic, Ukraine and the Republic of Croatia.

> To ensure the said integration the Commission actively cooperates with the relevant international bodies, involved in different aspects of inland waterway transport, such as UNECE, CCNR, EC, etc.

CESNI—European Committee for Drawing up Standards in the Field of Inland Navigation

CESNI (see CESNI 2020) adopts technical standards in various fields, in particular as regards vessels and crew to which the respective regulations at the European and international level, including the European Union and the CCNR, will refer with a view to their application.

The most relevant issues of CESNI are *ES-TRIN—European Standards for Technical Requirements for Inland Navigation*. Since 2015, the CESNI Committee has regularly updated and published the European Standard laying down Technical Requirements for Inland Navigation vessels. This standard lays down the uniform technical requirements necessary to ensure the safety of inland navigation vessels. It is available in four languages (German, English, French, Dutch) and brings together in a standardized way the requirements previously contained in directive 2006/87/EC and in the Rhine vessel inspection regulations RVIR. It contains provisions on inland navigation vessel construction and equipment, special provisions for certain categories of vessels such as passenger and container vessels, provisions on the model of inland navigation vessel certificate as well as for instructions on how to apply the technical standard.

References to ES-TRIN are now included in the legal frameworks of the EU and the CCNR (respectively directive (EU) 2016/1629 and Rhine vessel inspection regulations). The Danube Commission also decided in 2017 to recommend the standard in its international instruments and the International Sava River Basin Commission intends to create a reference to the standard in its legal framework.

ES-TRIN is not binding per se, the European Union (EU), the Central Commission for the Navigation on the Rhine (CCNR) and other international organisations or third countries can apply this Standard by referring to it in their respective legal frameworks. In other words, ES-TRIN has been established, with great effort, as the centerpiece and baseline for the technical requirements for inland navigation vessels in Europe and has contributed to the reinforcement of governance at the European level (Fig. 5.1).

As a reminder, a vessel operating on the EU waterways or the Rhine must carry either a Union inland navigation certificate or a Rhine vessel inspection certificate. Both certificates are issued by the competent national authorities (inspection bodies) and confirm the full compliance of the vessel with the technical requirements (in particular ES-TRIN). Once again, the objective of these technical requirements is

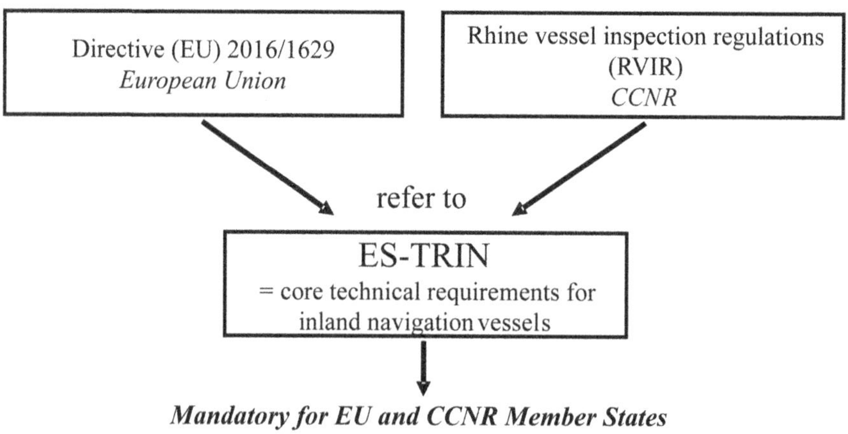

Fig. 5.1 Statutory regulations. *Source* CESNI 2020

to guarantee a high level of safety in inland navigation, thereby also protecting the environment and the people on board.

ES-TRIN contains provisions on the approval of installations and equipment, as well as installation firms and technical services. Such approvals are the responsibility of the national authorities under the ES-TRIN and associated regulatory frameworks (i.e. Rhine vessel inspection regulations or directive (EU) 2016/1629).

5.4 Other Noteworthy International Documentation on the Danube

The Danube is more than just a significant waterway. It is a major European phenomenon whose various aspects are subject of numerous studies, projects and research. Sometimes the final reports of those are rather eclectic—treating at the same time topics as diverse as agriculture, municipalities, invasive species and navigation. Conclusions, which had the power to grow into statutory regulations for IW vessels are implemented in documents described in the previous section. However, those reports also offer a wide range of data which may not be articulated as rules or regulations—but still may be extremely useful for ship designers. A shortlist of such documents and brief notes on their contents are given in the following. Some of them are already cited and used as a source in Chap. 2.

AGN—European Agreement on Main Inland Waterways of International Importance.

The document is a part of the so-called UN Treaty Collection, issued in 1996. Among other things, it contains explicit numbering of inland waterways of international importance with their technical and operational characteristics, as well as numbering of inland navigation ports of international importance.

Blue Book—Inventory of Main Standards and Parameters of the Waterway Network.

The document is an issue of UNECE; at present, the latest version is the third revised edition from 2017. It contains a handful of valuable data on the network of inland waterways of international importance, such as a definition and list of bottlenecks and missing links, navigation characteristics, parameters of locks and technical characteristics of ports.

ICPDR—International Commission for the Protection of Danube River (ICPDR 2020).

The Commission issues documents on highly heterogeneous aspects of the Danube river and Danube Basin, one of which is navigation. Linked up with the Danube Commission and the International Commission for the Protection of the Sava River Basin, they produced the *Joint Statement on Inland Navigation and Environmental Sustainability in the Danube River Basin*. Among other things, the *Joint Statement* provides principles and criteria for environmentally sustainable inland navigation on the Danube and its tributaries, including the maintenance of existing waterways and the development of future waterway infrastructure.

5.5 Concluding Remarks

A brief, itemized summary of the topics discussed in this chapter follows:

- Classification Societies' Rules focus on the structural strength of vessel hull.
- Class of IW vessel is not yet an obligation prescribed by international regulation and not mandatory in all European countries but is expected to be in the future.
- Classification Societies are most commonly empowered by flag state administrations to assess compliance with the statutory regulations also.
- Consistency and converging of the extensive library of EU statutory documentation is steered, pretty much as shown in Fig. 5.1. UNECE Resolution 61 expands the harmonization of regulations to non-EU countries also.
- The Danube Commission decided in 2017 to recommend the ES-TRIN standard in its international instruments.

Design and building of contemporary vessel for all European inland waterways, including the Danube, should be performed according to rules of a recognized classification society and in compliance with respective, updated ES-TRIN standards.

References

Bačkalov I (2017) International maritime regulations (in Serbian). Book. University of Belgrade, Faculty of Mechanical Engineering. ISBN 978-86-7083-945-8
BV (2020) https://group.bureauveritas.com. Accessed 14 Dec 2020
CCNR (2015) 200 years—bicentenary of the CCNR. https://www.200years-ccnr.org/?lang=en_US#book. Accessed 07 Feb 2021
CCNR (2021) https://www.ccr-zkr.org. Accessed 14 Jan 2021
CEMT (2020) https://www.itf-oecd.org/sites/default/files/docs/wat19922e.pdf. Accessed 14 Dec 2020
CESNI (2020) https://www.cesni.eu. Accessed 14 Dec 2020
DC (2020) https://www.danubecommission.org. Accessed 14 Dec 2020
DNV-GL (2020) https://www.dnvgl.com. Accessed 14 Dec 2020
EC (2020) https://www.consilium.europa.eu/en/. Accessed 14 Dec 2020
Grgić A (2017) Seminar for engineers of shipyard "Begej" Zrenjanin (in Serbian). Internal document
ICPDR (2020) https://www.icpdr.org. Accessed 14 Dec 2020
LR (2020) https://www.lr.org. Accessed 14 Dec 2020
UNECE (2020) https://www.unece.org/info/ece-homepage.html. Accessed 14 Dec 2020

Chapter 6
Transport Efficiencies and Performance Indicators

6.1 Background of Energy Efficiency Indicators

Maritime Transport—Recent IMO Activities

IMO's Marine Environment Protection Committee MEPC (IMO 2009a) developed measures to enhance energy efficiency in international shipping and thereby reduce greenhouse gas emissions. These are technical and operational measures as well as possible market-based instruments. The following indicators were introduced:

- **Energy Efficiency Design Index (*EEDI*)** for *new ships*, is a measure of the amount of CO_2 emitted by the ship per capacity mile (tonne-mile). *EEDI* is a theoretical index of ship performance at one draught and in ideal weather conditions. The *EEDI* has come into force in 2013 in order to limit maritime shipping CO_2 emissions mainly by application of more efficient designs, hull forms, cleaner engines and fuels etc. Introduced measures have already initiated various innovations and technical developments which increased energy efficiency of ships.

  According to IMO (2018a), *basic*[1] (or simplified) *EEDI* can be calculated from:

$$EEDI = \frac{\mathrm{CO_2}\ emissions}{Transport\ work} = \frac{P_{ME} \cdot SFC \cdot CF}{K \cdot V_{ref}}$$

[1] *Basic,* as only the first term in the numerator of *EEDI* formula, which represents the main engine, is shown here. The other terms, less significant, take into account the auxiliary engines (2nd term) and, if applicable, innovative electrical and mechanical technologies (4th and 5th term, respectively); the 3rd term represents shaft generator if present.

D. Radojčić et al., *Design of Contemporary Inland Waterway Vessels*,
https://doi.org/10.1007/978-3-030-77325-0_6

where

P_{ME} 75% MCR of main engine brake power [kW].
SFC Specific fuel consumption [g/(kWh)].
CF Carbon emission or conversion factor, depends on type of fuel. According to IMO (2009a) $CF = 3.206$, 3.000 and 2.750 [g CO_2/g fuel] for diesel oil, LPG (propane) and LNG, respectively.
K Ship capacity (t).
V_{ref} Ship speed corresponding to P_{ME} [kn].

The installed propulsion power, however, should be sufficient to enable safe manoeuvrability even under adverse conditions (see for instance IMO 2013).

Note that the most effective way to reduce EEDI is by a) reducing the design speed, thus the power too, and b) using alternative fuel types (for instance LNG).

MARPOL Annex VI, Chap. 4, Regulation 21 provides formula for the upper limit, i.e. required EEDI, depending on the type of seagoing ship:

$$Required\ EEDI = (1 - x/100) \cdot Reference\ EEDI$$

The coefficient x is a reduction factor that lowers the reference line through phases 1, 2 and 3, as shown in Fig. 6.1. This means that new designs are required to be more efficient in three phases, i.e. CO_2 emissions per transport work should be progressively reduced by 30% for ships delivered after 2024, compared to the 2013 level. A further ambition is to reduce CO_2 emissions in international shipping for 70% by 2050, compared to 2008 level (see IMO 2018b).

Reference EEDI represents the energy efficiency of a typical existing ship of the considered type above 400 GT and delivered in a period between 1998 and 2010. It can be evaluated as:

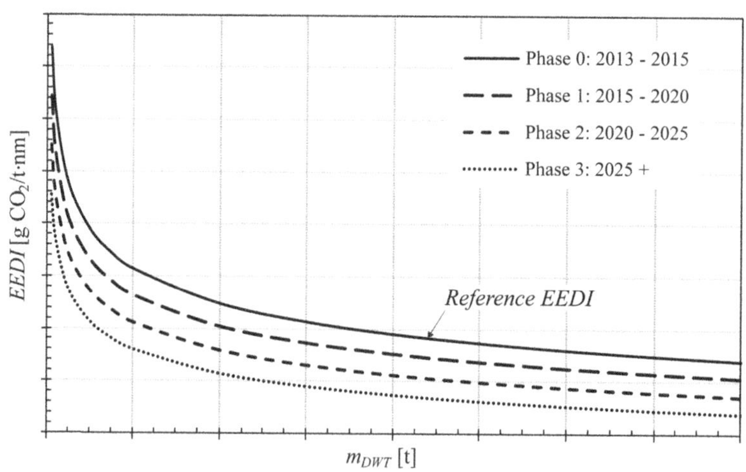

Fig. 6.1 Concept of required *EEDI* reduction

$$Reference\ EEDI = a \cdot K^{-c},\ \text{where}\ a,\ c = \text{f(ship type)}$$

Note that Attained *EEDI* may satisfy *EEDI* requirement at the time ship is built, but later can exceed it due to the phase reduction of Required *EEDI*.

- **Energy Efficiency Operational Indicator** *(EEOI)* enables ship operators to measure the fuel efficiency of an existing ship and to gauge the effectiveness of adopted measures for the reduction of energy consumption. Essentially it estimates total carbon emissions in a given time period per unit of revenue tonne-miles. According to IMO (2009b), *EEOI* can be calculated as:

$$EEOI = \frac{FC \cdot CF}{K \cdot s}$$

where

FC	Consumed fuel oil per voyage or in a given time period
s	Actual distance sailed per voyage or in a given time period [nm]
	Other symbols as above.

The *EEOI* provides a figure for the efficiency of particular ship and enables comparison of its energy or fuel efficiency with similar vessels.

- **Ship Energy Efficiency Management Plan (*SEEMP*)** incorporates guidance on best practices which include improved voyage planning, speed and power optimization, ship and cargo handling optimization etc. The purpose of *SEEMP* is to improve the *efficiency of an individual ship* by reducing its energy consumption, see IMO (2016).

The abovementioned regulations, amongst others, initiated application of already existing technologies within the international shipping industry, which would be neglected without proper incentives. As it turned out, even the existing technologies provided a significant increase of energy efficiency of a ship. Introduced measures also initiated the development of innovative energy efficiency technologies for the reduction of unnecessary energy dissipation during the ship navigation.

Actions of IMO's MEPC are essentially in the line with the Kyoto Protocol[2] to the United Nations Framework Convention on Climate Change (UNFCCC) which *concerns the seagoing ships*. Sooner or later similar measures will have to be applied to inland ships as well.

IW Vessels

Although more than 15,000 IW cargo vessels operate along the European waterways (see Sect. 1.1.3), strongly supporting European economic growth, regulations related to the energy efficiency for inland waterway ships still do not exist. Moreover, the

[2]Kyoto Protocol (1997) was improved and replaced by the Paris Agreement in 2016. Both treaties were aimed to curb the release of GHG.

absence of adequate goal-oriented energy and emission benchmarks for IW vessels is a large obstacle to performance improvements of these vessels.

However, several different performance indicators regarding energy efficiency and economic aspects of IW selfpropelled ships and pushed convoys were recently introduced within various research projects. Consequently, some of the methods proposed in those projects can be used for the assessment of existing and new designs, with respect to energy efficiency and economic performances of IW vessels. A brief list of those with short explanations is given in the following.

6.2 Energy Efficiency Indicators for IW Vessels

Unlike marine ships, IW vessels navigate in restricted waters, with limited and variable depth and/or breadth of fairway. The current speed varies along the river, depending on the cross-section, hence the hydrodynamic forces acting on a ship during navigation are changing too. Besides, river bends, bridges, the interaction with other ships, etc. make navigation of IW vessels even more complex.

Therefore, in order to maintain good manoeuvrability and obtain adequate stopping ability (i.e. increase safety) propulsion engines of IW vessels are usually considerably more powerful than it would be necessary just for achieving service speed in calm and unrestricted water. Furthermore, some IW selfpropelled vessels are designed to navigate with one or more coupled barges, hence are equipped with more powerful engines than would be needed for separate navigation. As a consequence of the abovementioned, the IMO concept of *EEDI* that is based on 75% of installed propulsive engine power and corresponding ship speed in calm and unrestricted water, is not suitable for IW vessels. Therefore, adequate energy efficiency indicators for IW vessels should be adjusted to these specific conditions.

6.2.1 Transport Efficiency Coefficient (TEC)

The Vienna Model Basin (VMB 2013) introduced a Transport Efficiency Coefficient (*TEC*) which is adjusted to IW vessels and is basically similar to *EEDI*, though is neglecting the environmental terms of *EEDI* equation. *TEC* is based on delivered power instead of engine break power, hence excludes the transmission (shaft) efficiency. Therefore, *TEC* cannot be used for exact estimation of the overall environmental impact of a ship, but may be useful for the assessment of the efficiency of IW vessels from the hydrodynamic point of view.

$$TEC = \frac{Delivered\ power\ at\ the\ propeller}{Transport\ work} = \frac{P_D}{tdw \cdot V_S}$$

where

P_D Delivered power [kW].
tdw Transported cargo [t].
V_S Ship speed over ground [km/h].

This coefficient is not dimensionless. It provides specific metrics of a ship that shows how much energy is needed to transport one ton the distance of one km [kWh/t·km]. Accordingly, the smaller the *TEC*, the more efficient is the vessel from the hydrodynamic point of view.

The *TEC* concept may be applied to selfpropelled IW vessels as well as to the pushed convoys.

6.2.2 Transport Efficiency Index (TEI)

Transport Efficiency Index (*TEI*), according to Hofman et al. (2013), can be calculated from:

$$TEI = \frac{Transport\ work}{Work\ done\ by\ propeller} = \frac{m_{DWT} \cdot V_S}{P_D}$$

where

m_{DWT} Deadweight [t].
P_D Delivered power [kW].
V_S Ship speed over ground [km/h], $V_S = V \pm V_C$.
V Ship speed through water [km/h].
V_C River current speed [km/h].

This index is similar to the Transport Efficiency Coefficient—*TEC* (Sect. 6.2.1). However, this non-dimensionless indicator is rising as ship efficiency increases. Accordingly, the higher the *TEI* the more efficient the vessel is from the hydrodynamic point of view.

This concept may be used for evaluation of hydrodynamic efficiency of IW selfpropelled vessels as well as for the pushed convoys.

6.2.3 Modified Energy Efficiency Design Index (EEDI*)

Modified Energy Efficiency Design Index (*EEDI**), which indicates both—the hydrodynamic and environmental performances, is proposed in Simić and Radojčić (2013) and Simić (2014). This index is applicable to typical IW ship navigation and is comparable to the IMO's *EEDI* concept. *EEDI** can be evaluated from:

$$EEDI^* = \frac{P_{Bref} \cdot SFC \cdot CF}{m_{DWT} \cdot V}$$

where

$EEDI^*$	Modified energy efficiency design index [g CO_2/t·km]
P_{Bref}	Reference engine power for achieving V [kW]
SFC	Specific fuel consumption [assumed 200 g/(kWh)]
CF	Carbon emission factor [3.206 g CO_2/g fuel, if diesel fuel is used]. Other symbols as above.

The proposed equation for the modified energy efficiency design index—*EEDI** has a similar structure as IMO *EEDI*, but some significant differences exist. For instance:

- *EEDI** depends on *actual ship speed through the water* (V), instead of speed over ground that corresponds to predetermined engine loading (75% MCR, according to IMO approach);
- *EEDI** is *based on reference power* (P_{Bref}), i.e. not on maximal installed power (MCR) as for IMO *EEDI*.

According to the proposed concept, *EEDI** should be evaluated for all considered ships for the same speed through the water and for the same fairway conditions (primarily fairway depths). This enables a fair comparison of various IW vessels with respect to energy efficiency and environmental performance.

Consequently, suggested *EEDI** depends on vessel type, deadweight and speed through the water, while the IMO *EEDI* depends on a ship type and deadweight (see Sect. 6.1). Therefore, the required value of *EEDI** of IW vessels is determined from a surface instead from a single curve. It can be expressed as:

$$EEDI^*_{Req} = a \cdot m^c_{DWT}$$

where coefficients a and c depend on:

(a) Ship type (selfpropelled dry cargo and tanker, or pushed convoy), and
(b) Speed through the water (for deep water navigation), or depth Froude number (in case of shallow water navigation).

These coefficients should be determined from the propulsive characteristics of similar ships, e.g. those that are built in last ten years. Coefficients a and c for deep and shallow water navigation (h = 5 m), for selfpropelled cargo vessels, are given in Table 6.1 (Simić 2014).

Coefficients a and c are determined by regression analysis and are valid for:

$$10km/h \le V \le 22km/h,\ 0.4 \le Fn_h \le 0.65 \text{ and } 100t \le m_{DWT} \le 3000t$$

Equations given in Table 6.1 are based on data for a limited number of selfpropelled cargo ships, the majority of which are older than ten years. External influences, such as wind, waves, etc. are neglected. Therefore, these expressions should be considered as preliminary (phase 0, resembling the IMO concept) and should be enhanced upon new data on similar ship types. Consequently, instead of artificially

Table 6.1 Coefficients for selfpropelled cargo vessels for the required *EEDI**

Deep water navigation
$a = 0.39554 \cdot V^2 - 11.27833 \cdot V + 111.69043$
$c = -0.00114 \cdot V^2 - 0.05177 \cdot V + 0.70843$

Shallow water navigation (h = 5m)
$a = 93.712 \cdot F_{nh}^{-3} - 516.38 \cdot F_{nh}^{-2} + 886.54 \cdot F_{nh}^{-1} - 414.86$
$c = -0.4181 \cdot F_{nh}^{-3} + 2.5716 \cdot F_{nh}^{-2} - 5.2767 \cdot F_{nh}^{-1} + 3.3485$

imposing reduction of required *EEDI* for 10, 20 or 30%, which is introduced by IMO concept for the seagoing ships, hereby the required *EEDI** is reduced intrinsically according to future technologically improved IW vessels (new designs), i.e. following realistic technological developments, and market demands. However, if needed, a goal-oriented concept, similar to IMO EEDI, may also be created.

Required *EEDI** surfaces, for selfpropelled cargo vessels, obtained by equations given in Table 6.1 are shown in Fig. 6.2.

These results can be converted into a 2D format, which is more convenient for validation of efficiency of IW vessels (see Figs. 6.3 and 6.4).

The proposed concept can also be applied to pushed convoys, but the required *EEDI** surfaces should be recalculated to match the pushed convoy configuration.

*EEDI** reference surfaces may be used during the design phase for benchmarking of a *new design*, see Fig. 6.5a. For required payload and ship speed, the hull form should be defined first. Follows evaluation of attained *EEDI** (based on principal hull form parameters) and comparison with the required value. If the attained *EEDI** is above the required *EEDI**, some of the parameters should be changed, otherwise proceed to the next design phase.

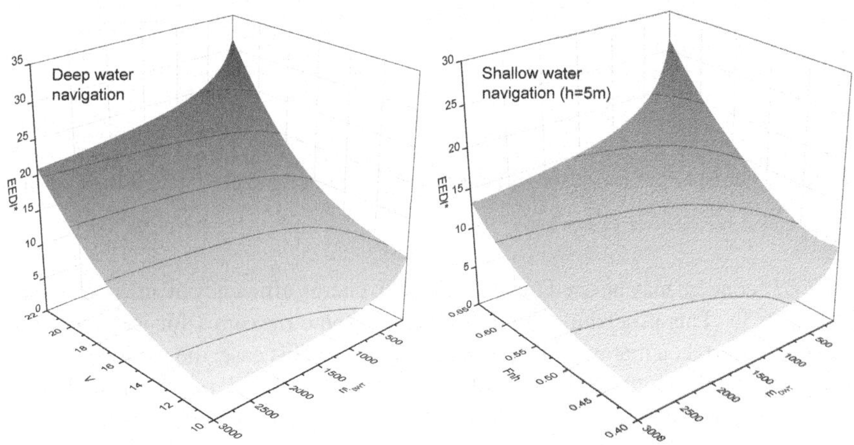

Fig. 6.2 Required *EEDI** surfaces for selfpropelled cargo vessels for deep and shallow water (h = 5 m) navigation

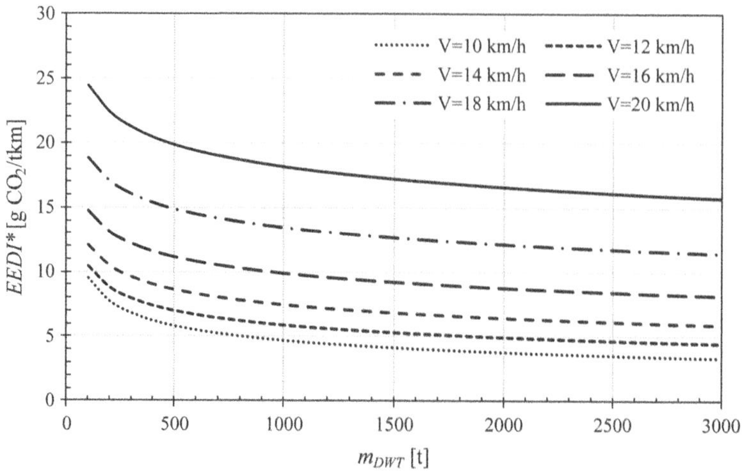

Fig. 6.3 Required *EEDI** curves for selfpropelled cargo vessels for deep water navigation

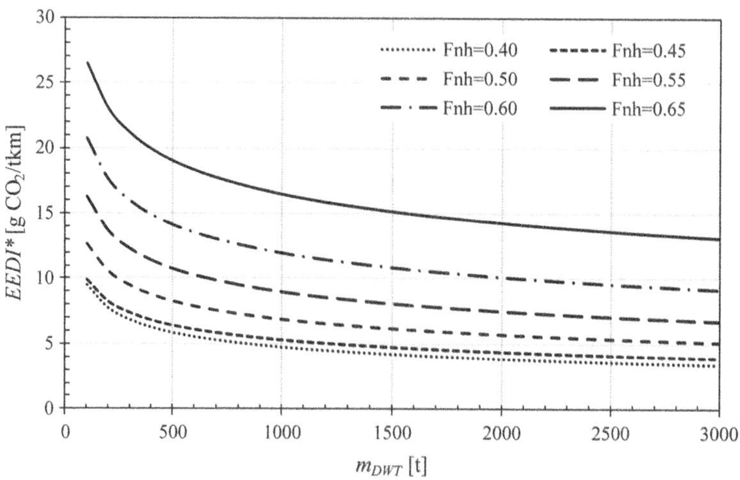

Fig. 6.4 Required *EEDI** curves for selfpropelled cargo vessels for shallow water navigation (h = 5 m)

*EEDI** concept may be used for assessment of energy efficiency of an *existing ship*, see Fig. 6.5b. This procedure should indicate if some measures for improvements could be undertaken that would increase efficiency. In this case, the payload, service speed, brake power and fuel consumption should be measured, paying special attention to external influences. If the attained *EEDI** is higher than the required value, something should be done. For instance:

- Payload and/or speed should be changed,
- Ship hull and/or propeller should be cleaned/repaired or optimized.

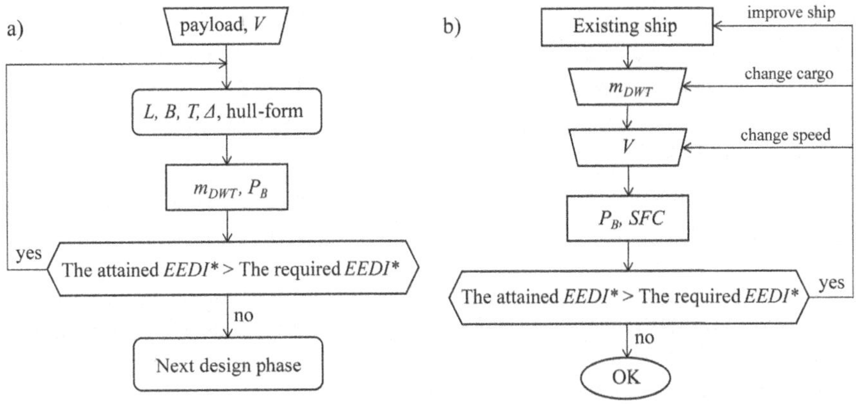

Fig. 6.5 Procedure for evaluation of energy efficiency of **a** a new ship during the design stage, or **b** of an existing ship. *Source* Radojčić and Simić (2014)

*EEDI** concept was tested for few existing IW vessels for which the full-scale and model-scale results were available (see Radojčić and Simić 2014). Regarding energy efficiency, different conclusions were derived from the full-scale and model-scale measurements. Accordingly, the conclusion was that the power margin[3] needs further clarification since it was not clear whether the attained *EEDI** was poor or the power margin was low. This, however, leads to discussion about the extrapolation of results from a model to full-scale (service conditions), which is beyond the scope of this book.

6.2.4 German Proposal for Energy Efficiency Evaluation (EED_{IIWV})[4]

The project FE (2019) was initiated by the German Federal Ministry of Transport and Digital Infrastructure and executed by DST. Within this project a new concept for evaluation of energy efficiency for IW vessels was proposed. This concept is still in the proposal phase, but will be presented here as it appears to be a promising one.

According to this proposal the IW vessels are categorized into four types: 1) dry cargo and container selfpropelled ships, 2) tankers, 3) pushed convoys and 4) passenger ships. Only types 1 and 3 are of interest for this book.

This concept has the following sub-concepts (classification according to navigation zone):

---

[3]Calm water power margin, according to ITTC, is the additional power above the tow tank prediction that ensures a calm water speed requirement is met. Sea margin (here river margin) is basically similar and accounts for average environmental conditions.

[4]The main source of information in this section is FE (2019).

- For deep and calm water,
- For the waterways of navigational Zone 3, i.e. shallow water (according to the European waterway classification, see Sect. 5.3 and EC 2006), and
- For the canals corresponding to Zone 4 waterways, including MD canal (see Sect. 5.3 and EC 2006).

Similar to IMO procedures, two options for further reduction of *EEDI* are considered: reduction for 15 and 25%, depending on ship type and navigation zone.

The *EEDI* for dry cargo and container selfpropelled IW vessels (type 1 ships) and pushed convoys (type 3), depending on the navigation zone, can be estimated according to the procedures given in Tables 6.2 and 6.3, respectively. Required coefficients denoted by Greek letters are given in Table 6.4.

In Tables 6.2 and 6.3, the following symbols are used:

B	Ship/convoy breadth [m].
h	Fairway depth [m]
SFC	Specific fuel consumption [assumed 220 g/(kWh)].
CF	Carbon emission factor [3.206 g CO_2/g fuel].
P_D	Delivered power [kW].
$EEDI_{IWV}$	Attained/estimated energy efficiency design index [g CO_2/t·km] $EEDI_{IWV} = CF \cdot SCF \cdot P_D/(V_S \cdot m_{DWT})$
$EEDI_{Req}$	Required energy efficiency design index [g CO_2/t·km] Other symbols as above.

Table 6.2 Procedure for calculation of *EEDI* for dry cargo and container selfpropelled IW vessels (type 1)

Deep water (h>7.5m)
$P_D = \alpha_1 \cdot m_{DWT}$ ➡ V_S ⇨ $EEDI_{IWV}$
$EEDI_{Req} = \alpha_4 + \beta_2 \cdot EXP\left(\frac{m_{DWT}}{-\gamma_2}\right) + \vartheta_1 \cdot EXP\left(\frac{m_{DWT}}{-\delta_1}\right)$
Zone 3 (shallow water)
$P_D = \left(\alpha_6 + \beta_4 \cdot EXP(-\gamma_4 \cdot B) - \delta_2 \cdot EXP\left(\frac{h}{-\varepsilon_1}\right)\right) \cdot m_{DWT}$ ➡ V_S ⇨ $EEDI_{IWV}$
$EEDI_{Req} = (\alpha_7 + \beta_5 \cdot V_C + \gamma_5 \cdot V_C^2) + (\delta_3 + \varepsilon_2 \cdot V_C - \zeta_1 \cdot V_C^2 + \eta_1 \cdot V_C^3) \cdot EXP\left(\frac{m_{DWT}}{-\theta_1}\right)$
Zone 4 (canal)
V_S ➡ P_D ⇨ $EEDI_{IWV}$
$EEDI_{Req} = \alpha_{12} - \beta_{11} \cdot m_{DWT}$

Note shaded arrow indicates measurement, while hollow arrow refers to the calculation

Table 6.3 Procedure for calculation of *EEDI* for pushed convoys (type 3)

Deep water (h>7.5m)

$$P_D = \left(\alpha_3 + \beta_1 \cdot EXP\left(\frac{B}{-\gamma_1}\right)\right) \cdot m_{DWT}$$ ➡ V_S ⇨ $EEDI_{IWV}$

$$EEDI_{Req} = \alpha_4 + \beta_2 \cdot EXP\left(\frac{m_{DWT}}{-\gamma_2}\right) + \vartheta_1 \cdot EXP\left(\frac{m_{DWT}}{-\delta_1}\right)$$

Zone 3 (shallow water)

$$P_D = \left(\alpha_9 + \beta_7 \cdot EXP(-\gamma_4 \cdot B) - \delta_5 \cdot EXP\left(\frac{h}{-\varepsilon_4}\right)\right) \cdot m_{DWT}$$ ➡ V_S ⇨ $EEDI_{IWV}$

$$EEDI_{Req} = (\alpha_8 - \beta_8 \cdot V_C + \delta_6 \cdot V_C^2) + (\gamma_7 + \varepsilon_5 \cdot V_C + \zeta_3 \cdot V_C^2) \cdot EXP\left(\frac{m_{DWT}}{-\eta_3}\right)$$

Zone 4 (canal)

V_S ➡ P_D ⇨ $EEDI_{IWV}$

$$EEDI_{Req} = \alpha_{13} - \beta_{12} \cdot m_{DWT}$$

Note shaded arrow indicates measurement, while hollow arrow refers to the calculation

Table 6.4 Coefficients required for equations given in Tables 6.2 and 6.3

Deep and calm water					
Coeff.	Value	Coeff.	Value	Coeff.	Value
α_1	0.262	β_1	0.25	γ_2	470
α_3	0.146	β_2	13	δ_1	4500
α_4	10	γ_1	11	ϑ_1	8
Zone 3					
Coeff.	Value	Coeff.	Value	Coeff.	Value
α_6	0.375	γ_4	0.13	ε_2	0.78
α_7	21	γ_5	0.28	ε_4	2.8
α_8	18	γ_7	8	ε_5	0.25
α_9	0.375	δ_2	0.5	ζ_1	0.46
β_4	0.0625	δ_3	11	ζ_3	0.375
β_5	0.7	δ_5	0.5	η_1	0.154
β_7	0.0625	δ_6	0.75	η_3	3100
β_8	2.5	ε_1	2.8	θ_1	800
Zone 4					
Coeff.	Value	Coeff.	Value	Coeff.	Value
α_{12}	22.4	β_{11}	0.004	-	-
α_{13}	24.9	β_{12}	0.004	-	-

Range of application of equations given in Tables 6.2 and 6.3 are the following:

- **Deep Water**

$$T = 1.5 \cdot D \quad \text{(where } D \text{ is propeller diameter)}$$
$$2.0\,m \leq T \leq 3.2m$$

.

- **Shallow Water—Zone 3**

$$T = 1.5 \cdot D$$
$$3.5m \leq h \leq 7.5m$$

Additional restrictions for Zone 3 only are given in Table 6.5.

- **Canals—Zone 4**

$$T = 1.7 \cdot D, \text{ but not exceeding } 2.5\text{ m}$$

Note that draught to propeller diameter ratios of 1.5–1.7 are suitable for the Rhine vessels, but may be too high for the Danube ships.

Table 6.5 Range of application of equations for navigation Zone 3

	App. range	Type 1	Type 3
T [m]	min.	2.0	2.0
	max.	2.8	3.0
L [m]	min.	40.0	76.5
	max.	135.0	230
B [m]	min.	5.0	9.0
	max.	17.0	34.4
m_{DWT} [t]	min.	250	500
	max.	6000	18000
V_C [km/h]	min.	2.0	2.0
	max.	8.0	6.0
h/T	min.	1.4	2.0

Deep and Calm Water Navigation

For deep water navigation, delivered power (P_D) for the estimation of energy efficiency of IW vessel is predetermined by the equations given in Tables 6.2 and 6.3, while speed over ground (V_S) should be measured during trials. In the case of full-scale trials, the delivered power can be controlled either via direct measurements on a propeller shaft, or based on fuel consumption measurements. In early phases of ship design, the empirical methods may be used for speed/power estimations instead of the measurements.

Consequently, the required *EEDI* of IW vessels, similarly to the IMO concept, is determined with a single curve and depends on m_{DWT}, see Fig. 6.6. Estimated $EEDI_{IWV}$ in case of a new design, or the attained $EEDI_{IWV}$ in case of an existing ship, should not exceed the required $EEDI_{Req}$ (identical for type-1 and -3 ships).

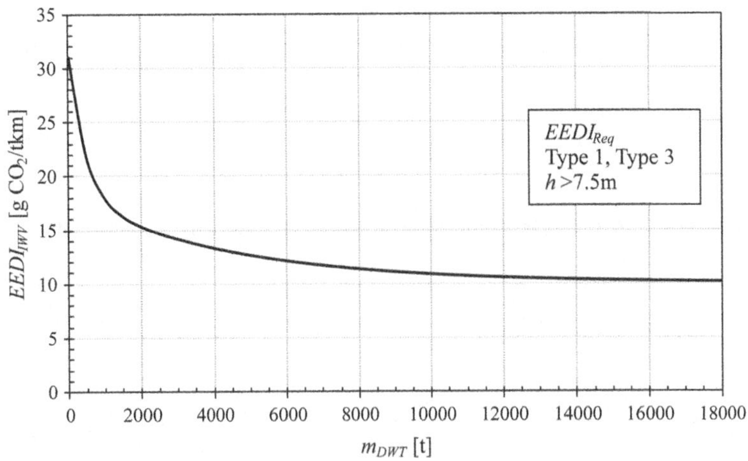

Fig. 6.6 Required *EEDI* for deep and calm water navigation

Shallow Water Navigation—Navigation Zone 3

The delivered power on waterways of navigational Zone 3 can be calculated from the equations given in Tables 6.2 and 6.3. As in the case of deep and calm water, the delivered power should be adjusted according to the predetermined value, while speed over ground should be measured during the trials, while sailing upstream. The river current velocity (V_C) is assumed to be known.

The river current velocity has a significant influence on ship speed over the ground; thus, instead of a single line for required $EEDI_{Req}$ (deep water case), there are now a number of lines, i.e. one for each current velocity (see equations in Tables 6.2 and 6.3. for Zone 3). Similarity with the modified *EEDI** concept (Sect. 6.2.3) should be noticed, where required *EEDI** is determined from a surface which depends on the ship's payload and ship speed through the water.

The estimated $EEDI_{IWV}$ of a new design should not exceed the required $EEDI_{Req}$, see Figs. 6.7 and 6.8.

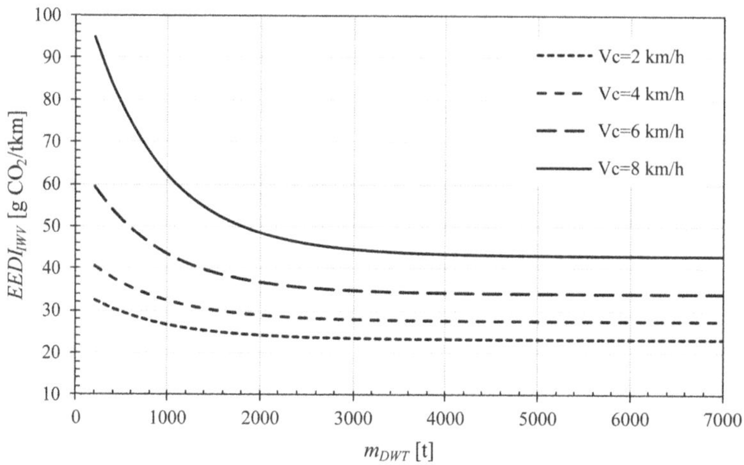

Fig. 6.7 Required *EEDI* for IW selfpropelled vessels of Type 1 and navigational Zone 3

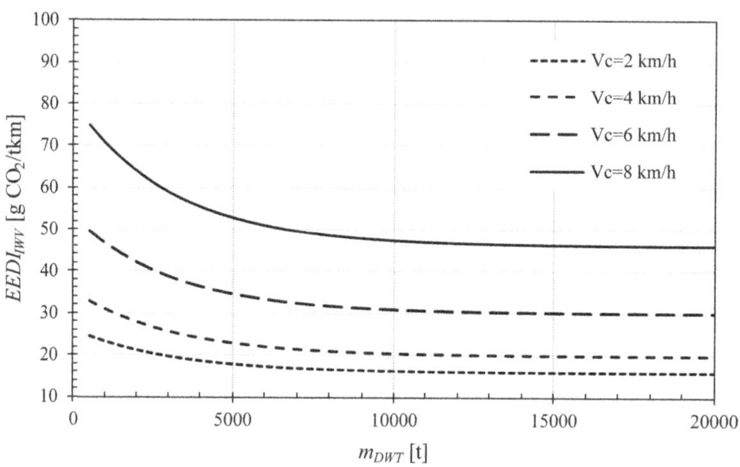

Fig. 6.8 Required *EEDI* for convoys (vessels of Type 3) and navigational Zone 3

Navigation on Zone 4—Canals

The proposed concept is further expanded for navigation on waterways of navigational Zone 4, including major canals with trapezoidal cross-section, amongst which is also the MD canal. Cross-section of the MD canal, for instance, with side grade of 1:3, is given in Fig. 6.9.

Delivered power should be measured during trials for predetermined ship speed, which depends on the vessel/convoy breadth according to:

(a) $V_S = 10 km/h$ for $B < 11.0m$, or
(b) $V_S = 9.4 km/h$ for $B \geq 11.0m$.

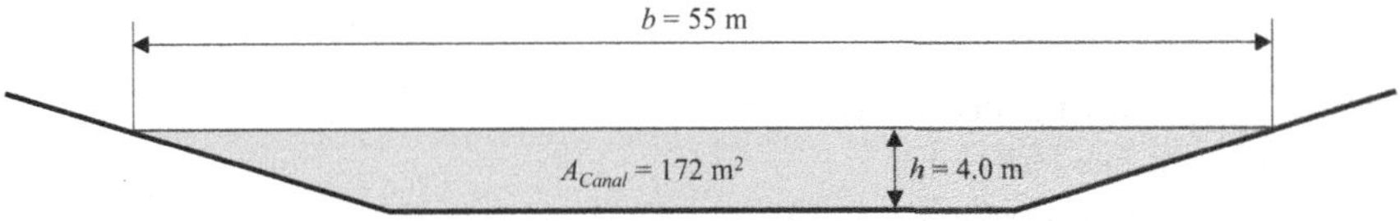

Fig. 6.9 Cross-section of a canal considered within this project

Unlike other considered cases, here the speed is predetermined with the aim of reducing hydrodynamic impacts typical for shallow and restricted waterway navigation.

The value of $EEDI_{Req}$, depending on a ship type, can be determined from the equations given in Tables 6.2 and 6.3. Note that the proposed acceptable limit of energy efficiency, i.e. $EEDI_{Req}$, is in a linear correlation with the m_{DWT}.

6.3 Economic Indicators[5]

The economic indicators for the assessment of cost-performances of IW cargo vessels introduced within the project Innovative Danube Vessel (IDV) are presented in this section. These indicators are primarily developed and analyzed for pushed convoys (see Hofman et al. 2013), but can also be applied for IW selfpropelled cargo ships.

The indicators can be used either during the design stage of a new ship, or to analyze some of the transportation dilemmas in the case of existing ships (e.g. the assessment of optimal barge formation for the given mass of cargo and pushboat power, or determination of optimal speed depending on ship loading and fairway conditions). Same efficiency assessment concepts may be applied to unrestricted as well as for restricted fairway conditions.

6.3.1 Economic Efficiency Coefficient (EEC)

A nondimensional Economic Efficiency Coefficient (*EEC*), as a function of *TEI* (see Sect. 6.2.2), is discussed in Hofman et al. (2013). It is evaluated as follows:

$$EEC = \frac{k_f}{k_c \cdot SFC} \cdot TEI$$

[5]The main source of information in this section is Hofman et al. (2013)

where

TEI	Transport Efficiency Index
k_f	Freight rate [€/(t·km)]
k_c	Cost of transport [€/t]
SFC	Specific fuel consumption [t/(kWh)].

Within this approach, only fuel costs are discussed. All other costs of transportation may be introduced in the denominator for enhanced analysis. This coefficient can be further transformed to:

$$EEC = \frac{k_f}{k_c \cdot SFC} \cdot \frac{Transport\ work}{Work\, done\, by\ propeller} = \frac{k_f}{k_c \cdot SFC} \cdot \frac{m_{DWT} \cdot V_S}{P_D}$$

Here, V_S refers to the designed ship speed over ground, while P_D is corresponding delivered power. Larger value of EEC indicates economically better convoy/vessel performance.

6.3.2 Profit Rate Indicator (Π_t)

The parameter that estimates cost performances of a ship, named the Profit Rate Indicator, is introduced in Hofman (2006) and Hofman et al. (2013). This indicator is defined by the following equation:

$$\Pi_t = \frac{Income - Costs}{t} = \frac{k_f \cdot m_{DWT} \cdot s - k_c \cdot SFC \cdot P_D \cdot t}{t}$$

where

Π_t	Profit rate indicator [€/h]
s	Trip distance [km]
t	Trip time [h]
	Other symbols as above.

Presented parameter can be transformed as follows:

$$\begin{aligned}\Pi_t &= k_f \cdot m_{DWT} \cdot V_S - k_c \cdot SFC \cdot P_D \\ &= k_f \cdot m_{DWT} \cdot (V \pm V_C) \cdot \left(1 - \frac{1}{EEC}\right)\end{aligned}$$

where signs '+' and '−' refer to downstream and upstream navigation, respectively.

EEC explained in Sect. 6.3.1, as a dimensionless indicator of efficiency, can be considered as an intermediate step required to determine the expected profit rate. From the given equations, it can be noticed that the higher Economic Efficiency Coefficient shows a higher Profit Rate Indicator.

The presented approach enables the determination of a technically realistic speed of the vessel which results in maximal profit for given convoy/ship loading and fairway conditions. Essentially that is the optimal or economic speed of transportation (V_{EC}). According to Hofman et al. (2013), from the definition of profit rate (profit per hour) one could find the profit rate maximum as:

$$\frac{d\Pi_t}{dV_S} = k_f \cdot m_{DWT} - k_c \cdot SFC \cdot \frac{dP_D}{dV_S}$$

If the profit rate indicator tends to its maximum:

$$\frac{d\Pi_t}{dV_S} = 0 \rightarrow \frac{dP_D(V_{EC})}{dV_S} = \frac{k_f \cdot m_{DWT}}{k_c \cdot SFC} \rightarrow V_{EC}$$

from which, as indicated, follows the economic speed of transportation.

6.3.3 Profit Coefficient (p)

The Profit Rate Indicator can be transformed into another dimensionless form, suitable for the analysis of off-design conditions (partial loading), called Profit Coefficient, see Hofman et al. (2013):

$$p = \frac{\Pi_t}{k_f \cdot m_{ref} \cdot V_{ref}}$$

where

m_{ref} Reference deadweight [t]
V_{ref} The reference speed [km/h]
Other symbols as above.

This coefficient is derived to enable further cost analysis corresponding to the transportation of any cargo weight (m_{ref}), at the reference speed. It can also be used to determine the economic speed for given IW vessel, loading condition and fairway restrictions, similarly as in case of Profit Rate Indicator (Sect. 6.3.2). Concerning the pushed convoys, profit coefficient approach may be used for finding optimal barge formation for given mass of cargo and pushboat power.

For example, the influence of the pushed convoy draught on economic speed and profit coefficient is given in Fig. 6.10, see Hofman et al. (2013). Here the convoy that consists of a pushboat and 6 barges in the formation P + 2 + 2 + 2 is considered. Within this example the upstream navigation is assumed, while the river current speed and water depth are $Vc = 3$ km/h and $h = 7.5$ m (other data required for the evaluation of profit rate are same as in Sect. 8.8.2, see Table 8.34). Each line of the diagram corresponds to different barge draught. Presented results confirm that the increase of draught is always beneficial from the profit rate point of view. However,

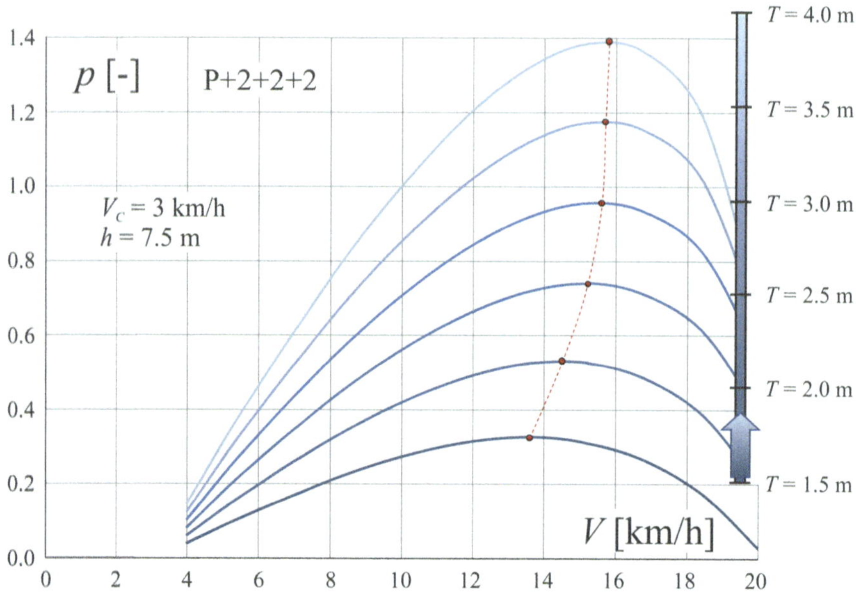

Fig. 6.10 Profit coefficient versus convoy speed through water, for different convoy draughts (Hofman et al. 2013)

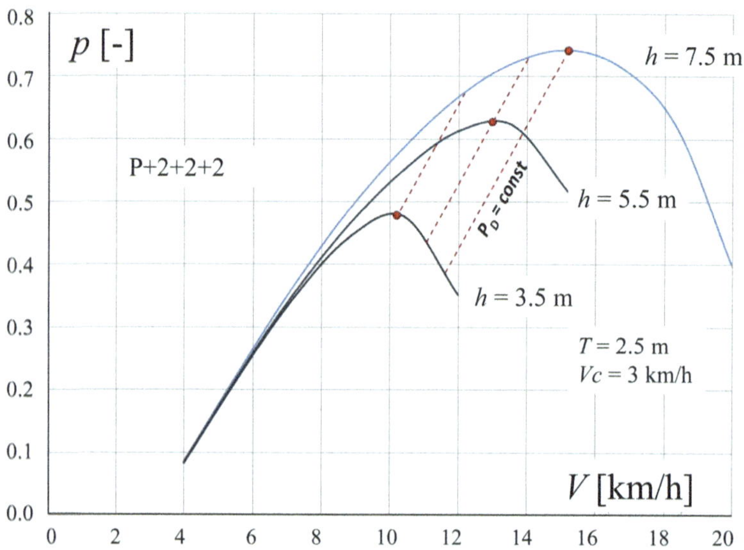

Fig. 6.11 Profit coefficient as a function of convoy speed through water, for three different water depths (Hofman et al. 2013)

the main feature is the occurrence of profit coefficient maximums, indicating that there exists a technically realistic speed of navigation that gives the maximal profit. The values of maximal profit and the corresponding economic speeds are marked on the diagram, and connected by the dashed line.

A similar analysis, taking into account the influence of the fairway depth, is given in Fig. 6.11. In this example the convoy draught is 2.5 m and the river current speed is 3 km/h. As expected, with the decrease of water depth, not only the maximal profit p_{max}, but the corresponding economic speed and optimal power are also reduced.

6.4 Concluding Remarks

Mandatory requirements concerning the energy efficiency of IW vessels still do not exist. However, few recently developed concepts for assessment of energy efficiency of IW vessels are presented in this chapter; their summary is given in Table 6.6. Although elaborated concepts are essentially based on IMO *EEDI* approach, i.e. maritime environment, they are all adjusted to typical IW navigation imposed by the waterway conditions.

The primary goal of the abovementioned concepts is reduction of CO_2 emissions. However, their side effect is to encourage the application of new technologies within the IW shipping industry (these issues are further elaborated in Chap. 9).

As the economic speed varies during the voyage due to variations of fairway conditions (e.g. water depth) it is not possible to keep the maximal profit per hour of the selfpropelled ship or a convoy without adjusting the engaged engine power. The best way to keep the maximal profit rate through the whole voyage would be the real-time measurements of power, speed and other vital parameters,[6] in order to evaluate the economic speed in real time, and accordingly adjust the ship/convoy speed. The economic indicators suitable for the assessment of cost-performances of commercial IW vessels and the estimation of the so-called economic speed of IW vessels are summarized in Table 6.7.

Summarising, to select the best transportation solution, from the viewpoint of the maximal profit rate and the lowest environmental impact, methods for estimation of the energy efficiency of IW vessels (presented in Sects. 6.2.1–6.2.3) may be done for the economic speed (estimated from Sects. 6.3.2 and 6.3.3).

[6]Something similar was done within the MoVeIt! project, see Godjevac and Drijver (2014)

Table 6.6 An overview of presented concepts for the evaluation of energy efficiency of IW vessels

Concept (Sect.)	Hydrodynamic assessment	Environmental assessment	Goal-based Criterion	Comments
TEC (6.2.1)	yes	no	no	General concept that should be further developed.
TEI (6.2.2)	yes	no	no	General concept that should be further developed.
*EEDI** (6.2.3)	yes	yes	yes	Predetermined V^1, P_B is measured. Based on ship speed through the water. Applicable for European IW vessels. Required criteria for deep and shallow water (h=5m) are based on a limited number of selfpropelled IW vessels.
$EEDI_{IWV}$ (6.2.4)	yes	yes	yes	Predetermined $P_D{}^2$, V_S is measured. Based on ship speed over ground. Applicable for vessels typical for the Rhine. Available required criteria for deep, shallow water and canals for IW selfpropelled dry cargo vessels and pushed convoys.

Notes:

[1] Speed depends on fairway conditions; the economic speed may also be considered

[2] For navigation through canals the speed over ground is predetermined

Table 6.7 An overview of presented concepts for the assessment of cost-performances of commercial IW vessels/convoys

Concept (Sect.)	Comments
EEC (6.3.1)	Only fuel costs are considered; may be enhanced with other costs.
Π (6.3.2)	Applicable for cost analysis (design conditions – full load). Enables determination of the economic speed.
p (6.3.3)	Applicable for cost analysis (off-design conditions – reduced load). Enables determination of the economic speed.

References

EC (2006) Council Directive 82/714/EEC, Laying down technical requirements for inland waterway vessels

FE (2019) Beurteilung des Energiebedarfs von Binnenschiffen mit Hilfe von Energie-Effizienz Indizes, Kurzfassung zum Endbericht Nr. 2252, FE-Projekt 40.0399/2017, April 2019

Godjevac M and Drijver M (2014) Power configuration of an inland pusher. ICMT 7th-9th July, Glasgow 2014

Hofman M (2006) Inland container vessel: optimal characteristics for specified waterway. RINA International Conference Coastal Ships & Inland Waterways, London 2006

Hofman M, Krstović M, Kalajdzic M (2013) WP3, Performance indicators and transport efficiency of pushed convoys. Innovative Danube Vessel Project

IMO (2009a) MEPC.1/Circ.681: Interim Guidelines on the method of calculation of the Energy Efficiency Design Index for new ships

IMO (2009b) MEPC.1/Circ.684: Guidelines for voluntary use of the Ship Energy Efficiency Operational Indicator (EEOI)

IMO (2013) MEPC.232(65): 2013 Interim guidelines for determing minimum propulsion power to maintain the manoeuvrability of ships in adverse conditions

IMO (2016) MEPC.282 (70): Guidelines for the development of a Ship Energy Efficiency Management Plan (SEEMP)

IMO (2018a) MEPC.308 (73): 2018 Guidelines on the method of calculation of the Attained Energy Efficiency Design Index (EEDI) for New Ships

IMO (2018b) MEPC.304 (72): Initial IMO strategy on reduction of GHG emissions from Ships

Kyoto Protocol (1997) Protocol to the United Nations Framework Convention on Climate Change. UN December

Radojčić D, Simić (2014) WP2, A hydrodynamic improvements - energy efficiency of IWW self-propelled ships. FP7-Move It Project

Simić A (2014) Energy efficiency of inland waterway self-propelled cargo ships. In: RINA International conference: Influence of EEDI on Ship Design, London

Simić A, Radojčić D (2013) On energy efficiency of inland waterway self-propelled cargo vessels. FME Transactions 41(2)

VMB (2013) WP2, Innovative Danube Vessel, Analysis of technical solutions

Part II
Design Guidelines with the Basics on Shallow Water Hydrodynamics

Chapter 7
Basics on Inland Waterway Vessel Hydrodynamics

7.1 Introduction

From the hydrodynamic viewpoint, the two main differences between the IW vessels and conventional seagoing ships should be emphasized:

- *Fairway differences:* Fairway restrictions are different, i.e. IW vessels are usually sailing in shallow waters, often restricted by the width too, while the seagoing ships do not have those limitations.
- *Design differences:* The hull form is different because the IW vessels are tailored to navigate in the shallow water. Limited draught influences propulsors and steering devices as well.

Both affect the flow around the hull, thus their hydrodynamics too. These differences are briefly discussed in the following. Note that only shallow water effects important for IW ship design will be discussed here. General ship hydrodynamics is skipped; for that, the other sources should be consulted, for instance, Molland et al. (2017), Carlton (2012) etc.

7.2 Shallow Water Resistance

According to ITTC (2017a), shallow water effects on ship resistance should be taken into consideration if:

- $h/T < 4$, or
- $Fn_h > 0.5$, where $Fn_h = V/\sqrt{(g \cdot h)}$ is the Froude number based on water depth (h).

Note that most vessels navigating on the Danube will fulfil at least one, if not both criteria, indicating that special attention should be paid to shallow water effects on ship resistance.

D. Radojčić et al., *Design of Contemporary Inland Waterway Vessels*,
https://doi.org/10.1007/978-3-030-77325-0_7

Empirical and Numerical Methods

The most reliable way for the determination of resistance is by model tests, but in the early design stages application of *empirical methods* might be more convenient. Nevertheless, only a few empirical methods are available for IW vessels. For instance, Marchal et al. (1996) which is applicable for pushed convoys, and the modified Guldhammer and Harvald method (see Georgakaki and Sorenson 2004), for selfpropelled IW vessels. The first method takes into account the depth of the waterway, while the second one is appropriate for deep water navigation, thus additional correction is required for shallow water navigation.

Traditional empirical methods for estimation of resistance of seagoing ships in shallow water apply some kind of speed corrections, i.e. modify estimated ship speed due to shallow water effects. Thus, the accuracy of these methods depends not only on the recommended correction itself but also on the prediction of the ship resistance in deep water. Although the most empirical methods based on ship speed corrections due to shallow water effects are relatively old, e.g. Schlichting (1934), Karpov (1946), Apukhtin (1947), Lackenby (1963), Jiang (2001), etc., they can still be taken as valid. Similar methods applicable for canal navigation are Landweber (1939) and Xu-tao (1990). If the shallow-water effects are expected during the trials of the seagoing ships, the Lackenby's method is recommended (ITTC 2014).

Recent developments of *numerical methods* enabled better insight into the physical details of water flow around the ship. Also, the influence of shallow water effects on various resistance components is better explained. Consequently, instead of speed corrections used by the traditional empirical methods mentioned above, the new methods are focused on the increase of a particular *resistance component* in shallow water, which is explained in the following text.

Resistance Components

The total ship resistance (R_T) can be divided into two major components (ITTC, 2017b): (a) the *viscous resistance* (R_V) and (b) the *wave-making resistance* (R_W). The viscous resistance can account up to 85% of total resistance at low speed (Fn < 0.12), and up to 40% for fast ships at higher speeds. Several other resistance components are also existing, for instance, hull roughness, air resistance etc., but are less important for this analysis. Accordingly, the total ship resistance coefficient in calm and deep water, in dimensionless form, can be expressed as:

$$C_T = C_V + C_W + \Delta C_F + C_{AA} + C_a$$

where

C_T Total resistance coefficient [−], $C_T = 2 \cdot R_T / (\rho \cdot S \cdot V^2)$.
C_V Viscous resistance coefficient [−], $C_V = 2 \cdot R_V / (\rho \cdot S \cdot V^2) = C_F \cdot (1 + k)$.
C_F Frictional resistance coefficient, according to the ITTC-1957 model-ship correlation line [−], $C_F = 0.075/(\log Re - 2)^2$

C_W Wave-making resistance coefficient [−], $C_W = 2 \cdot R_W / (\rho \cdot S \cdot V^2)$
ΔC_F The roughness allowance coefficient [−].
C_{AA} The air resistance coefficient [−].
C_a The correlation allowance [−].
k Form factor [−].
S Wetted surface area [m^2].
Re Reynolds number [−], $Re = (V \cdot L)/\nu$.
R_T Total ship resistance [kN].
R_V Viscous ship resistance [kN].
R_W Wave-making ship resistance [kN].
V Ship speed through water [m/s].
L Ship waterline length [m].
ρ Water density [t/m^3].
ν Kinematic viscosity [m^2/s].

During navigation in shallow water, the clearance between the ship hull bottom and fairway bed is reduced, compared to deep water navigation. Consequently, according to the continuity equation and the Bernoulli principle, the flow speed below the ship will increase and the local pressure will decrease. Besides the wave-making resistance, these will affect frictional resistance, thus viscous resistance too.

7.2.1 Wave Making Resistance in Shallow Water

During navigation, due to the local pressure distribution around the hull, two wave systems are formed: *transverse* and *divergent* waves. At low speeds the depth of water doesn't affect the wave pattern (Fig. 7.1a), hence the wave-making resistance too. The wave pattern is similar as in the case of deep water navigation, and the so-called *Kelvin wedge* is identical in both cases.

By increasing the ship speed above that which corresponds to $Fn_h > 0.5 \div 0.7$,[1] the local pressure distribution around the hull starts being directly affected by the water depth. Namely, the wave amplitudes increase, as well as the angle of divergent wave system propagation, i.e. the Kelvin wave angle increases gradually till the ship speed approaches the critical speed, see Fig. 7.1b. The *critical speed* theoretically corresponds to $Fn_h = 1$, but effectively it is around $Fn_h \approx 0.96$. This leads to significant growth of the bow wave, and consequently to, an often drastic, increase of the wave-making resistance (R_W).

[1] This speed range is above the feasible (operational) IW vessel speed range but it is treated here as is essential for shallow water hydrodynamics in general.

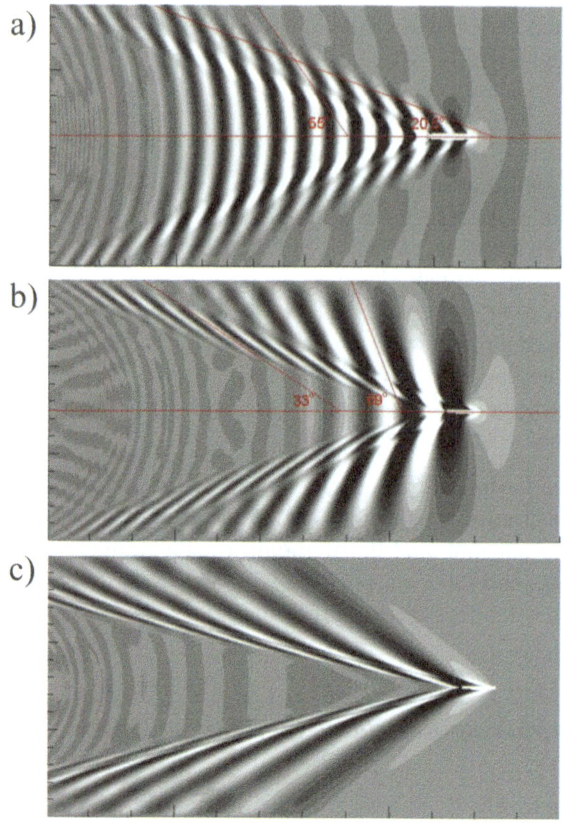

Fig. 7.1 Wave pattern at **a** sub-critical speed, **b** critical speed and **c** super-critical speed. *Source* Nwogu 2003

When the depth Froude number exceeds 1.0, the diverging wave angle decreases. Transverse waves diminish and the wave system is made up entirely of the diverging waves, see Fig. 7.1c, leading to a decrease of wave-making resistance. A direct correlation between the bow wave angle and depth Froude number is shown in Fig. 7.2.

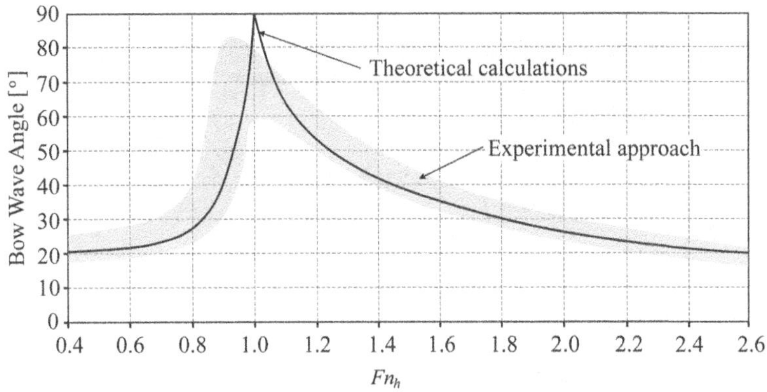

Fig. 7.2 Impact of depth Froude number on diverging bow wave angle. *Source* Radojčić and Bowles 2010

The typical wave-making resistance in deep and shallow water, with respect to the depth Froude number and ship speed, is shown in Figs. 7.3 and 7.4, respectively. Figure 7.3 clearly illustrates three characteristic regions:

- *Sub-critical* ($Fn_h < \sim 0.65$) where the effects of the water depth are negligible or small,
- *Critical* ($\sim 0.65 < Fn_h < \sim 1.25$ with maximum at $Fn_h \approx 0.96$) where the effects of water depth can be extremely pronounced and negative, and
- *Super-critical* ($Fn_h > \sim 1.25$) where the effects are relatively small but generally favourable, i.e. $R_{Wh} < R_{W\infty}$.

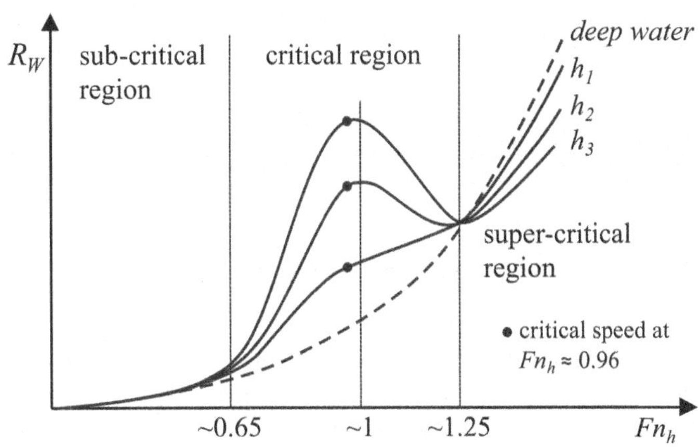

Fig. 7.3 Wave-making resistance versus Fn_h in deep (dashed line) and shallow water ($h_1 > h_2 > h_3$)

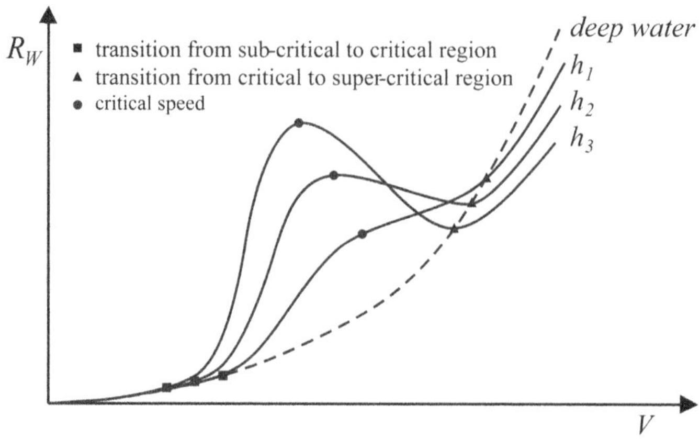

Fig. 7.4 Wave-making resistance versus ship speed in deep (dashed line) and shallow water ($h_1 > h_2 > h_3$)

Accordingly, the *transitional speed* (the speed which defines the transition from one region to another) depends on the waterway depth. Thus, as the depth decreases, the transitional speeds decrease too, see Fig. 7.4. The relationship between the waterway depth and transitional speeds is vividly depicted in Fig. 7.5.

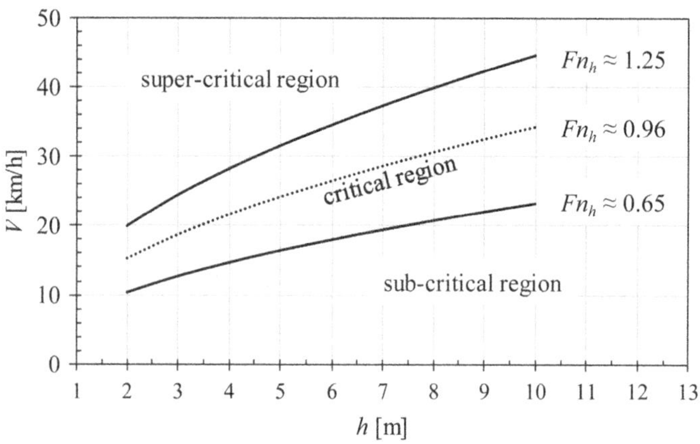

Fig. 7.5 Transitional speed with respect to waterway depth

Wave-making resistance curves show a pronounced peak at the critical depth Froude number, i.e. at the critical speed which depends on the water depth. This phenomenon may be well expressed through the ratio of shallow water wave-making resistance to deep water wave-making resistance $r = Rw_h/Rw_\infty$. Thus, in the sub-critical region $r \approx 1$, in the critical region $r > 1$, while in the super-critical region $r < 1$.

The increase of wave-making resistance, i.e. resistance ratio r, in the critical region is of *primary importance for fast vessels* and depends mainly on the ratio of h/L. This is well depicted by the so-called shallow-water resistance charts, shown in Fig. 7.6. These charts are obtained by relatively complicated theoretical calculations (see Hofman and Radojčić 1997, and Hofman and Kozarski 2000), but are essentially simple for the application and therefore very useful, as the only influential parameters that are tied together are L, h and V; the other ship parameters (ship form and dimensions) are practically not important and may be neglected. In those charts, black and dark grey zones indicate an increase of wave-making resistance. These zones should be avoided. Furthermore, it can be noticed that the only way to avoid the negative shallow water effects is to avoid the critical region itself.

Commercial IW vessels usually navigate at relatively low speeds, in sub-critical or low critical region (i.e. up to $Fn_h < 0.65$ and in most cases below $Fn_h = 0.55 \div 0.60$). Note, however, that the speed through the water is relevant for the explained hydrodynamic phenomena. Note also that other shallow water side effects, such as squat (dynamic sinkage and trim, see Sect. 7.5), enhance the increase of wave-making resistance.

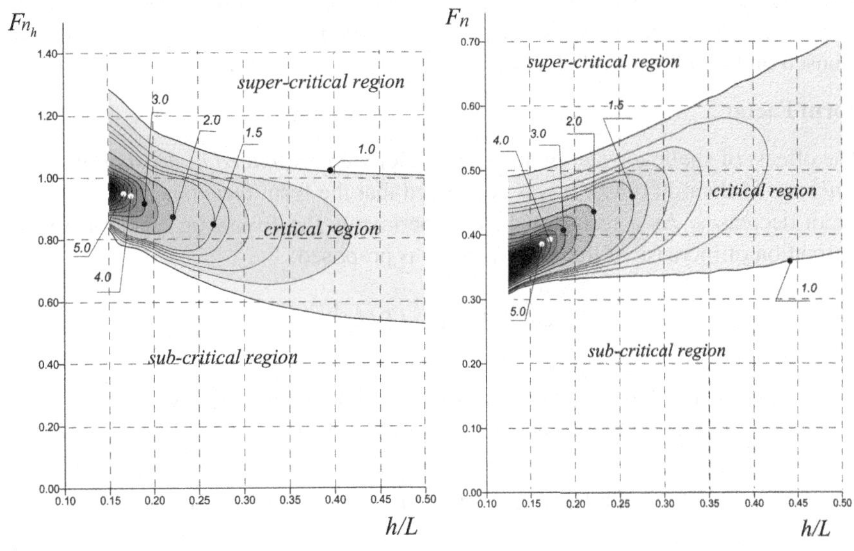

Fig. 7.6 Shallow water resistance charts—resistance ratio $r = Rw_h/Rw_\infty$ as a function of h/L and Fn_h or Fn. *Source* Hofman and Radojčić 1997, and Hofman and Kozarski 2000

7.2.2 Viscous Resistance in Shallow Water

For commercial IW vessels, the viscous resistance dominates over other resistance components as they sail at low speeds. Consequently, to minimize the negative hydrodynamic effects of shallow water on ship resistance, the accurate prediction of viscous resistance in shallow water is very important. The influence of fairway depth on the viscous resistance was, however, often neglected in the past:

(a) As the effects on wave-making resistance are more obvious and pronounced, and
(b) Due to the absence of appropriate engineering methods for reliable estimations.

According to the ITTC-1978 (see for instance Molland et al. 2017), the viscous resistance consists of *frictional resistance* and *viscous pressure resistance*. In nondimensional form, this can be presented as:

$$C_V = 2 \cdot R_V / \left(\rho \cdot S \cdot V^2\right) = C_F \cdot (1 + k)$$

where C_F and k are frictional resistance coefficient and form factor, respectively. The form factor determines viscous pressure resistance. Recent research is therefore focused on the influence of shallow water on these two coefficients.

Form Factor

The effects of shallow water on the form factor were *examined in the shallow water tank*, see Millward (1989). It was concluded that the form factor increases as water depth decreases. As a result of those experiments the following equation for the estimation of increase of the form factor was proposed:

$$\Delta k = 0.644 \cdot (T/h)^{1.72}$$

where T and h are ship's draught and water depth [m], respectively.

Another similar equation stems from Kamar (1996):

$$\Delta k = \left(80.967 \cdot C_B \cdot B \cdot \sqrt{B \cdot T} \cdot (T/h)^{1.845}\right) / L^2$$

where B, L and C_B are ship breadth, length and block coefficient, respectively. This equation takes into account principal ship dimensions, thus enabling more detailed analysis than the previous one. Note that both equations are based on towing tank tests of 7 different seagoing ships.

Numerical calculations are based on CFD methods which enable detailed analysis of fluid flow around the ship, including the interaction of fluid and relevant surfaces (ship hull, waterway bed etc.), through definition of the boundary conditions. Viscous-flow computations give an almost complete picture of the actual flow field in the vicinity of the ship's hull, but give frictional and pressure resistance instead of viscous and wave-making resistance. To obtain the viscous pressure resistance from the total pressure resistance, and to determine a form factor (k), Raven (2012, 2016) applied double-body computations, in which the water surface was treated as a symmetry plane. As a consequence, the viscous pressure resistance was evaluated, enabling the determination of the form factor.

Based on numerical simulations done for 17 different ship types (from slender- to full-form seagoing ships, including a typical Rhine IW self-propelled vessel), Raven (2012, 2016) proposed an equation for prediction of the form factor for shallow water navigation:

$$k_h = (1 + k_\infty) \cdot \left(1 + 0.57 \cdot (h/T)^{-1.79}\right) - 1$$

where

k_h Form factor in shallow water [−]
k_∞ Form factor in deep water [−].

Given equation is valid for $h/T \geq 2$. For $h/T = 2$ the increase of viscous resistance in shallow water is up to 17%. Due to the specified boundary, this equation is not applicable for extremely shallow water which can be found on the Danube.

Note that the main parameter in the equation is the ratio of h/T while the ship speed is irrelevant. Similar conclusions were obtained by the aforementioned experimental approach.

Frictional Resistance

The commonly used frictional resistance coefficient (C_F), evaluated according to ITTC-1957 recommendations, is based on the assumption that the waterway is not restricted, i.e. is infinitely deep and wide. However, in the case of shallow water navigation, the speed of water flow around the ship increases, thus the frictional resistance will be affected too.

To investigate the influence of water depth on frictional resistance, a series of *numerical tests* were performed, see Zeng et al. (2018) and Zeng (2019). Within these simulations, the flow between two flat plates was analysed at different Reynolds numbers ($10^5 < Re < 10^9$) and different distances between the plates D ($0.01 < D/L < 1.00$). Based on this research, a modification of the well known ITTC-1957 correlation line was proposed, with the aim to improve the prediction of the frictional resistance and include shallow water effects. The following equation was proposed:

$$C_F = \frac{0.08169}{(\log Re - 1.717)^2} \cdot \left(1 + \frac{0.003998}{\log Re - 4.393} \cdot \left(\frac{D}{L}\right)^{-1.083}\right)$$

where

D is the distance between plates [m] and
L is the length of the flat plate [m].

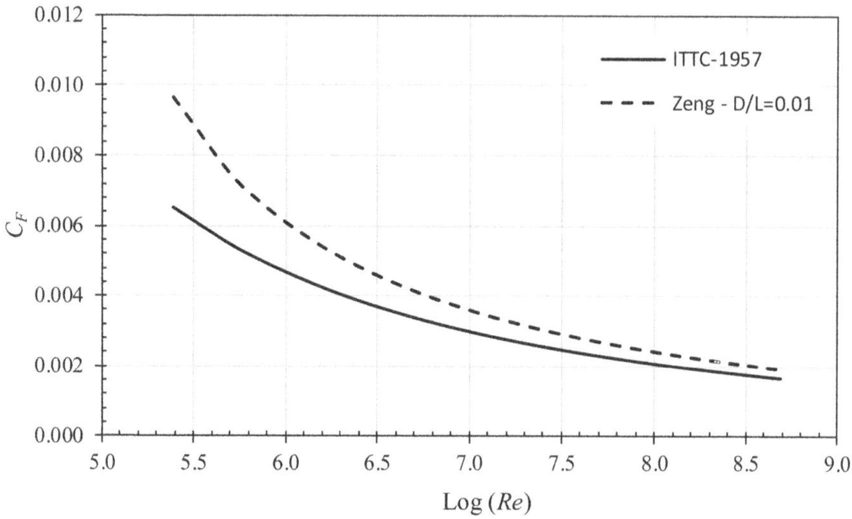

Fig. 7.7 Comparison of the frictional coefficient curves in deep (ITTC-1957) and shallow water ($D/L = 0.01$)

For deep water ($D \approx L$), results obtained by given equation are close to the ITTC-1957 equation, but as the ratio D/L decreases that difference increases. The ITTC-1957 equation is unresponsive to water depth, thus underestimates frictional forces in shallow water. For $D/L = 0.01$, which corresponds to typical navigation in shallow water, the increase of frictional resistance (thus viscous resistance too) of up to 50% compared to deepwater may be expected for low *Re*, see Fig. 7.7. With the increase of *Re* the influence of shallow water on viscous resistance reduces and vice versa. For a ship speed that corresponds to $Fn_h = 0.65$, the expected increase of C_F is about 8.3%.

Summarizing, it should be noted that the abovementioned methods for evaluation of the form factor and the coefficient of frictional resistance in shallow water result in an increase of the viscous resistance. Nevertheless, these methods have been recently developed, and require further validation.

7.3 Propulsive Efficiency in Shallow Water

The hydrodynamic propulsive efficiency depends on the propulsive coefficients which will be briefly examined in the following. The propulsive coefficients are a quantitative measure of propulsor-hull interaction. By far the most influential coefficient is the propeller open water efficiency, thus it will be examined separately from the other propulsive coefficients. Besides, it essentially depends on the propulsor (see Sect. 9.4) which is usually treated as an independent subject.

7.3.1 Deep Water—An Overview

Open Water Efficiency

The open water efficiency (η_O) represents the hydrodynamic propeller efficiency determined in the laboratory conditions (*homogeneous wake field with no hull in front of it*). It is determined from the *open water test*, which is by definition conducted in conditions that correspond to deep water. The open water efficiency depends on the propeller geometry and loading conditions and can be expressed as:

$$\eta_O = (J/2\pi) \cdot \left(K_T/K_Q\right),$$

where

η_O Propeller open-water efficiency [−]
J Advance coefficient [−]
K_T Thrust coefficient [−]
K_Q Torque coefficient [−].

Thrust (K_T) and torque (K_Q) coefficients may also be obtained from the expressions which are based on model experiments of ducted propeller[2] series and are functions of their main characteristics. Typical polynomial equations for K_T and K_Q, as functions of advance coefficient and propeller pitch (i.e. *P*/*D*), follow:

$$K_T = \sum C_T(P/D)^x(J)^y$$
$$K_Q = \sum C_Q(P/D)^x(J)^y$$

The regression coefficients for Ka series for propellers (3–65,[3] 4–70 etc.) and nozzle 19A (or some other nozzle type) can be found in Kuiper (1992) for instance; for the IW ships more appropriate are the Meyne propellers (7–85), see Meyne and Guesnet (1996). Note that ducted propellers and nozzles are discussed in Sect. 9.4.3.

Propulsive Coefficients

The hydrodynamic propulsive efficiency (η_D), or just propulsive efficiency, often also called quasi propulsive efficiency, is defined as a ratio of effective (towing) power (P_E) and the power delivered to the propeller (P_D). Total or overall propulsive efficiency (η_P) takes into account the transmission efficiency too and is a product of propulsive efficiency and transmission efficiency (η_S):

$$\eta_D = \frac{P_E}{P_D} \quad \eta_P = \eta_D \cdot \eta_S$$

[2]A pair that consists of a nozzle and a propeller (ducted propeller) is considered as an undividable propulsor.

[3]Propellers with 3 blades and expanded blade area ratio 0.65.

where

η_P Total propulsive efficiency [−].
η_D Propulsive (hydrodynamic) efficiency [−]
η_S Transmission or shaft efficiency (includes the gearbox efficiency) [−].

The transmission efficiency is a measure of the power loss on its way between the prime mover (engine) and the propeller and is discussed in Sect. 9.4.1. The hydrodynamic propulsive efficiency η_D consist of the following components:

$$\eta_D = \eta_O \cdot \eta_R \cdot \eta_H = \eta_O \cdot \eta_R \cdot \frac{1-t}{1-w}$$

where

η_O Propeller efficiency in open water [−]
η_R Relative rotative efficiency [−]
η_H Hull efficiency [−]
t The thrust deduction coefficient [−], $t = 1 - R_T/T$
w Wake fraction coefficient [−], $w = 1 - V_A/V$
V_A Propeller speed of advance [m/s], $V_A = V \cdot (1 - w)$
T Propeller thrust [kN].
Other symbols as above.

$\eta_D, \eta_O, \eta_R, \eta_H, t$ *and* w are called *propulsive coefficients.*

The most influential coefficient is open water efficiency η_O, which is separately elaborated above. The relative rotative efficiency η_R and hull efficiency η_H describe the interaction between the ship hull and propeller. They directly influence the propeller efficiency when it operates behind the ship (*non-homogeneous wake field with the hull in front of it).* Coefficients η_R and η_H are determined experimentally from the so-called *self-propulsion test* which should be preceded with separate tests of resistance and propeller open water tests; the first one without the propeller and the second without the hull in front of the propeller, see Molland et al. (2017) for instance.

In deep water conditions, the propulsive hydrodynamic efficiency is close to propeller open water efficiency, as the product of $\eta_R \cdot \eta_H$ is close to one. However, in the case of IW vessels navigating in shallow water, the propeller inflow conditions are considerably different.

7.3.2 Shallow Water Operation

From the hydrodynamic viewpoint, shallow water navigation means that the clearance between the bottom of a ship and the waterway bed is reduced. Due to this, the water inflow to the propeller is changed, thus all propulsive coefficients are affected too (for the speed range of interest, w and t are higher, while η_R and η_O are lower than in deep water). This results in lower propulsive efficiency η_D.

In the shallow water, and particularly *for the pushboats* (see Bilen and Žerjal, 1999), the relative rotative efficiency η_R and open water efficiency η_O are reduced due to:

- Non-axisymmetric (i.e. oblique) water inflow caused by the propeller tunnels (see Sect. 8.2);
- Non-uniform water inflow to the propeller, caused by the nozzle integrated into the hull (to facilitate installation of the larger propeller, see Sects. 8.2.2 and 8.2.3), and struts and flanking rudders if present;
- Overloading of a propeller as ship resistance increases in shallow water;
- Potential cavitation (as inflow to the propeller is insufficient, which leads to a decrease of pressure in the surrounding water).

As a consequence, the propulsive coefficients for a *pushboat,* obtained through the classical model testing procedure, might be misleading, see Bilen (1997) and Bilen and Žerjal (1999). Namely, the self-propulsion tests (needed for determination of relative rotative efficiency η_R) are conducted under the so-called atmospheric conditions, inherently meaning that the effects of cavitation are ignored. Therefore, to harmonize model test results with the reality, a redefinition of η_R is suggested in a way that it should represent the cavitation effects too. Thus, for the series of Danube pushboats, η_R was recalculated and determined from the full-scale measurements instead experimentally from the self-propulsion tests. This resulted in considerably *lower relative rotative efficiency values of 0.7 ÷ 0.8* (for seagoing ships the usual range is 0.95 ÷ 1.1). Adding to this a *low open water efficiency of 0.4 ÷ 0.5* (for seagoing ships the range is 0.5 ÷ 0.7), results in a relatively low pushboat *propulsive efficiency of 0.3 ÷ 0.4*, rarely above that (for seagoing ships the range is 0.50 ÷ 0.70). Without η_R correction, the pushboat speed was overpredicted and the propellers were too "heavy".[4]

The abovementioned discussion practically reveals an indirect impact of shallow water from *design differences*, mentioned in Sect. 7.1. Namely, due to navigation in shallow water, the propellers are often overloaded as they are limited in diameter, which results in the necessity for "unconventional" hull forms with tunnels, affecting the propulsive coefficients.

The propulsive coefficients of *selfpropelled vessels* have values somewhere between those of pushboats and seagoing ships. The coefficients w, t and η_R required for evaluation of the propulsive efficiency, may be estimated from the empirical methods applicable to IW vessel hull forms, see for instance Rotteveel et al. (2014), Pompée (2015) etc. Only few of them take into account the influence of water depth.

Heuser (1986) gives the propulsive efficiency η_D directly from the model tests (resistance and self-propulsion tests of IW vessels; note that the first provides R_T, thus $P_E = R_T \cdot V$, while the second provides the required delivered power P_D for a selected

[4]Good results, concerning cavitation impacts, were obtained with medium skewed 5-bladed propellers instead of usual 3 and 4 bladed Ka series propellers.

propeller). The results are given in graphical format as $\eta_D = f(V, P_D/(B \cdot D))$ and are valid for selfpropelled IW ships with two propellers in nozzles, see Fig. 8.10, Sect. 8.3.1. For the naked propellers the propulsive efficiency (η_D) is approximately 5% lower.

All methods mentioned above should be used cautiously, and only for approximate estimation of propulsive efficiency. For more precise analysis the towing tank tests are recommended.

Summarizing, the propulsive efficiency of IW vessels in shallow water is usually between 40 ÷ 50% but can be as low as 20 ÷ 30% (Pompée 2015). The majority of the Danube pushboats have a relatively low propulsive efficiency of around 30 ÷ 40% (Bilen and Žerjal 1999), though this can also be lower and depends on several factors.

The previous discussion is focused on the subcritical region. In the *critical region*, the situation is different as the variation of propulsive coefficients is much more pronounced, see Radojčić and Bowles (2010). Specifically, the propulsive efficiency hollow (η_D reduction) exists and mirrors the resistance hump around the critical depth Froude number, except that its minimum corresponds to $Fn_h \approx 0.9$ (while the resistance hump is around $Fn_h \approx 0.96$, see Figs. 7.3, 7.4 and 7.5). This hollow is explained by the increase of propeller loading due to a significant increase of resistance in the vicinity of the critical speed.

7.4 Power Evaluation

The energy efficiency of IW vessel (see Chap. 6) primarily depends on the power needed to propel the vessel at a certain speed (fuel consumption of generating sets and other minor consumers onboard are neglected in this analysis). That power is proportional to GHG emissions (if engine type variations are ignored). Consequently, it is of primary importance to reduce the power needed for moving the ship. This power, called the engine brake power (P_B), depends on the vessel speed (V), resistance (R_T), transmission efficiency (η_S) and propulsive efficiency (η_D):

$$P_B = \frac{V \cdot R_T}{\eta_S \cdot \eta_D} = \frac{P_D}{\eta_S}$$

The transmission efficiency depends on the adopted concept of ship propulsion, which is discussed later in Sect. 9.4.1. As said before, delivered power (P_D) is the power demanded by the propulsors for achieving the required speed. It depends on the hydrodynamic characteristics of the ship and propulsor. Thus, the evaluation of delivered power implies the determination of a) ship resistance, and b) propulsion coefficients (already discussed in Sects. 7.2 and 7.3, respectively).

A complex interaction between the ship, propeller and engine is depicted in Fig. 7.8. and is generally valid for both the seagoing and the shallow draught vessels. The *fairway and design differences* mentioned at the very beginning of this chapter are reflected only through the input, i.e. specific values of propulsive coefficients (η_R, η_S, w and t). Other input values (V and D) are essentially predetermined, while a pair *P/D* and n can vary, but only specific values will give optimal result, i.e. a high value of propulsive efficiency η_D.

The fragmented approach (Fig. 7.8) is not convenient for application in the early design stages due to unavailable or unreliable propulsive coefficients for IW vessels navigating in shallow water. Thus, to enable estimation of delivered power for IW selfpropelled ships and pushed convoys in the early design stages, the methods (Mathematical Models—MMs) presented in Sects. 8.4.1 and 8.4.2 are proposed. Presented are Simple and Complex methods. The second are developed by Artificial Neural Networks (ANN) and are based on the various systematic and unsystematic model- and full-scale measurements. Note, however, that presented MMs are based on a limited number of data and thus should be used with caution.

A typical influence of water depth and speed on delivered power is shown in Fig. 7.9. This particular example is obtained by Complex MM (suitable for P_D evaluation of IW selfpropelled ships, see Sect. 8.4.1) and is done for an existing Danube selfpropelled ship Deligrad, see worked-out example in Sect. 8.8.1.

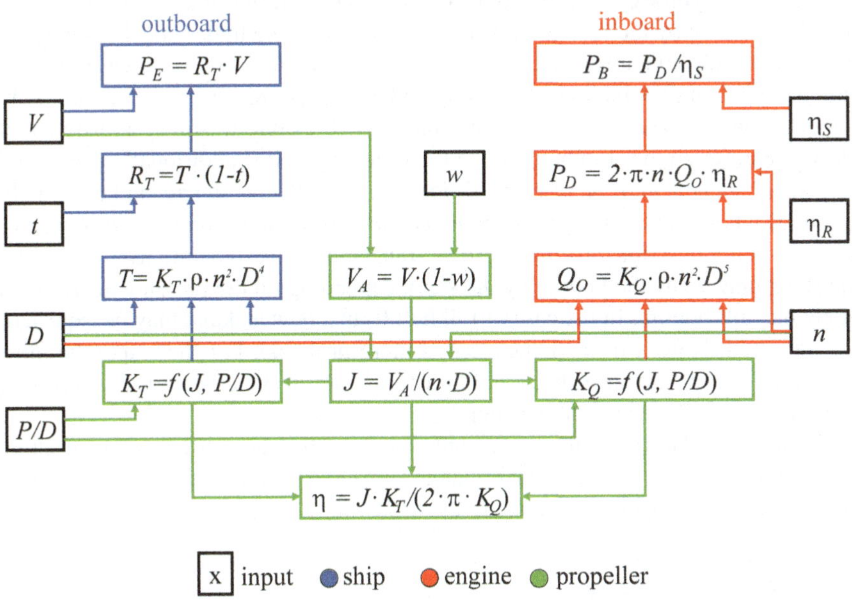

Fig. 7.8 Interaction between the ship, propeller and engine. *Source* Ren et al. 2019

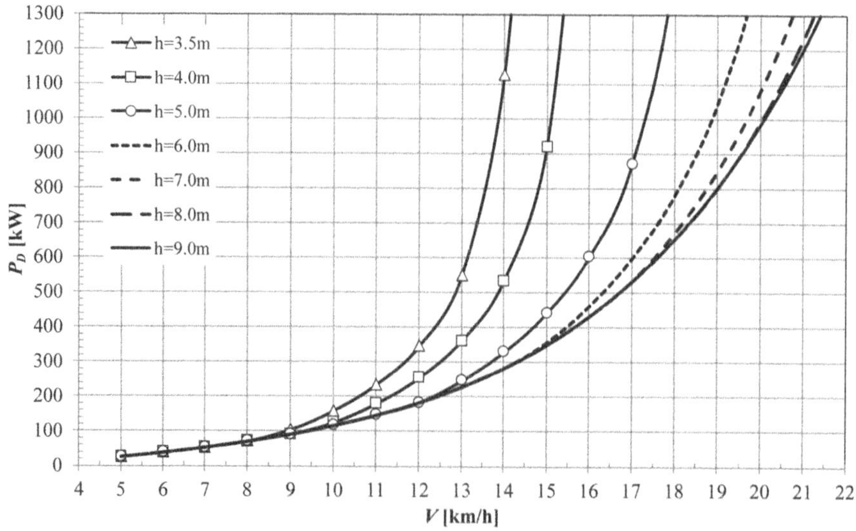

Fig. 7.9 Power-speed diagram of the IW selfpropelled ship Deligrad with respect to the fairway depth

7.5 Squat

The squat is a hydrodynamic phenomenon that appears during shallow water navigation. It is caused by the increased water velocity under the hull. Namely, according to the Bernoulli principle, the increase of water speed leads to a decrease of pressure below the hull, which results in ship sinkage. This phenomenon is particularly important in the case of shallow water navigation, as due to squat, already small clearance between the ship bottom and the waterway bed is additionally decreased. The squat is a *combination of vertical sinkage and a change of trim* and essentially is a *maximal* dynamic draught increase, as opposed to *mean running (or dynamic) sinkage* which describes *average* vertical translation. The dynamic trim caused by the squat may be by the bow or by the stern, depending on the pressure distribution along the hull bottom. In the case of full hull forms, bow sinkage may be expected; for finer hull forms a trim by the stern is more probable. For the squat, the most influential factors are ship speed and hull form, and in the case of navigation in a restricted waterway, the waterway depth.

For low-speed navigation, even in confined water, the squat effects are negligible, but with the increase of speed, the dynamic sinkage is increasing. The most notable consequences of this phenomenon are the following:

(a) Possible grounding (bow or stern),
(b) Change of resistance,
(c) Change of propulsive efficiency and
(d) Effect on manoeuvrability.

The most effective technique for reducing the danger of grounding and other side effects of squat is to reduce the ship's speed. That is, the squat effect is approximately proportional to the square of the speed.

Having in mind the restricted waterway depth, a proper estimation of squat is important for IW vessels. Unfortunately, most of the existing empirical squat prediction methods have been developed for seagoing ships. Few methods, however, may be used for the estimation of bow/stern sinkage of IW vessels and are given in Sect. 8.5.

7.6 Wash

The wash problems are usually associated with navigation of high-speed craft, see Radojčić and Bowles (2010). Similar hydrodynamic effects can arise when IW vessels are navigating in a confined waterway, particularly if relatively high propulsion power is used at lower speeds. Faster IW vessels with full forms (high C_B and B/T) when navigating in shallow water, can generate large waves. This is followed by the increase of wave-making resistance, i.e. power demand (see Sect. 7.2.1). However, manoeuvrability, vibrations and noise can also be affected. Described hydrodynamic effects are known as *wave-wash or just wash*. Dynamic sinkage and trim, as well as the influence of the return flow field, superimposed on the primary ship waves, enhance these effects.

The divergent waves propagating from the ship towards the river banks, see Fig. 7.10, can:

(a) Endanger other users of the waterway which are anchored or are passing by at a relatively close distance
(b) Cause environmental problems—bank erosion, and jeopardise the local ecology.

During navigation in confined waterways, ship propulsors can generate a significant load on (a) the river bed (which depends on its morphology), and (b) river banks. The second is emphasized when the vessel is manoeuvring with deflected rudders (the loads are similar to those produced by the bow thruster, see VBW 2016). These hydrodynamic effects are known as *propeller wash*. Note that the wave wash effects are highly dependent on the shape and slope of the river bank. Especially, in shallow inner bends of free-flowing rivers, even moderate stern waves in the deeper fairway, can lead to bank erosion due to wash (i.e. steep and breaking waves, see Fig. 7.10).

The potential consequences of the mentioned hydrodynamic phenomena depend on the ship speed, displacement, geometry and waterway restrictions (in depth and breadth). Consequently, the wash is more noticeable when the ship speed and/or displacement are large. This generates large waves, thus the wave energy increases too.

Fig. 7.10 Typical shallow water wave system. *Source* DST

The most effective way to minimize or avoid wash is to (a) design a slender ship, and/or (b) reduce ship speed. The speed reduction, however, seems to be a more feasible solution.

Consequently, various mandatory IW navigation rules impose the speed limit to avoid excessive wash which causes damage to stationary objects and structures or moving vessels. In particular, the ship speed must be reduced, but not below the speed required for safe steering. Acceptable, as well as the safe speed, depends on ship size and waterway restrictions. Appropriate navigation signs/signals which indicate that special attention should be paid during navigation to avoid excessive wash are specified in the Chaps. 3 and 5 of UNECE (2018).

7.7 Manoeuvrability

The IW vessels should satisfy high manoeuvring requirements due to specific navigational conditions which may be the consequence of the strong river current, sharp and narrow bends, shallow water sectors, locks and bridges, relatively dense traffic, etc. These requirements concern:

- Minimum speed relative to the water in all water conditions in ahead navigation,
- Change of course,

- Stopping manoeuvre (ability to stop within a certain distance) and
- Astern sailing (steering control when running astern).

The Specificity of IW Vessel Maneuvering

Due to complex hydrodynamic forces acting on a ship or a convoy while turning under rudder and moving ahead, the actual vessel/convoy heading direction (tangent to its path) and vessel/convoy axis, differ. This angle is called the *drift angle* (β). The drift angle defines the width of the vessel's/convoy's path which is considerably wider than its real breadth, see Fig. 7.11. Consequently, IW manoeuvres should be considered as very demanding, particularly in the case of large pushed convoys.

Moreover, due to restricted waterway breadth, overtaking and encounter manoeuvres of IW vessels require utmost precision. The pressure distribution in the water surrounding the hull influenced by the ship motion tends to move the vessel transversally by shifting the bow or stern towards or away from another vessel coming from the opposite direction, see Fig. 7.12. In narrow waterways, high pressure in the bow region tends to move the vessels away from each other (Fig. 7.12b), while reduced pressure in the stern region (caused by the propeller suction), tends to move the vessels towards each other (Fig. 7.12d). Similar hydrodynamic phenomena occur during the overtaking manoeuvre, or in case of a vessel passing close to banks.

Summarizing, the vessel cannot keep the course, thus the *straight course* becomes a *zig-zag course* with continuous course corrections. For instance, for a 100 × 11.4 m ship, the path is 9 ÷ 10 m and 12 ÷ 13 m wider for, respectively, upstream and downstream navigation. A minimal safety distance between the vessels (10 m) and the

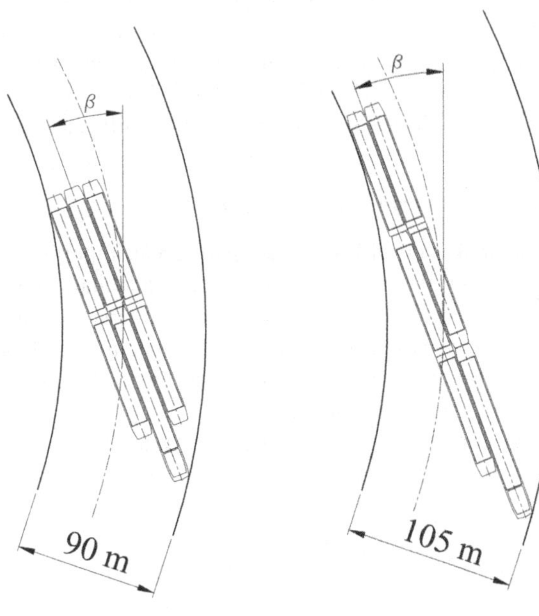

Fig. 7.11 A minimal fairway width during navigation of pushed convoy through the river bends. *Source* EUDET 1998

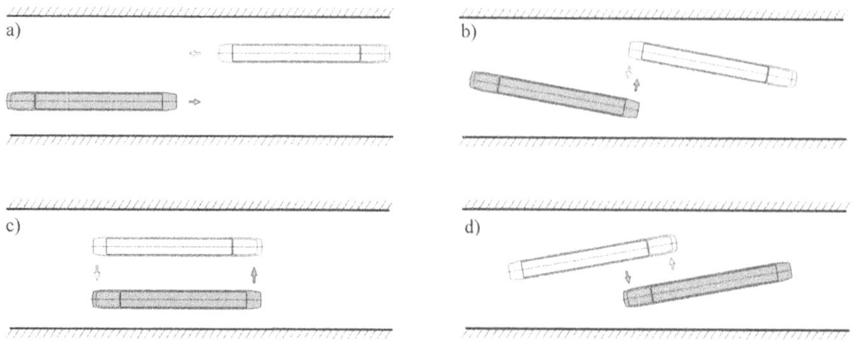

Fig. 7.12 Encounter manoeuvre of two ships heading in opposite directions on the waterway with a restricted width

banks (5 m), for a two-way navigable waterway should be accounted for. However, for *recommended unrestricted navigation* (ignoring the water depth) a minimal distance between the ship side and the banks should be approximately *half of the vessel length*, see EUDET (1998).

Shallow Water Effects

Regardless of specificities usually connected with the IW vessel manoeuvrability (river width restrictions, dense traffic, river current etc.), shallow water effects on manoeuvrability are interesting and unclear. Namely, it is well known that the turning circle is larger in shallow water than in deep water, but it is less known that this, to some extent, depends on the ship type. For instance, for twin-propelled, wide-beam ships (which in a way resemble IW vessels) this may be quite opposite (see Liu et al. 2015). Also, a pushboat with a barge convoy may have a smaller turning circle (and worse course keeping) in shallow than in deep water. These effects become noticeable when $1.5 < h/T < 3.0$, and significant for extremely shallow water when $1.2 < h/T < 1.5$.

The Maneuvering Requirements

Safe navigation of IW vessels is of great importance. Hence, according to CESNI (2019), all IW vessels must display adequate manoeuvrability, which must be confirmed through several tests: stopping capacity, capacity for going astern, capacity for taking evasive action and turning capacity. The detailed procedures for the implementation of mentioned tests, including required criteria for each test, are given in the abovementioned document.

Important manoeuvres for IW ships are "crash-stop", turning, flanking (moving sideward) and running astern. For instance, all ships operating on the Danube have to fulfil tight and precisely defined requirements for stopping. These requirements set up the maximum distances in relation to the ground (after previously running downstream at a certain speed) after which the ship has to stop in relation to water.

Steering Devices

Full control of course and trajectory during navigation is extremely important. However, it is often not so easy to satisfy those requirements due to limited ship draught and specific form of IW vessels.

The rudders on IW vessels differ from those on seagoing ships, e.g. have the end-plates, fish-tail etc. (see Sect. 9.4.2). Their aspect ratio has to be close to one or even less, which means that the maximal side force can be obtained at a higher rudder deflection angles, in some cases of more than 60° (compared to 35° which is typical maximum for the seagoing ships). Moreover, to obtain sufficient manoeuvring forces, the increase of the total rudder area is often necessary. This requires complex multiple-rudder configurations, such as twin- or even triple-rudder arrangements per propeller.

The pushboats may also have flanking rudders located forward of the propellers (see Fig. 9.30). They significantly enhance manoeuvrability during astern operations. Note that IW vessels (selfpropelled and pushboats) are often equipped with bow thrusters which further improve their manoeuvrability. Moreover, according to CESNI (2019), IW vessel whose length exceeds 110 m must be equipped with a bow thruster that is controlled from the wheelhouse and is also effective when the craft is in an unladen state.

7.8 Concluding Remarks

Amongst all principal elements of the design[5] of IW vessels, which differ from seagoing ships as they are tailored for restricted waterways, an issue of fairway restrictions is essential only for hydrodynamics.

The most important specifics concerning the hydrodynamics in shallow water are the following:

- *Shallow water effects resistance* when $h/T < 4$ or $Fn_h > 0.5$. The wave amplitude and the Kelvin wave angle increase, thus resistance too (see Figs. 7.1 and 7.2).
- Concerning the *wave-making resistance*, three distinctive regions exist (see Figs. 7.3, 7.4 and 7.5):
  - *Sub-critical*—where shallow water does not impact resistance or just starts to be influential (below $Fn_h \approx 0.65$),
  - *Critical*—with a pronounced negative impact on resistance (i.e. resistance is larger than in deep water with a maximum at $Fn_h \approx 0.96$, theoretically $Fn_h = 1$), and
  - *Super-critical*—where effects are small but positive (i.e. resistance is smaller than in deep water, above $Fn_h \approx 1.25$).

[5]Hydrostatics, hydrodynamics, structures etc.

Note that the critical and super-critical regions are *important only for high-speed crafts*, but are given here as they are an essential part of shallow water hydrodynamics.

- For commercial IW vessels only the *sub-critical* region ($Fn_h < 0.65$) is of interest.
- The *viscous resistance* is larger in the shallow than in deep water (for the region of interest of considered IW vessels).
- The *propulsive efficiency* of selfpropelled ships and pushboats depends on many factors and is considerably lower than for the seagoing ships; usually around 40 ÷ 50% and 30 ÷ 40% respectively.
- Procedures for the *evaluation of required power* in shallow water are similar as in deep water, except that propulsive coefficients for IW vessels are less explored. The mathematical models for power evaluation, presented in Sect. 8.4, may be used in the preliminary design phases.
- The *squat* is an unavoidable effect of shallow water navigation. The most effective way to reduce it is by speed reduction. Approximate methods for squat evaluation are given in Sect. 8.5.
- The *wash* is more important for restricted than for shallow water navigation, though its effects endanger other users of the waterway and cause bank erosion in both cases.
- The IW vessel *manoeuvrability* is inherently endangered due to the nature of the fairways (river bends and the width restrictions in the first place). Besides, the manoeuvrability is generally worse in shallow than in the deep water.

Only the essentials of shallow water hydrodynamics, needed to fully comprehend other book's subjects, are presented in this chapter. Per se, that is not a standalone knowledge, but complementary with the textbooks already mentioned in the introduction.

References

Apukhtin P (1947) Influence of shallow water on resistance of passenger vessels (in Russian). Sudostroeniye, No. 4

Bilen B (1997) Innovations in pushboat convoy technology (in Serbian). Institute of Technical Sciences of Serbian Academy of Sciences and Arts, Belgrade, ISBN 86–80321–01-X

Bilen B, Žerjal M (1999) An optimized propulsive and manoeuvring system for river pushboats. In: RINA International conference on coastal ships and inland waterways, London

Carlton JC (2012) Marine propellers and propulsion. Butterworth-Heinemann. ISBN:9780080971230

CESNI (2019) European Standard laying down Technical Requirements for Inland Navigation Vessels (ES-TRIN). Edition 2019/1

EUDET (1998) Evaluation of the Danube Waterway as the key European Transport Resource

Georgakaki A, Sorenson SC (2004) Report on collected data and resulting methodology for inland shipping. The ARTEMIS Project. Department of Mechanical Engineering, Section for Energy Engineering, Technical University of Denmark

Heuser H (1986) Verdrängungsschiffe auf flachem Wasser. Teil II. Schiffstechnik Band 33, Heft 1

Hofman M, Kozarski V (2000) Shallow water resistance charts for preliminary vessel design. International Shipbuilding Progress 47(449)

Hofman M, Radojčić D (1997) Resistance and propulsion of fast ships in shallow water (in Serbian). Monograph, University of Belgrade, Faculty of Mechanical Engineering, ISBN 86–7083–297–6

ITTC (2014) Recommended procedures and guidelines—analysis of speed/power trial data, 7.5–04–01–01.2. Specialist Committee on Performance of Ships in Service, 27th ITTC 2014

ITTC (2017a) Recommended procedures and guidelines—captive model test, 7.5–02–06–02. Propulsion Committee of the 28th ITTC 2017

ITTC (2017b) Recommended procedures and guidelines—1978 performance prediction method, 7.5–02–03–01.4. Propulsion Committee of the 28th ITTC 2017

Jiang T (2001) A new method for resistance and propulsion prediction of the ship performance in shallow water. In: Proceedings of 8th PRADS Symposium, Shanghai

Kamar L (1996) Wassertiefe—Ihr Einfluss auf den Formfaktor von Seeschiffen. Schiff & Hafen 6

Karpov A (1946) Calculation of ship resistance in restricted waters (in Russian). Trudy GII. T. IV 2

Kuiper G (1992) The wageningen propeller series. MARIN Publication 92–001, ISBN 90–900 7247–0

Lackenby H (1963) The effect of shallow water on ship speed. Shipbuilder and Marine Engineer 70

Landweber L (1939) Tests of a model in restricted channels. Nav. Exp. Model Basin, Rep. 460

Liu J, Hekkenberg R, Rotteveel E, Hopman H (2015) Literature review on evaluation and prediction methods of inland vessel manoeuvrability. Ocean Eng 106. https://doi.org/10.1016/j.oceaneng

Marchal JLJ, Shen YD, Kicheva D (1996) An empirical formula to estimate the resistance of a convoy in a restricted waterway. J Ship Res 40(2)

Meyne K, Guesnet T (1996) Untersuchung einer Serie von Dusen-Skew-Propellern mit hoher Flugelzahl (Meyne-VBD-Serie). Sommertagung der Schiffbautechnischen Gessellschaft in Berlin

Millward A (1989) The effects of water depth on hull form factor. International Shipbuilding Progress 36(407)

Molland AF, Turnock SR, Hudson DA (2017) Ship resistance and propulsion—practical estimation of ship propulsion power. Cambridge University Press, ISBN, p 9781316494196

Nwogu O (2003) Boussinesq modeling of ship-generated waves in shallow water. PIANC USA Annual Meeting, Portland 2003

Pompée PJ (2015) About modelling inland vessels resistance and propulsion and interaction vessel-waterway—Key parameters driving restricted/shallow water effects. PIANC Smart Rivers, Buenos Aires

Radojčić D, Bowles J (2010) On high speed monohulls in shallow water. In: The second chesapeake power boat symposium, Annapolis

Raven H (2012) A computational study of shallow water effects on ship viscous resistance. In; 29th Symposium on naval hydrodynamics, Gothenburg 2012
Raven H (2016) A new correction procedure for shallow water effects in ship speed trials. In: Proceedings of PRADS, Copenhagen
Ren H, Ding Y, Sui C (2019) Influence of EEDI (Energy Efficiency Design Index) on ship–engine–propeller matching. J Mar Sci Eng 7(425). https://doi.org/10.3390/jmse7120425
Rotteveel E, Hekkenberg R, Liu J (2014) Design guidelines and empirical evaluation tools for inland ships. In; European inland waterway navigation conference, Budapest
Schlichting (1934) Ship resistance in water of limited depth resistance of seagoing vessels in shallow water. Jahrbuch der STG 35
UNECE (2018) European Code for Signs and Signals on Inland Waterways, Resolution No. 90
VBW (2016) Driving dynamics of inland vessels—vessel behaviour on European inland waterways and waterway infrastructure with special respect to German waterways. Published by VBW, Karlsruhe, ISBN 978-3-939230-49-6 (Online)
Xu-tao Q (1990) Study on the optimal coordination between ship and restricted waterway. International Shipbuilding Progress 37(411)
Zeng Q (2019) A method to improve the prediction of ship resistance in shallow water. PhD dissertation, Delft University of Technology
Zeng Q, Thill C, Hekkenberg R, Rotteveel E (2018) A modification of the ITTC57 correlation line for shallow water. J Marine Sci Technol

Chapter 8
Essential Expressions and Design Guidelines for the Preliminary Design of IW Vessels

8.1 Introduction

This chapter is unalike to others as it is conceived to be *a kind of a manual or a handbook* consisting of some essential expressions and design guidelines specific for IW vessels. In depth explanations are given in other chapters (Chap. 7 for instance) or in general naval architecture textbooks. It thus contains specific expressions and instructions needed in the early design phases which are not easy to find in ordinary textbooks. Consequently, this chapter is organized as follows:

- Section 8.2 gives design guidelines for hull form of selfpropelled IW ships and pushboats (Sects. 8.2.1 and 8.2.2, respectively). Follows a brief discussion on effects of propeller diameter that influence propeller tunnel design.
- Section 8.3 is supposed to give the essentials on stability and safety, but as these issues are strictly regulated by various rules, these subjects are abridged – see Note on Stability and Safety in the following.
- Section 8.4 gives MMs for *power* evaluation of IW selfpropelled ships and pushed convoys (Sects. 8.4.1 and 8.4.2).
- Section 8.5 presents methods for evaluation of *squat*.
- Section 8.6 gives several methods for estimation of *steel weight and lightweight mass* of selfpropelled cargo vessels (Sects. 8.6.1 and 8.6.2, respectively).
- Section 8.7 gives design guidelines for evaluation of *principal dimensions* of conventional selfpropelled container ships.
- Section 8.8 gives *typical worked examples* for selfpropelled vessel and pushed convoys (Sects. 8.8.1 and 8.8.2, respectively) showing detailed procedures for assessment of power, squat, energy and economic indicators.

D. Radojčić et al., *Design of Contemporary Inland Waterway Vessels*,
https://doi.org/10.1007/978-3-030-77325-0_8

8.2 Hull Form—Design Guidelines[1]

The economic factors and fairway restrictions (concerning the Danube, limited waterway depth, air draught, and size of lock chambers are already discussed in Chap. 2) dictate the main dimensions and hull form of IW vessels. Two IW vessel types will be treated in this section—the selfpropelled ships and the pushboats (see also Sects. 4.2.1, 4.2.2 and 4.3.1). The design guidelines that follow are a kind of the benchmark for these particular IW vessels types and stem from the Rhine experience (see Heuser 1986). Hull forms for extremely shallow water are discussed in Sect. 9.3.2; see also VEBIS project in Sect. 10.1.3 and Zibell and Müller (1996).

As a propeller diameter is directly connected to tunnel design (stern shape), Sect. 8.2.3 discusses the effects of a propeller diameter.

8.2.1 *Selfpropelled Ships*

IW vessels have to navigate efficiently and safely in shallow and restricted waterways. To enhance economic efficiency, the freight capacity should be increased and the power requirement reduced. This results in a full hull form with a relatively long parallel middle body. The geometry of the bow and the stern are important for the resistance and propulsion of the ship.

The sectional areas curve of a typical IW vessel is shown in Fig. 8.1. The length of the stern, the parallel middle body and the bow part of the vessel, are marked with $L_{WL\text{-}S}$, L_{PMB} and $L_{WL\text{-}B}$, while the volumetric displacements are marked with ∇_S, ∇_{PMB} and ∇_B respectively. Each part will be analysed separately.

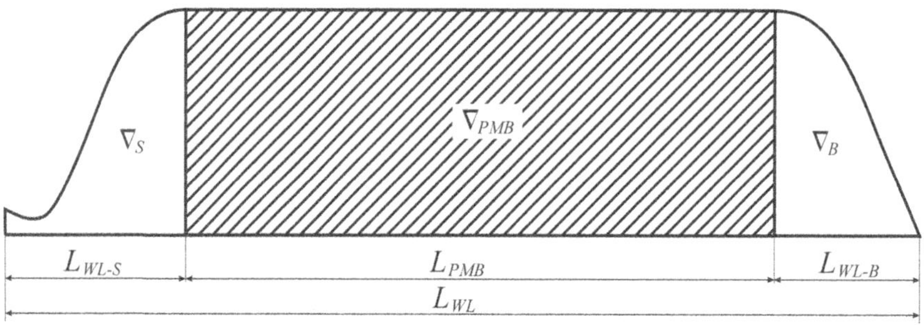

Fig. 8.1 Sectional areas curve of IW vessel. *Source* Heuser (1986)

[1]The main source of information in this section is Heuser (1986).

Parallel Middle Body

The parallel middle body (PMB) is similar to a cuboid ($L_{PMB} \times B \times T$) with rounded bilges (the transition between the bottom and the sides). The full form of the PMB provides maximal volumetric displacement, i.e. cargo-carrying capacity. The length of the parallel middle body is usually up to 70% of L_{WL}, with block coefficient of close to 0.995. Consequently, the block coefficient of IW vessel can be estimated from:

$$C_B = \frac{L_{WL-S} \cdot C_{B-S} + L_{PMB} \cdot 0.995 + L_{WL-B} \cdot C_{B-B}}{L_{WL}}$$

where

C_{B-S} block coefficient of the stern part of the vessel,
C_{B-B} block coefficient of the bow part of the vessel.

Bow Shape

The bow is a front part of the ship where the cross-sectional area starts to decrease (Fig. 8.1). Four typical bow types exist, see Fig. 8.2:

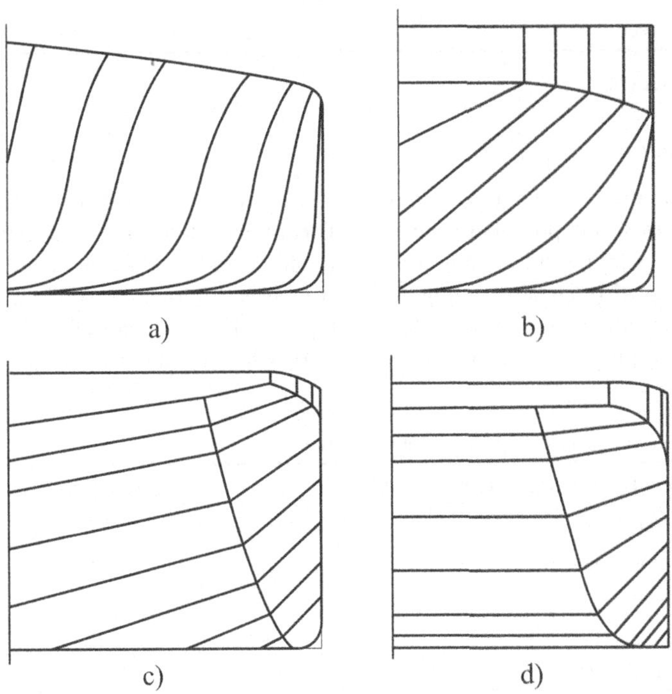

Fig. 8.2 Typical bow types of IW vessels. *Source* Heuser (1986)

- *U-shaped bow* sections (Fig. 8.2a) for slower (relative to water) IW vessels. Compared to other bow types, U-shaped bow offers higher displacement, thus deadweight tonnage too, for the identical shape of the rest of the vessel. However, it also creates a higher resistance.
- *V-shaped bow* sections (Fig. 8.2b) is better for higher speeds (relative to water) as it requires less propulsion power. This bow type is convenient for the selfpropelled vessels designed to push additional barges (coupled convoys).
- *Wedge-shaped bow* (Fig. 8.2c) is mainly used for the barges, due to the simple construction and lower building cost. This bow type requires more power for the same speed, compared to U- and V-shaped bows.
- *Pontoon bow* (Fig. 8.2d) with a flat bottom is an even simpler and less efficient type of wedged-shaped bow. It can be found on the Europa IIa barges, while the wedge-shaped bow is a feature of Europe IIb barges, see Sect. 4.2.1.

Recommendations based on the experience, regarding the waterline length and the block coefficient of the bow part of an IW ship are given in Table 8.1.

Table 8.1 Recommendations for IW vessel's bow

Bow type	Bow length	Block coefficient
Conventional bow, "U" or "V"	$L_{WL-B} \geq 1.55 \cdot B$	$C_{B-B} \leq 0.78$
Wedge-bow	$L_{WL-B} \geq 5.0 \cdot T$	$C_{B-B} \leq 0.75$

Source Heuser (1986)

Stern Shape

Special attention should be paid to the geometry of the ship stern to facilitate good water inflow to the propeller/propellers. This directly impacts the propulsive efficiency. The width of the stern is equal to the ship breadth, while the length depends on the adopted propulsion configuration (which is different for single and double propeller installations), see Table 8.2 and Fig. 8.3. The stern part of a ship is longer for a single-propeller installation (Fig. 8.3a), while in the case of two-propeller

Table 8.2 Recommended length of the stern of IW vessels

Propulsion configuration	Stern length	Other data
One propeller	$L_{WL-S} \geq 1.9 \cdot D + 1.85 \cdot B$	-
Two propellers with parallel shafts	$L_{WL-S} \geq 3.2 \cdot D + 1.45 \cdot B$	$d \approx 0.45 \div 0.50 \cdot B$ $\beta \approx 0°$
Two propellers with convergent shafts	$L_{WL-S} \geq 2.5 \cdot D + 1.65 \cdot B$	$d \approx 0.28 \cdot B$ $\beta \approx 5° \div 7°$

Source Heuser (1986)

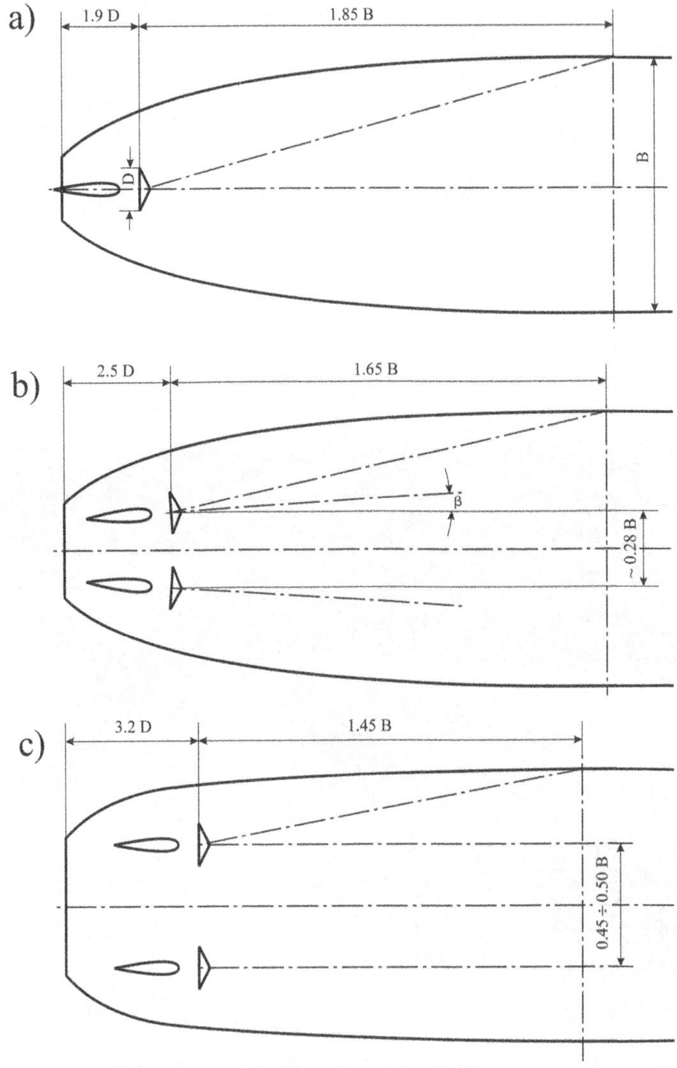

Fig. 8.3 Typical propulsion configurations of IW vessels. *Source* Heuser (1986)

configurations the length depends on the distance between the propeller shafts (d), see Fig. 8.3b, c.

To avoid unfavourable interaction between the propellers (when the distance between the shafts is reduced due to other design constraints) the angle between the shaft axis and a centerline (β) should be in the range between 5° and 7°, as shown in Fig. 8.3b. In that case, the ship can still maintain a straight course, even if one drivetrain is out of order (Heuser 1986).

Smaller IW vessels, with displacements of up to 1000 t, are usually equipped with a single-propeller. The larger ones may have a single propeller, but more

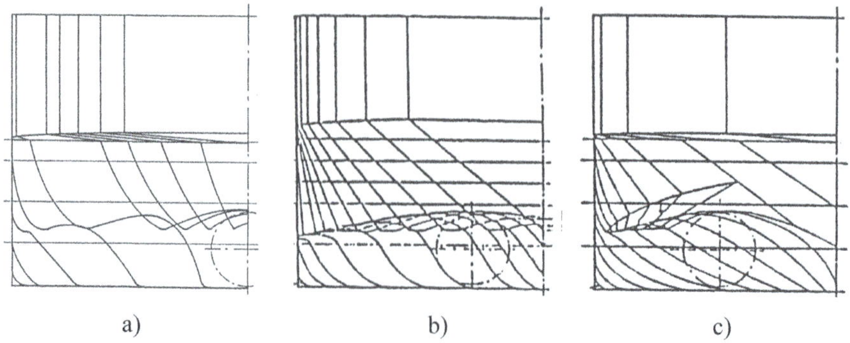

Fig. 8.4 Typical stern sections of IW vessels. *Source* Heuser (1990)

Fig. 8.5 The stern of a single propeller ship with a skirt. *Source* Radojčić (2009)

often are equipped with two propellers. Some of the advantages of double-propeller configurations specific to IW ships are:

- Reduced propeller load and uniform water inflow (resulting in higher propulsive efficiency and reduced probability of cavitation, thus reduced vibrations too).
- Shallow water operation (due to smaller propellers).
- Better manoeuvring performances (enabling also multiple rudder installation).
- Lower immersed transom area (thus reduced ship resistance).

Fig. 8.6 The stern of twin-propeller selfpropelled IW vessel, without propeller tunnels (Selfpropelled ship Deligrad—see Fig. 8.26)

As expected, the main disadvantages of multi-propeller installations are increased investment and maintenance costs.

Maximal recommended values of block coefficient of the stern part are given in Table 8.3.

Table 8.3 Recommended block coefficient of IW vessel's stern

Number of propellers	Block coefficient
One propeller	$C_{B-S} \leq 0.31 + 0.15 \cdot \frac{L_{WL-S}}{B}$
Two propellers	$C_{B-S} \leq 0.32 + 0.17 \cdot \frac{L_{WL-S}}{B}$

Source Heuser (1986)

Propeller Tunnels

IW vessels usually have propeller tunnels, which enables installation of the propellers with diameters larger than the nominal ship draught, thus enabling more efficient propulsion in lower water depths. A propeller tunnel in combination with a skirt also prevents propeller ventilation (sucking the air from the surface) when the vessel operates at a low draught. In brief, the propeller tunnels increase the energy efficiency of the vessel. However, the propeller tunnels have to be carefully designed to enable a good water inflow to the propeller. Typical stern sections of IW vessels are shown in Fig. 8.4. Figure 8.4a, b show a typical form of a propeller tunnel, while a tunnel with skirts is shown in Figs. 8.4c and 8.5. Note however that the selfpropelled ships may also have a simple stern without a tunnel or a skirt, as shown in Fig. 8.6. This kind of stern can be effectively improved (for navigation at reduced draught) with adjustable retractable tunnels, see Sect. 9.4.8 and Fig. 9.47.

110 m IW ships usually have a propeller diameter of around 1.75 m. Rarely more than 1.8 m even in cases where draughts are large and larger propellers could be installed. For smaller draughts (below 2 m) somewhat deeper tunnels are required to accommodate large enough propeller diameters. Newer designs have even smaller propellers (1.4 m $\div$ 1.6 m) to improve resilience in low water periods.

8.2.2 *Pushboats*

The pushboats, as a rule, have deep propeller tunnels due to small draught and inherently heavily loaded propellers (see Sect. 4.2.2 as well). Recommendations for the tunnel geometry, based on model experiments performed in VBD (today DST), are given in Fig. 8.7 and Table 8.4.

In Bilen and Bilen-Katić (1997) for typical 2-propeller Danube pushboats of around 2000 kW with $D \approx 1.7$ m (other dimensions are $L = 32 \div 42$ m, $B = 10 \div 13$ m, $H = 2.6 \div 2.8$ m, $T = 1.6 \div 2.1$ m, $\Delta = 370 \div 525$ t), a bit different tunnel dimensions are recommended. Another school of design is given in Latorre (1994); presented is a 4-step procedure for the design of a river pushboat for operation in a water depth of 2.5 $\div$ 3.5 m. See also Mulder and Williams (2006).

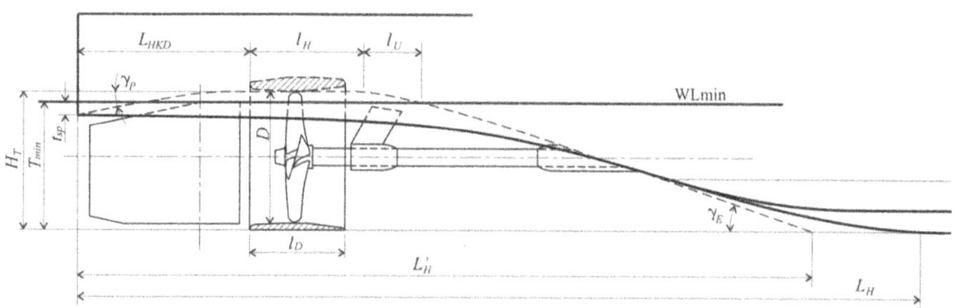

Fig. 8.7 The geometry of a propeller tunnel of IW pushboat with respect to the minimal draught (T_{min})

The Heart of a Pushboat—Tunnel Stern Arrangement.

A pushboat is basically an engine room, steering compartment and a wheelhouse of the whole convoy. Its performance without the convoy is irrelevant. Therefore, from a naval architect's viewpoint the essence of a river pushboat design lies in a tunnel stern arrangement, particularly concerning the following aspects:

Table 8.4 Recommended dimensions of a propeller tunnel for a pushboat (relates to minimal draught)

Characteristics		Recommended Value		Characteristics		Recommended Value	
L_H/L_{WL}	[-]	=	$0.42 \div 0.58$	$t_{sp\ min}$	[m]	=	$0.05 \cdot D$*
L_H/D	[-]	=	$5.4 \div 7.3$	γ_P	[°]	<	12
L'_H/L_H	[-]	=	$0.80 \div 0.90$	γ_E	[°]	=	$15 \div 25$
l_H/l_D	[-]	≥	1.3	H_T	[m]	=	$1.008 \cdot D + 0.08$
l_U/D	[-]	≥	0.75	L_{HKD}	[m]	=	$2.6 \cdot T_{min}$* - 0.2
l_D/D	[-]	=	$0.50 \div 0.75$	D	[m]	=	$1.38 \cdot T_{min}$ - 0.11

Note: The tip clearance between the propeller blades and a nozzle should be up to $0.01 \cdot D$
* vessel ready for navigation with minimal supplies

Source Heuser (1986)

- The tunnel should accommodate a ducted propeller of an appropriate diameter that can absorb necessary power needed for pushing a given tow at a certain speed. Propeller diameter may be up to 20% larger than is the vessel's draught. Moreover, to facilitate accommodation of as large as possible propeller, nozzles are often deformed a) on the top as are countersunk into the tunnel (for some 90 deg. or 25% of a circle), and b) on the bottom as the segment of an outer diameter of a nozzle is cut off. The distribution of total power needed for pushing a convoy is often the only solution, resulting in twin and triple, or even quadruple propeller installations (see Hagesteijn et al. 2015 and shallow draught concepts, Sects. 10.1.3 and 10.3.3).
- Ducted propeller enables up to $15 \div 20\%$ better efficiency and $20 \div 30\%$ higher bollard pull compared to naked propellers. The drawbacks of ducted propellers are suction of stones and sand in very shallow water, and ice floes during the ice periods (see Sect. 9.4.3).
- The design of a propeller and a nozzle directly influences the propulsive efficiency η_D which is generally low (between $0.30 \div 0.40$, rarely more, see Sect. 7.3. Amongst others, it depends on the water depth, the number of barges in the tow, their formation and loading. The same ducted propeller must also satisfy the stopping requirements of the whole convoy.
- Tunnel/tunnels should enable a good water inflow to ducted propellers. For shallow water operation, the water inflow from the sides (instead under the keel) is essential. Note that the central propeller/propellers are particularly sensitive as they have to "steal" water from the side propellers. Inadequate water inflow leads to low efficiency and excessive vibrations. The tunnel geometry should prevent ventilation a) from the sides over the bilges in ahead sailing, and b) from the stern in astern or backing condition, usually during the crash-stop manoeuvres of a heavy convoy. Note that inappropriate tunnel design can cause tunnel losses,

thus its geometry is important. Moreover, due to small hydrostatic pressure and overloading of the propeller and nozzle's inner ring cavitation is also possible.

- The manoeuvrability of a convoy is achieved with 1 or 2 (rarely 3) ahead rudders and 2 flanking rudders (both per propeller). Note that a large number of rudders is necessary as a) draught is restricted, thus they cannot be deep enough, and b) steering of the whole convoy (in some cases 300+ m long) in a confined waterway must be facilitated (see Sect. 9.4.2). A large number of rudders requires independent steering gears.
- Flanking rudders are used only for astern operations and are positioned forward of the propeller. When not in use they disturb the water inflow, thus their angle (opening) is important as it influences the propulsive efficiency. Contemporary pushboats have bow thrusters (or three-channel thrusters, see Fig. 9.44), or innovative retractable flanking rudders (see Fig. 9.30) instead of the conventional flanking rudders.
- If rudder propellers are used the tunnels should allow rotation of the propulsors around their vertical axis. This usually results in the employment of considerably smaller propellers than it would be the case with a conventional installation.
- The pushboat's stern draught shouldn't vary much, regardless of its displacement. Thus spending of the consumables should influence only the bow draught. This, however, may jeopardise the efficiency of a bow thruster and ballasting may be necessary.

8.2.3 *Positive Effects of Large Propeller Diameter*

Ideal Propeller Efficiency

For the *first and very approximate estimates* of propeller efficiency, which should stress the importance of a large propeller diameter for a given propeller thrust (the ship speed is assumed to be given in advance), a well-known momentum theory may be used according to which the (open water) propeller efficiency η_O is:

$$\eta_O = \eta_I \cdot FM,\ \eta_I = 2/\left(1 + \sqrt{1 + C_{Th}}\right),$$

$$C_{Th} = T/\left(0.5 \cdot \rho \cdot V_A^2 \cdot D^2 \cdot \pi/4\right).$$

where:

η_I ideal efficiency (of an actuator/propulsor with an infinite number of blades and no frictional or rotational losses), [-].
C_{Th} thrust loading coefficient, [-].
T required propeller thrust, [N].
ρ water density [kg/m^3].

V ship speed [m/s].
V_A speed of advance of the propeller, [m/s], $V_A = V \cdot (1\text{-}w)$, where w is the wake fraction coefficient
D propeller diameter [m].
FM Figure of Merit.

The equation for ideal efficiency (see also Fig. 8.8) indicates that the larger propeller loading (C_{Th}), the smaller is the ideal efficiency. Since $C_T = \text{f}\ (D)$ it follows that η_I increases as D increases (for $V_A = \text{const.}$). Note that the propeller efficiency η_O is considerably smaller than η_I.

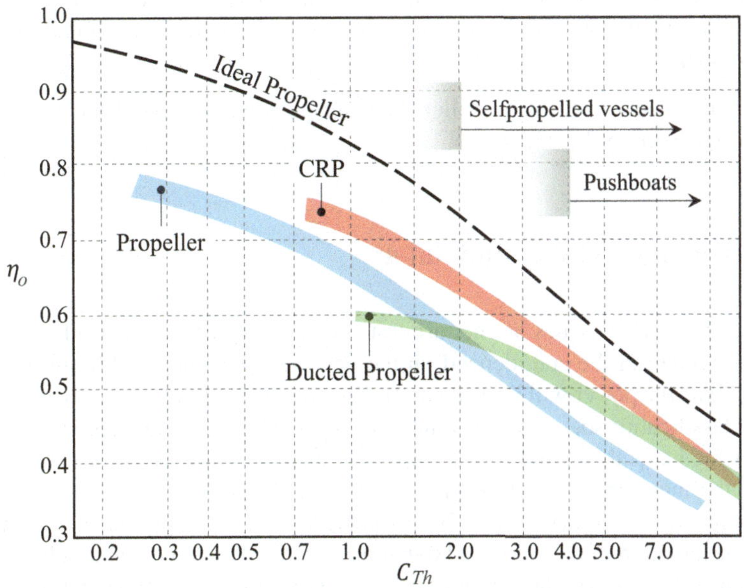

Fig. 8.8 Ideal efficiency and maximal achievable open water efficiencies of some propeller types

FM and the wake fraction (w) vary to a great extent, for seagoing ships they may be assumed to be around $0.7 \div 0.8$ and $0.15 \div 0.20$, respectively; for IW vessels these values are different. However, as the abovementioned expressions may be used only for comparative purposes to indicate the relation between D and T, assuming V is predetermined, exact values of FM and w are not important as long as they are assumed to be constant throughout the comparative process.

Propeller Cavitation

In previous, propeller cavitation was not considered. One of the simplest recommendations for IW cargo vessels, to avoid cavitation, is the following: $P_D/(D^2\pi/4) \leq 350 \div 400\ \text{kW/m}^2$. Consequently, the larger the propeller diameter is the lower the propeller cavitation will be. More about cavitation can be found in Carlton (2012) and Molland et al. (2017).

Right Choice of Propeller Diameter

Concerning the right choice of a propeller diameter, as once chosen it hardly can be changed later, two different situations occasionally occur:

(a) "Pushboat case"—The diameter is chosen in the early design stage according to the initially designed draught. When the vessel is finished the draught happens to be larger than initially anticipated, while the propeller kept its initially determined diameter (hence the benefits of potentially larger propeller diameter were not fully exploited).
(b) "Selfpropelled vessel case"—The propeller diameter is chosen according to the minimal draught requirements, in the first place to prevent ventilation (thus a lower efficiency may be expected when the vessel is exploited at larger draughts).

Note that the choice of propeller diameter goes hand in hand with the tunnel design (see Sects. 8.2.2 and 9.4.3).

8.3 Note on Stability and Safety

Stability and safety issues are not arbitrary. Unlike other topics considered in this chapter, which, in the end, depend on the designer (e.g. choice of a hull form, or method how the power will be evaluated), quite opposite, they are strictly prescribed by the classification societies' rules and statutory regulations (see Chap. 5). Moreover, besides the general demand for all ships that "their stability has to be satisfactory" and/or the metacentric height positive, for most ship types stability requirements defined in those documents have to be proved by *lege artis* stability calculations which include evaluation of the static and dynamic stability. However, some rather simple constraints that can be checked purely upon ship dimensions, type and only basic hydrostatic calculations, can be extracted from the abovementioned regulations and serve in very early design stages. As an example, two of such, related to freeboard and metacentric height, are presented here. Other safety issues, e.g. safety requirements considering fire protection, visibility from the wheel house etc. are beyond the scope of this book and the reader is advised to consult the relevant rules and regulations listed in Chap. 5. Accordingly, the reader is notified to have in mind the following:

(a) The expressions which are given in the following are among some *sine qua non* conditions; i.e. accomplishment of those does not absolve the designer from conducting further detailed stability calculations.
b) To be in line with the main objective of this book, quoted are requirements for IW dry cargo, open deck ships (container and bulk carriers, often referred as Type C) and sailing in Zone 3; requirements for passenger ships and tankers, as well for other Zones, may significantly differ.

Freeboard

In the case of IW vessels, in addition to *freeboard* (vertical distance from the water line to the deck at ship side amidship), a *safety distance* (*safety clearance*) is introduced. It is measured as the vertical distance from the waterline corresponding to the maximal draught to the *lowest point* above which the ship hull cannot be considered as watertight. For ship types considered here, that point is usually the upper edge of hatch coaming.

According to UNECE 2006 Resolution No 61[2] (see Chap. 5) for Ship type C in Zone 3:

- Minimal freeboard is 150 mm and
- Minimal safety distance is 500 mm.

Note that this implicitly defines the minimal hatch coaming height of 350 mm. Commonly, due to longitudinal strength requirements, container stuck arrangement and desire that the coaming functions also as a hatch railing—the hatch coaming is much higher.

Metacentric Height of Container Ship

According to EC (2006), see Chap. 5, the minimal metacentric height of a container ship should be:

- $GM_{min} = 0.5$ m when carrying secured containers
- $GM_{min} = 1.0$ m when carrying non-secured containers

Note that, as explained at the beginning of the section, these are only "the first minimum" values. The same statutory documents require additional calculation of the static heel of a vessel exposed to simultaneous action of heeling moments caused by beam wind force and the centrifugal force due to turning—which should not exceed 5 degrees or the angle at which the deck enters the water. This requirement can lead to significantly (somethimes unproperly) higher values of GM_{min}, especially in the case of shallow draught vessels (see Bačkalov 2015 and Bačkalov et al. 2016).

Probabilistic Approach

Needless to say, the above rules are *deterministic*. A *probabilistic* approach, already present in the maritime sector, is not yet applied in IW stability regulations. To put it simple, instead of prescribing explicit minimal values of metacentric height or freeboard, probabilistic rules would specify *acceptable probabilities* of occurrence of unwilling events (incidents). Note that certain incidents are common for various types of ships (e.g. flooding of cargo hold) while some are quite specific (e.g. sliding of containers). Consequently, a probabilistic approach can provide rules that are better tailored for particular ships. Equally important is the fact that a probabilistic approach inherently leads to harmonization of the rules (for instance, required freeboard or safety distance may be different on the Rhine and on the Danube, but the

[2]Note that different rules may apply in the same area of navigation which subsequently leads to unequal safety levels (see Bačkalov 2015).

probability of cargo hold flooding should not be different). Therefore, an introduction of a probabilistic approach in IW vessels stability assessment is expected.

A safety assessment procedure, based on probabilistic analysis, is discussed in Bačkalov (2015). Amongst the conclusions of that research is that the deterministic rules are unable to reflect atypical designs. Specifically, the minimal stability requirements of the Directive 2006/87/EC (see Chap. 5) are too low, at least for the examined Danube shallow draught container vessel navigating in zone 3. It is therefore proposed that the freeboard and hatch coaming height should be based on a desired *range of safe navigation*,[3] instead on the waterway navigation zone.

Judicious and farsighted discussion of those issues for container ships can also be found in Hofman and Bačkalov (2010). Some of that is applied for a conceptual design of a vessel presented in Sect. 10.3.2.

8.4 Power Evaluation

Four empirical methods or Mathematical Models (MMs) for the estimation of delivered power, developed specifically for IW selfpropelled vessels and pushed convoys, are presented in this section. These are Simple and Complex MMs for both vessel types (Sects. 8.4.1 and 8.4.2 respectively). Complex MMs are developed using the Artificial Neural Network (ANN). Amongst the input parameters are the vessel's principal dimensions or pushed convoy configuration and main waterway characteristics. These MMs may be used in the early design stages.

8.4.1 Delivered Power of IW Selfpropelled Vessels

Simple Method

A simple, hence a very approximate method for estimation of influence of shallow water effects on delivered power of IW selfpropelled vessels, developed through broad investigation done in VBD (now DST), is presented in Heuser (1986). According to this method, to determine the total resistance of a particular IW selfpropelled vessel, the resistance of the *reference vessel* that has a ratio $L/B = 10$ has to be modified. Necessary corrections require evaluation of the specific resistance ratio (r) which depends on L/B and Fn_h, see Fig. 8.9.

To estimate the *delivered power*, the propulsive efficiency η_D of a vessel should be known. It can be evaluated from the diagram given in Fig. 8.10 (η_D is a function of V and p; the diagram is valid for *two-propeller vessels with propellers in nozzles*). The coefficient p can be calculated from:

---

[3]Combinations of influential parameters (wind, speed, number of container layers etc.) for which the probability of an unwilling event is still satisfactory.

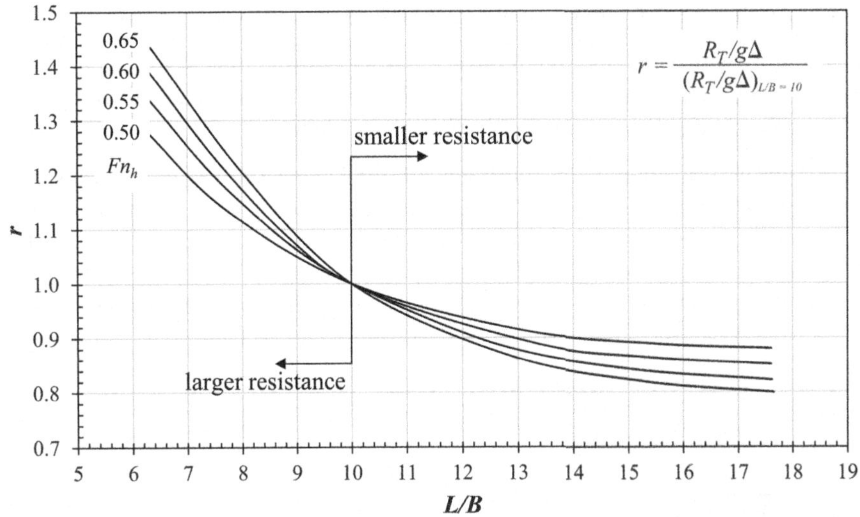

Fig. 8.9 Specific resistance relative to $L/B = 10$ vessel. *Source* Heuser (1986)

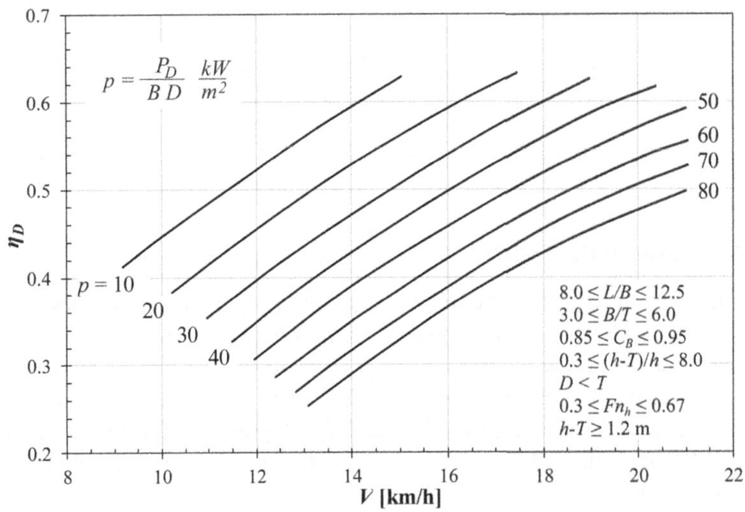

Fig. 8.10 Efficiency η_D for a two-propeller vessel with nozzles.[4] *Source* Heuser (1986)

$$p = \frac{P_D}{B \cdot D}\left[\frac{kW}{m^2}\right]$$

where D is the propeller diameter.

Concerning the diagram given in Fig. 8.10 note the following:

- The waterway depth does not influence η_D (which is not true, see Sect. 7.3)
- Due to the applicability limit $(h - T) \geq 1.2$ m, the given method cannot be applied to some critical shallow water sectors on the Danube.

For power evaluation of IW selfpropelled ship having $L/B \neq 10$, in addition to the diagrams given in Figs. 8.9 and 8.10, the diagram for a *reference* or *standard* two-propeller Rhine cargo vessel is also needed (for instance the one shown in Fig. 8.11, from Zöllner 2003); presented values are for water depth $h = 5$ m and ship of $L = 110$ m, $B = 11.4$ m, $T = 2.5$ m, $\Delta = 2752$ t, two propellers, $D = 1.76$ m in nozzles.

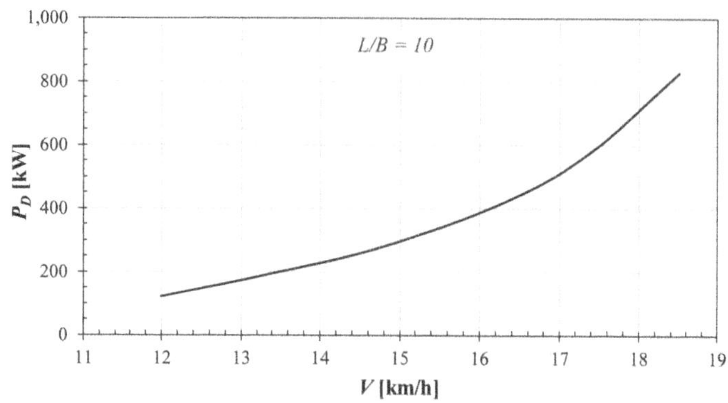

Fig. 8.11 $P_D - V$ diagram for "standard" 110 m vessel, $h = 5$ m. *Source* Zöllner (2003)

Complex Mathematical Model

The Mathematical Model (MM) for the estimation of the delivered power of IW selfpropelled cargo vessels, presented in Simić (2012), is based on the results of tests conducted on 50 IW selfpropelled cargo vessels consisting of more than 600 data points. Artificial Neural Network (ANN) with a sigmoid activation function ($sig = 1/(1 + e^{-x})$) was used for the MM development, resulting in a relatively complex MM, i.e. equation with 144 polynomial terms. Goodness-of-fit of the MM, expressed through the root-mean-square deviation (RMSD),[5] is around 3.7%.

Necessary input and output parameters or, respectively, independent and dependent (target) variables, for MM, are the following:

[4]For naked prropellers, the efficiency η_D would be approximately 5% lower.

[5]The root-mean-square deviation (RMSD) is the statistical measure of the differences between values predicted by a MM and the values actually observed (measured). Lower RMSD indicates better accuracy of the MM.

INPUT	OUTPUT
$X_1 = L/\nabla^{1/3}$	C_D
$X_2 = B/T$	
$X_3 = b$	
$X_4 = h$	
$X_5 = V/\sqrt{g \cdot R_h}$	

where

∇ volumetric displacement (m^3).
b waterway breadth (m).
V ship speed through water (m/s).
R_h hydraulic radius (m), $R_h = (b \cdot h - B \cdot T)/(2 \cdot h + 2 \cdot T + b + B)$.
C_D coefficient of delivered power [-], $C_D = P_D/(\rho \cdot g \cdot \nabla \cdot V)$.
Other symbols as above.

Database distribution and boundaries of applicability for the MM are presented in graphical format and as a set of equations suitable for programming, see Fig. 8.12 and Table 8.5 respectively.

From the database, used for the development of MM (Fig. 8.12), an *uneven distribution of data* can be noticed, particularly concerning the fairway restrictions.

Namely, most of the measurements were conducted in shallow water, while deep water (actually deeper as is not indefinitely deep)[6] was not covered enough. To overcome this, and enable a smooth transition between deep and shallow water hydrodynamic effects, additional conversion of some input parameters is necessary, in particular of X_4 and $X_5 = f(X_4)$.

In other words, if $V < V_{TR}$ the equivalent depth (h_{EQ}) should be evaluated from V_{TR} [km/h] $= 3.6 \cdot V$ (m/s), and h_{EQ} should then be used for evaluation of equivalent hydraulic radius (R_{hEQ}). Thus, new independent variables $X_4 = h_{EQ}$ and $X_5 = f(V, R_{hEQ})$ are obtained.

The *transitional speed* (V_{TR}), depending on fairway breadth and depth, can be estimated from the following equation:

$$V_{TR}(\text{km/h}) = h \cdot (0.0000023 \cdot b^2 + 0.0013493 \cdot b) + 2.3236268) \\ + 0.000009 \cdot b^2 + 0.004332 \cdot b - 3.600712$$

[6]Essentially, the real deep-water depth, whichever, is recalculated to depth of water which corresponds to the transitional speed V_{TR}. This is a kind of boundary line between the subcritical and critical regimes. In the subcritical region, from the hydrodynamic point of view, any depth larger than that evaluated from V_{TR} is irrelevant (see Sect. 7.2.1). For shallow water navigation, the ship's speed is always higher than the transitional speed, thus there is no need for conversion of input parameters.

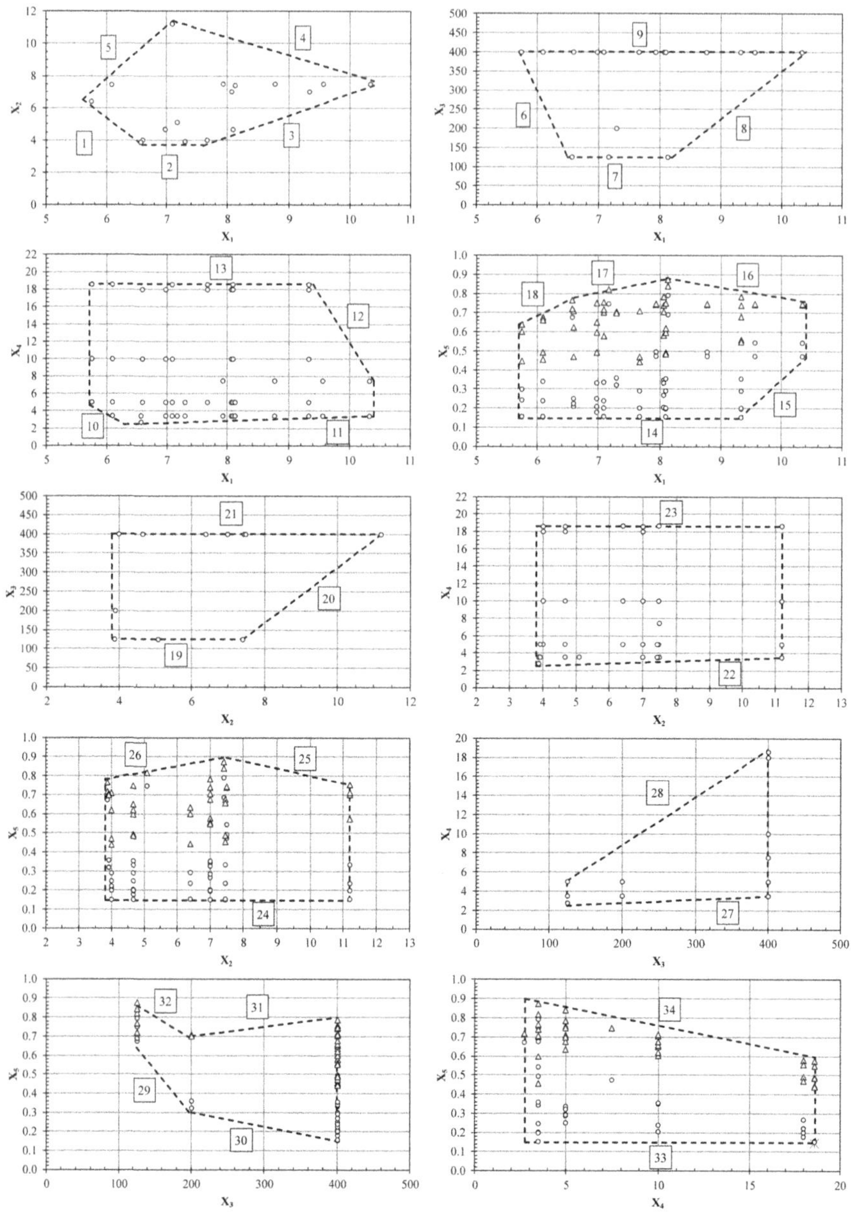

Fig. 8.12 Database distribution and boundaries of applicability of a MM for IW selfpropelled vessels

Table 8.5 Boundaries of applicability suitable for programming of a MM for IW selfpropelled vessels

No.	Boundaries	
1	$5.6 \leq X_1 < 6.6$	$X_2 \geq -2.8 \cdot X_1 + 22.18$
2	$6.6 \leq X_1 < 7.6$	$X_2 \geq 3.7$
3	$7.6 \leq X_1 \leq 10.4$	$X_2 \geq 1.34 \cdot X_1 - 6.52$
4	$7.1 \leq X_1 \leq 10.4$	$X_2 \leq -1.12 \cdot X_1 + 19.35$
5	$5.6 \leq X_1 < 7.1$	$X_2 \leq 3.27 \cdot X_1 - 11.79$
6	$5.7 \leq X_1 < 6.5$	$X_3 \geq -343.75 \cdot X_1 + 2359.375$
7	$6.5 \leq X_1 < 8.2$	$X_3 \geq 125$
8	$8.2 \leq X_1 \leq 10.4$	$X_3 \geq 125 \cdot X_1 - 900$
9	$5.7 \leq X_1 \leq 10.4$	$X_3 \leq 400$
10	$5.7 \leq X_1 \leq 6.3$	$X_4 \geq -3.67 \cdot X_1 + 25.6$
11	$6.3 \leq X_1 \leq 10.4$	$X_4 \geq 0.24 \cdot X_1 + 0.96$
12	$9.4 \leq X_1 \leq 10.4$	$X_4 \leq -11.1 \cdot X_1 + 122.94$
13	$5.7 \leq X_1 \leq 9.4$	$X_4 \leq 18.6$
14	$5.7 \leq X_1 \leq 9.3$	$X_5 \geq 0.147$
15	$9.3 \leq X_1 \leq 10.4$	$X_5 \geq 0.29 \cdot X_1 - 2.55$
16	$8.1 \leq X_1 \leq 10.4$	$X_5 \leq -0.053 \cdot X_1 + 1.31$
17	$6.6 \leq X_1 < 8.1$	$X_5 \leq 0.065 \cdot X_1 + 0.35$
18	$5.7 \leq X_1 < 6.6$	$X_5 \leq 0.153 \cdot X_1 - 0.233$
19	$3.8 \leq X_2 < 7.4$	$X_3 \geq 125$
20	$7.4 \leq X_2 < 11.2$	$X_3 \geq 72.37 \cdot X_2 - 410.53$
21	$3.8 \leq X_2 \leq 11.2$	$X_3 \leq 400$
22	$3.8 \leq X_2 \leq 11.2$	$X_4 \geq 0.135 \cdot X_2 + 1.98$
23	$3.8 \leq X_2 \leq 11.2$	$X_4 \leq 18.6$
24	$3.8 \leq X_2 \leq 11.2$	$X_5 \geq 0.147$
25	$7.5 \leq X_2 \leq 11.2$	$X_5 \leq -0.038 \cdot X_2 + 1.18$
26	$3.8 \leq X_2 \leq 7.5$	$X_5 \leq 0.032 \cdot X_2 + 0.66$
27	$125 \leq X_3 \leq 400$	$X_4 \geq 0.0036 \cdot X_3 + 2.045$
28	$125 \leq X_3 \leq 400$	$X_4 \leq 0.05 \cdot X_3 - 1.18$
29	$125 \leq X_3 \leq 200$	$X_5 \geq -0.0047 \cdot X_3 + 1.23$
30	$200 \leq X_3 \leq 400$	$X_5 \geq -0.00075 \cdot X_3 + 0.45$
31	$200 \leq X_3 \leq 400$	$X_5 \leq 0.0005 \cdot X_3 + 0.6$
32	$125 \leq X_3 \leq 200$	$X_5 \leq -0.0023 \cdot X_3 + 1.15$
33	$2.75 \leq X_4 \leq 18.6$	$X_5 \geq 0.147$
34	$2.75 \leq X_4 \leq 18.6$	$X_5 \leq -0.019 \cdot X_4 + 0.95$

After the input variables X_4 and X_5 are re-evaluated as explained above, follows the evaluation of a coefficient of delivered power, thus delivered power too, according to the MM given in Table 8.6 and flow chart presented in Fig. 8.13. An example

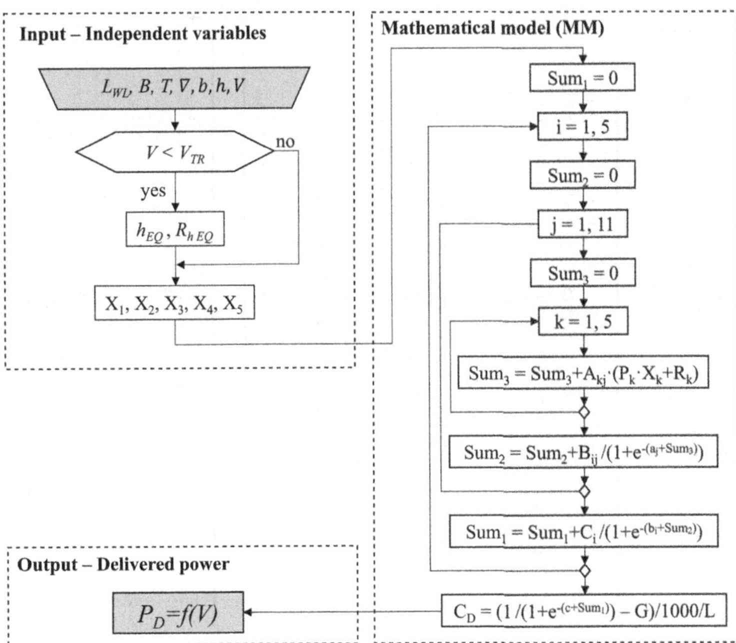

Fig. 8.13 The Flow chart for the evaluation of the delivered power of IW selfpropelled vessels

Table 8.6 Mathematical model and corresponding ANN coefficients

$$C_D = \frac{sig\left\{c + \sum_{i=1}^{5}\left\{C_i \cdot sig\left[b_i + \sum_{j=1}^{11}\left[B_{ij} \cdot sig\left(a_j + \sum_{k=1}^{5}\left(A_{kj} \cdot (P_k \cdot X_k + R_k)\right)\right)\right]\right]\right\}\right\} - G}{L}$$

A_{1j}	A_{2j}	A_{3j}	A_{4j}	A_{5j}	a_j
-2.62973	1.03914	0.54515	-5.22671	0.54363	0.83240
-2.45093	0.02407	1.29390	-0.82279	1.99099	-2.12594
0.34665	1.10963	-1.53864	-10.63455	6.50289	-4.54323
3.07244	-1.76541	-0.19061	-1.25267	1.60924	1.20133
4.38140	3.56541	0.95153	-0.78922	2.46340	2.86918
-2.47526	-0.00234	1.22394	-5.83226	-0.86820	0.14227
-1.50350	0.78361	-2.33393	-0.89590	2.91578	2.17729
3.30634	2.21796	2.03846	5.48810	-1.39568	-2.34102
0.80852	0.32919	-4.21854	6.72520	14.27642	-15.68451
0.57896	-5.87356	0.64051	-9.73264	8.95593	-6.73497
-0.41949	1.62357	0.51128	5.05424	2.56605	-4.88347

C_1	C_2	C_3	C_4	C_5	c
-4.85348	0.74726	-4.54208	-3.74272	-6.29805	13.10090

P_k	R_k	G	L
0.19603	-1.07549	0.04841	0.03232
0.12305	-0.42813		
0.00327	-0.35909		
0.05678	-0.10615		
1.24647	-0.13955		

B_{j1}	B_{j2}	B_{j3}	B_{j4}	B_{j5}	B_{j6}	B_{j7}	B_{j8}	B_{j9}	B_{j10}	B_{j11}	b_i
1.86458	-1.60346	-1.39233	-0.78252	3.80301	0.99054	-1.53327	-3.80687	-10.45898	-6.49082	-1.31314	5.06873
1.52137	-0.20169	-0.13242	-1.16487	2.29852	0.50053	-0.41515	-3.10789	-4.87319	-3.18205	-1.42125	3.84613
2.17873	-1.42456	1.97323	-1.15006	1.31352	-2.00239	-2.37647	-2.83132	-2.35419	1.10889	-0.36411	2.79079
3.21958	-2.71332	-6.61645	-4.34655	0.89775	-3.86923	-2.40985	3.89709	-0.87180	0.45096	-3.47135	4.27677
1.00840	-1.17829	-1.29786	-0.89079	3.88267	-0.25514	-1.01372	-1.92177	-8.20337	-12.13766	-0.51855	4.56944

of results obtained by the presented MM is shown in Fig. 7.9 and the calculation procedure is explained within an example given in Sect. 8.8.1.

8.4.2 *Delivered Power of Pushed Convoys*

Two Mathematical Models (MMs) are given, Simple and Complex, for estimation of the delivered power of pushed convoys. Both are based on the results of systematic towing tank tests performed in VBD (now DST), see VBD (1985). MMs cover common pushed convoy configurations shown in Fig. 8.14. They have, however, a serious shortcoming as are based on the measurements performed with just *one pushboat model*, thus the pushboat configuration (including propeller characteristics) can not be varied in the numerical analysis. Therefore, the characteristics of a particular pushboat ($L_{WLp} = 33.9$ m, $B_p = 14$ m, $T_p = 1.75$ m, $\nabla_p = 525$ m^3, $S_p = 545$ m^2, two propellers in nozzles, $D = 2.1$ m) are hidden within the MM.[7] Nonetheless, obtained results are useful in the *early design stages*, as well as for the *qualitative analysis* of various convoy configurations.

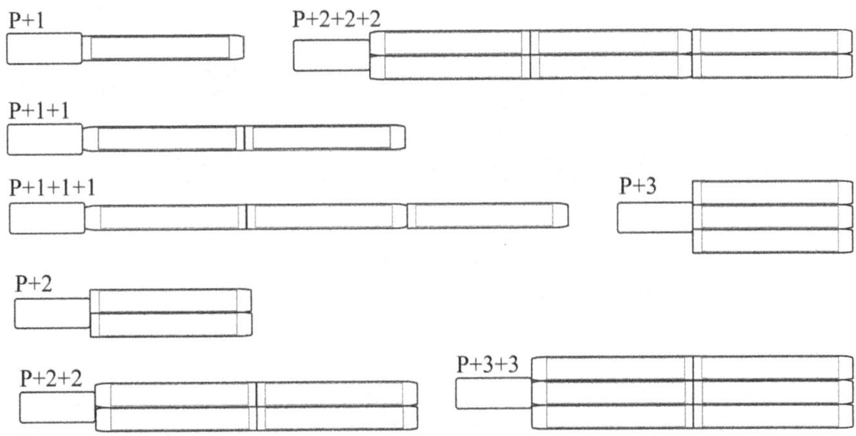

Fig. 8.14 Configuration of IW pushed convoys

[7]Concerning the mathematical modeling, the pushboat characteristics are comparable to secondary hull form parameters of conventional vessels. Secondary hull form parameters are usually not amongst the MM's input parameters. They should rather be considered as *important characteristics of the MM*.

In the IDV study (Hofman et al. 2013) the discrepancies between model tests and results evaluated based on MM, for identical push convoy configurations and water depths, were justified with dissimilarities of the pushboats. Namely, MM's pushboat and particular Danube model-tested pushboat had propellers of *2.1* m and *1.59* m (when extrapolated to full scale), which corresponds to propulsive efficiency (depending on the convoy configuaration) of *0.45* ÷ *0.55* and *0.35* ÷ *0.40*, respectively. Thus, there is no doubt that the MM's results are heavily influenced by the predetermined pushboat characteristics.

See also MM in Marchal et al. (1996) for evaluation of convoy resistance, which is formed on another database.

Simple Mathematical Model

This method, presented in Bilen (1997), enables the evaluation of resistance and delivered power for pushed convoy configurations depicted in Fig. 8.14 (except configuration P + 1 + 1 + 1). MM and corresponding regression coefficients are given in Tables 8.7 and 8.8, respectively. MM is valid *only* for Europe IIb barges, draught $T = 2.5$ m and water depth $h = 7.5$ m, while convoy speed may be varied; various sets of regression coefficients are corresponding to different convoy configurations given in each row in Tables 8.7 and 8.8.

Table 8.7 Mathematical model and corresponding regression coefficients for evaluation of resistance

$$R_T = A_0 + A_1 \cdot V + A_2 \cdot V^2 + A_3 \cdot V^3$$

Convoy Config.	A_0	A_1	A_2	A_3
P+1	-43.862	14.511	-1.281	0.050
P+1+1	-29.020	9.735	-0.790	0.037
P+2	-56.654	20.298	-1.794	0.0852
P+2+2	-60.381	21.666	-1.878	0.092
P+2+2+2	40.018	-8.465	1.419	0.000
P+3	-33.380	13.932	-1.175	0.094
P+3+3	-113.800	44.163	-4.434	0.208

where

V Convoy speed through the water (km/h).
R_T Convoy resistance (kN).

Table 8.8 Mathematical model and corresponding regression coefficients for evaluation of power

$$P_D = a_0 + a_1 \cdot V + a_2 \cdot V^2 + a_3 \cdot V^3 + a_4 \cdot V^4$$

Convoy Config.	a_0	a_1	a_2	a_3	a_4	V_{min} - V_{max} [km/h]
P+1	4205.55	-1365.50	162.27	-8.22	0.158	8 - 22
P+1+1	1782.50	-603.40	76.05	-4.04	0.089	8 - 21
P+2	1055.83	-433.60	65.71	-3.99	0.105	6 - 18
P+2+2	1498.97	-676.67	113.95	-7.93	0.233	6 - 15
P+2+2+2	1006.50	-448.83	75.10	-5.15	0.149	6 - 17
P+3	294.68	-114.87	18.46	-0.78	0.044	5 - 15
P+3+3	1017.11	-467.79	80.58	-5.47	0.179	6 - 14

where

V Convoy speed through the water (km/h).
P_D Delivered power (kW).

Complex Mathematical Model

Complex MM is the core of a computer code "CONVOY ", developed at the University of Belgrade, see Kalajdžić and Hofman (2009). This MM enables examination of several important issues of pushed convoys, which are based on the power evaluation for various speeds, convoy formations, deadweights, water depths etc.

Artificial Neural Network (ANN) with a sigmoid activation function $sig = 1/(1+e^{-x})$ was used for MM development, resulting in a relatively complex MM, i.e. equation with 198 polynomial terms. The database consisted of more than 1000 data points, 30% of which were used for checking of the MM. The goodness of fit of the MM, expressed through the root-mean-square deviation (RMSD), is around 3%.

MM's input and output parameters or, respectively, independent and dependent (target) variables, are the following:

INPUT	OUTPUT
$X_1 = L/B$	C_D
$X_2 = B/T$	
$X_3 = L/h$	
$X_4 = L/\nabla^{1/3}$	
$X_5 = Fn = V/\sqrt{gL}$	

where

L	the length of convoy (m), $L = L_{WLp} + nL_{WLb}$.
L_{WLp}	the length of the pushboat (m).
L_{WLb}	the length of the barge (m).
B	the breadth of convoy (m), $B = mB_b$.
B_b	the breadth of the barge (m).
T	the draught of the convoy (m).
∇	volumetric displacement of the convoy (m^3), $\nabla = \nabla_p + nm\nabla_b$.
n	number of barges in the longitudinal direction of the convoy [-].
m	number of barges in the transverse direction of the convoy [-].
V	speed through the water of the convoy (m/s).
C_D	the coefficient of delivered power [-], $C_D = 1000P_D/(\rho g\nabla V)$.
P_D	the delivered power (kW).
	Other symbols as in the rest of the book.

Database distribution and boundaries of applicability of the MM are presented in a graphical format in Fig. 8.15. MM and its ANN coefficients are given in Table 8.10.

If the pushed convoy *consists exclusively of Europe IIb barges,* the estimation of the input parameters may be simplified by evaluating L_{WLb}, ∇_b and T_b from the simple polynomial equations given in Table 8.9.

Table 8.9 Mathematical model and corresponding ANN coefficients

L_{WLb}	[m]	$L_{WLb} = 0.0653 \cdot T^3 - 0.638 \cdot T^2 + 2.9487\ T + 70.452$
∇_b	[m^3]	$\nabla_b = -2.9333 \cdot T^3 + 36.4 \cdot T^2 + 701.73 \cdot T + 40.4$
T_b	[m]	$T_b = 7 \cdot 10^{-12} \cdot \Delta^3 - 7 \cdot 10^{-8} \cdot \Delta^2 + 0.0014 \cdot \Delta - 0.042$

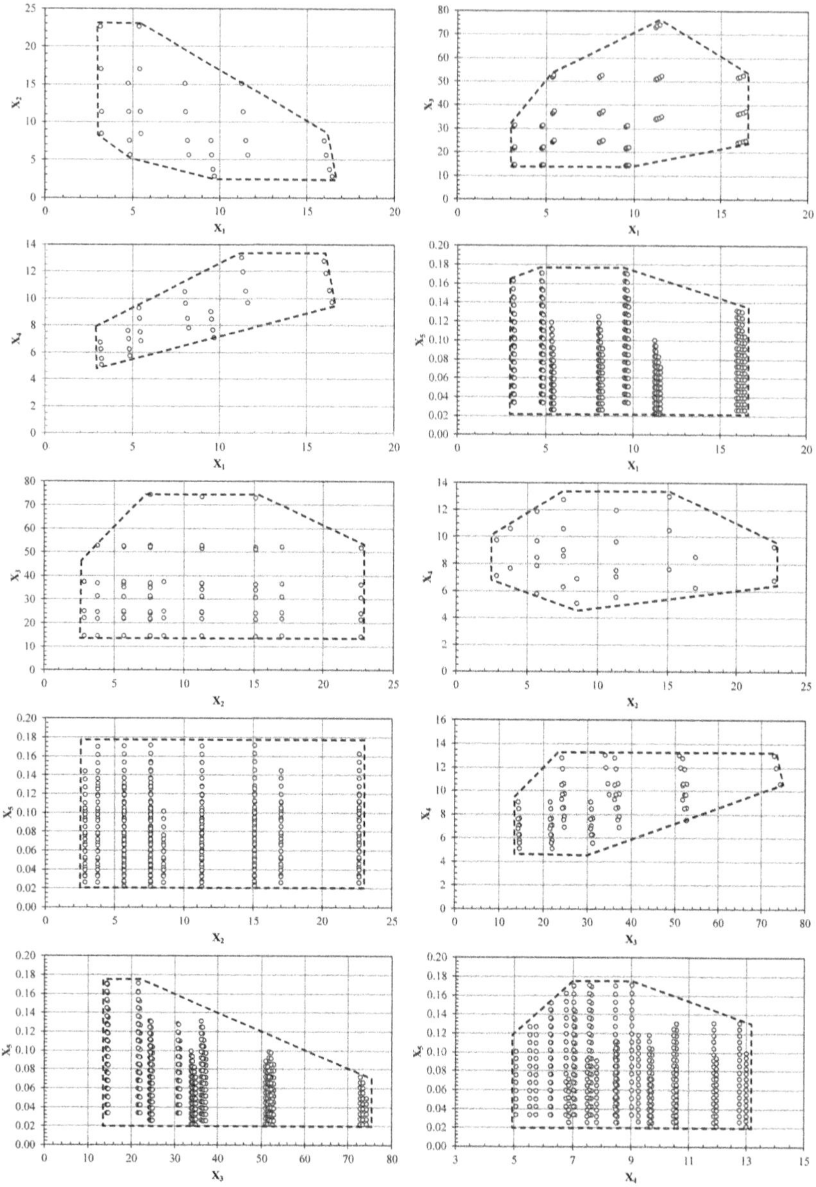

Fig. 8.15 Database distribution and boundaries of applicability of a MM for IW pushed convoys

Table 8.10 Polynomials based on the data for Europe IIb barge (valid for $T = 1.5 \div 4$ m)

$$C_D = \frac{sig\left(e_1 + \sum_{v=1}^{3}\left(E_{v1} \cdot sig\left(d_v + \sum_{w=1}^{5}\left(D_{wv} \cdot sig\left(c_w + \sum_{i=1}^{7}\left(C_{iw} \cdot sig\left(b_i + \sum_{j=1}^{9}\left(B_{ji} \cdot sig\left(a_j + \sum_{k=1}^{5}\left(A_{kj} \cdot (P_k X_k + R_k)\right)\right)\right)\right)\right)\right)\right)\right)\right)\right) - G}{L}$$

j	A_{1j}	A_{2j}	A_{3j}	A_{4j}	A_{5j}	a_j
1	-1.166348	-2.783221	-3.604903	-2.025712	-5.378228	6.417914
2	-2.648328	-0.713384	-0.180779	0.157007	4.189291	0.953689
3	0.842080	1.899729	-10.271500	4.530565	-3.682956	2.315466
4	-0.933936	-5.329044	-13.345910	-5.704818	1.905937	1.413591
5	5.992681	1.701130	2.617187	-17.344370	-1.330412	1.742030
6	-1.935798	-0.692122	-15.717640	-12.453900	1.275389	9.342219
7	-10.479590	8.390247	2.027237	0.315596	-7.363225	4.065030
8	13.841040	1.287127	7.118299	5.920812	-3.080441	-7.198208
9	1.218522	-2.802006	-1.881928	3.217824	2.582319	-0.506870

i	B_{1i}	B_{2i}	B_{3i}	B_{4i}	B_{5i}	B_{6i}	B_{7i}	B_{8i}	B_{9i}	b_i
1	2.712726	1.264464	-3.492853	0.789727	5.207363	-2.257747	-0.168757	-4.172150	2.229724	-1.433187
2	-1.740363	5.351151	-2.874480	-2.259038	7.908931	-8.963451	-1.649598	-2.299474	-2.080463	1.154354
3	1.794098	-3.577066	0.697624	2.222907	0.355194	0.560344	0.813611	1.011585	-1.190070	-1.969467
4	-7.676115	1.832165	-0.434195	0.718745	10.241510	1.154937	0.163853	1.979787	3.163420	-6.834247
5	-2.302156	2.406005	-5.800941	-5.075598	0.595637	-0.495264	-0.142848	-1.081158	-0.145598	3.183046
6	-6.502975	5.968524	-0.966492	-2.967121	2.640382	-1.374954	-4.790840	-2.574104	-1.896178	-0.075202
7	-2.656193	7.951061	-1.275850	-3.758516	-0.196248	-1.019759	-1.727610	-2.572931	0.960570	-3.360709

(continued)

Table 8.10 (continued)

w	C_{1w}	C_{2w}	C_{3w}	C_{4w}	C_{5w}	C_{6w}	C_{7w}	c_w
1	2.251047	2.372722	-0.140841	3.422574	2.082446	0.829893	1.029840	-3.716167
2	0.675065	2.014420	-6.406669	4.896750	-0.441944	-2.526683	2.273639	-0.942570
3	1.137268	3.613458	1.593404	4.092658	3.239221	4.495337	3.249649	-6.202133
4	1.297363	1.347450	-0.363973	3.401099	2.977446	3.609143	2.903754	-5.768052
5	0.134916	0.360245	-0.415023	3.027792	3.087563	2.090854	3.675019	-4.890333

v	D_{1v}	D_{2v}	D_{3v}	D_{4v}	D_{5w}	d_v
1	0.067516	-6.312922	3.287876	1.558534	0.770429	0.090030
2	-3.527841	-1.776605	-4.781667	-4.530158	-3.815368	9.381804
3	-2.545620	-4.272006	-0.169602	-0.117120	-1.678473	3.862690

	E_{11}	E_{21}	E_{31}	e_1
1	-3.561975	-4.996552	-2.092212	5.940332

k	P_k	R_k
L/B	0.067756	-0.164379
B/T	0.045393	-0.078597
L/B	0.015024	-0.165456
$L/\nabla^{1/3}$	0.113307	-0.524355
Fn	6.040268	-0.082886

	L	G
C_D	0.0322370	0.0399740

Note that presented MM can be used for the *estimation of the delivered power regardless of the type of barges*, as long as the input variables are within the boundaries of applicability (Fig. 8.15).

8.5 Squat

Three methods for evaluation of squat, applicable to selfpropelled IW ship forms, are given in the following.

Romisch Method (Romisch 1989)

This method is based on model experiments and enables evaluation of *bow (Sb) and the stern (S_S) squat*:

$$S_b = C_V \cdot C_F \cdot K_{\Delta T} \cdot T$$

$$S_S = C_V \cdot K_{\Delta T} \cdot T$$

where

C_V is a correction factor for ship speed,
$C_V = 8 \cdot \left(\frac{V}{V_C}\right)^2 \cdot \left[\left(\frac{V}{V_C} - 0.5\right)^4 + 0.0625\right]$.

C_F is a correction factor for the ship shape, $C_F = \left(\frac{10 \cdot C_B}{L_{PP}/B}\right)^2$.

$K_{\Delta T}$ is a correction factor for squat at ship critical speed, $K_{\Delta T} = 0.155 \cdot \sqrt{h/T}$.

V_C is the ship's critical speed, $V_C = 0.58 \cdot \sqrt{gh} \cdot \left[\left(\frac{h}{T}\right)\left(\frac{L_{WL}}{B}\right)\right]^{0.125}$.

This method is applicable for $h/T = 1.19 \div 2.25$. The formula for evaluation of critical speed is valid for the unrestricted (by width) but shallow waterway. According to Pompée (2015), this method is probably the only one that is suitable for IW vessels. Note that the same source gives a formula for the restricted waterway, i.e. canal, but it is not presented here.

Barrass Method (Briggs et al. 2009)

Barrass proposed a simple formula for *maximal ship squat* estimations. It is based on full-scale measurements:

$$S = \frac{K \cdot C_B \cdot V^2}{100}$$

where

K is the coefficient depending on the blockage factor, $K = 5.74 \cdot C^{0.76}$, $1 < K < 2$.

C is the blockage factor. It is a ratio of the cross-section area of a ship to that of the channel, $C = 0.98 \cdot B \cdot T/\left(Wh + nh^2\right)$.

W canal width measured at the bottom

n inverse side grade of the canal, n = 1, 2 or 3 representing grades of 1:1, 1:2 or 1:3, respectively (see Fig. 6.9).

Other symbols as above.

The abovementioned equation is applicable for $C_B < 0.9$ and for navigation in extremely shallow water $1.1 < h/T < 1.5$. Note that the full-form ships ($C_B > 0.7$, which is a typical value for IW vessels) tend to squat by the bow.

Raven Method (Raven 2016)

Dynamic sinkage of a ship, navigating in shallow water, can be estimated from the following equation:

$$S = L_{WL} \cdot C_Z \cdot \frac{\nabla}{L_{WL}^3} \cdot \left[\frac{Fn_h^2}{\sqrt{1 - Fn_h^2}} - \frac{Fn_{hd}^2}{\sqrt{1 - Fn_{hd}^2}} \right]$$

where

C_Z is a coefficient depending on a hull form. An average value is 1.46.

Fn_{hd} is Froude number based on corresponding 'deep water' depth, $hd = 0.3L_{WL}$, $Fn_{hd} = V/\sqrt{g \cdot (0.3L_{WL})}$.

Other symbols as above.

8.6 Lightship Weight Estimation

Approximate formulae for lightweight estimation of inland waterway vessels are rare in literature. Most of the ones presented in the following are based on the investigations published in Heuser (1986), Hofman (2006) and Hekkenberg (2012, 2014). Given are equations for the estimation of steel weight, lightweight, and some other local masses, for conventional selfpropelled IW vessels.

Note that in the remainder of the section all weights are given in tons, while the vessel's length (L), breadth (B), height (D) and design draught (T) are given in meters. The coefficient of determination[8] (R^2) in all presented mathematical models is in the range of $0.98 \div 0.99$.

[8]Statistics that demonstrate how well the mathematical model fits the data. R^2 can have a value between 0.0 and 1.0, where a value of 1.0 indicates a perfect fit (reliable model), while a value of 0.0 shows extremely bad modeling.

8.6.1 Steel Weight Mass

Simple and easy to use trend lines for the estimation of steel weight of dry bulk cargo IW vessels depending on the *LBT* are given in Hekkenberg (2014). A distinction is made between longitudinally and transversely framed hulls, and dry bulk and container vessels. A mathematical model and corresponding coefficients are presented in Table 8.11. The equation obtained is based on data for the lengths up to 135 m, $L/B = 6 \div 12$ and $T = 1.5 \div 4.5$ m.

Calculated steel weight of dry bulk vessels as a function of *LBT* and for a range of design draughts is presented in Fig. 8.16. Trend lines for container vessels are rather similar. One should be aware of the constraints regarding the range of *L*, *B* and *T* stated above.

Enhanced mathematical models for W_{steel} estimation—valid for enlarged ranges of ship principal dimensions: $L/B = 4 \div 20$, $L = 40 \div 180$ m, $B = 5 \div 25$ m and

Table 8.11 Mathematical model and corresponding coefficients.
$W_{steel} = c_1(LBT)^2 + c_2(LBT)$

	Dry bulk vessels				Container vessels			
	Transverse framing		Longitudinal framing		Transverse framing		Longitudinal framing	
***T* [m]**	c_1	c_2	c_1	c_2	c_1	c_2	c_1	c_2
1.5	1.80E-05	2.37E-01	2.62E-05	2.11E-01	3.00E-05	2.39E-01	3.27E-05	2.13E-01
2	8.33E-06	1.89E-01	1.49E-05	1.67E-01	1.36E-05	1.95E-01	1.83E-05	1.70E-01
2.5	7.18E-06	1.51E-01	9.86E-06	1.37E-01	8.10E-06	1.61E-01	1.14E-05	1.39E-01
3	3.00E-06	1.37E-01	6.04E-06	1.22E-01	5.59E-06	1.38E-01	7.97E-06	1.21E-01
3.5	1.73E-06	1.25E-01	3.72E-06	1.14E-01	3.66E-06	1.26E-01	5.20E-06	1.12E-01
4	1.78E-06	1.16E-01	2.58E-06	1.07E-01	3.45E-06	1.14E-01	3.10E-06	1.09E-01
4.5	1.61E-06	1.09E-01	1.50E-06	1.03E-01	3.49E-06	1.05E-01	1.91E-06	1.04E-01

Source Hekkenberg (2014)

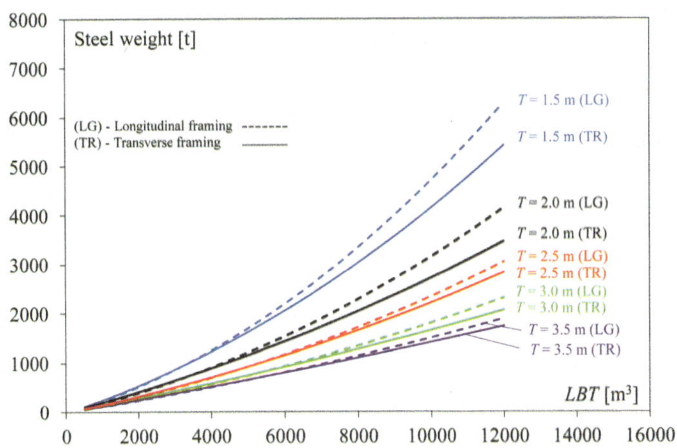

Fig. 8.16 Steel weight of dry bulk vessels

$T = 1.5 \div 4.5$ m are also developed in Hekkenberg (2014) and given in Table 8.12. Deviation from the actual (measured) vessel masses is said to be between $5 \div 10\%$.

Figure 8.17 presents the steel weight mass of dry bulk vessels, calculated according to Table 8.12 as a function LBT for $T = 2.5$ m. Trendlines for container ships are similar.

Table 8.12 Mathematical model and corresponding coefficients.

$$W_{steel} = c_1 + c_2 LB + c_3 L^2 B + c_4 LBT + c_5 L^{3.5} B + c_6 \frac{L^{1.3} T^{0.7}}{B} + c_7 \frac{1}{B^2 T^{1.5}}$$

	Dry bulk vessels		Container vessels	
***T* [m]**	Transverse framing	Longitudinal framing	Transverse framing	Longitudinal framing
c_1	-2.597E+01	4.985E+01	-2.200E+01	5.107E+01
c_2	2.320E-01	2.290E-01	2.540E-01	2.440E-01
c_3	-1.552E-03	-1.234E-05	-1.975E-03	-1.772E-04
c_4	4.444E-02	1.910E-02	4.473E-02	1.588E-02
c_5	8.134E-07	9.584E-07	1.059E-06	1.100E-06
c_6	1.024E+00	2.880E-01	9.600E-01	3.120E-01
c_7	7.691E+02	-1.066E+03	6.676E+02	-1.164E+03

Source Hekkenberg (2014)

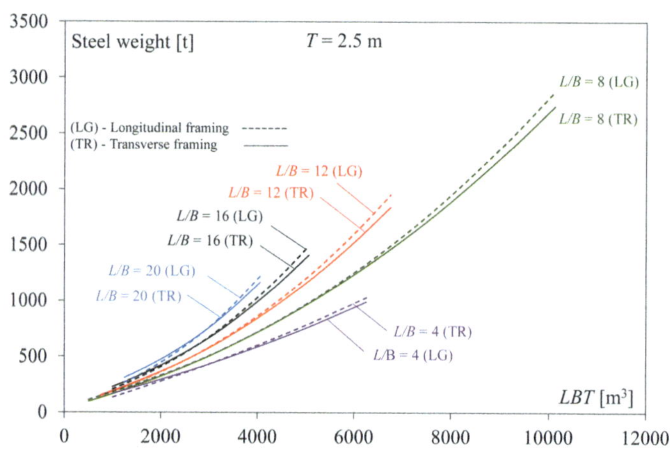

Fig. 8.17 Steel weight of dry bulk vessels—$T = 2.5$ m

8.6.2 *Lightweight Mass*

Ship *lightweight mass* includes steel weight mass and other "local" masses, such as wheelhouse, winches, masts, outfitting, etc.

One general formula for the lightweight mass of IW vessels can be found in Heuser (1986):

$$W_{lightweight} = -6.6607 \times 10^{-6}(LBD)^2 + 0.21822(LBD) - 4.1$$

The formula is based on data more than 30 years old and thus is a bit dated. It is revised upon additional data on container vessels on the river Rhine in Hofman (2006) as given below:

$$W_{lightweight} = -4.44 \times 10^{-6}(LBD)^2 + 0.195(LBD)$$

An illustrative comparison between Heuser and Hofman estimations of lightweight, depending on *LBT*, is presented in Fig. 8.18.

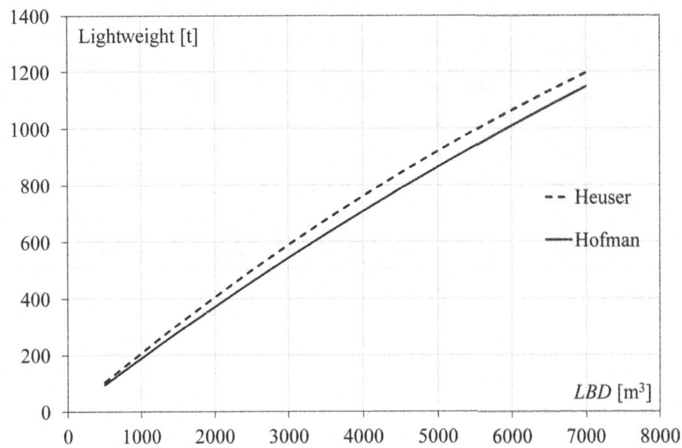

Fig. 8.18 Lightweight according to Heuser and Hofman

The approximative formula for ship lightweight is also derived in Hekkenberg (2012) and given here in Table 8.13. Lightweight is expressed as a function of *LBT* valid for: 1.5 m < T < 4.5 m, L < 135 m and 6 < L/B < 12. As previously, a distinction is made between longitudinally and transversely framed hulls, dry bulk and container vessels. Results of the function given in Table 8.13, for dry bulk vessels, are presented in Fig. 8.19. Trend lines for container vessels are similar.

Formulas considering a more detailed break of lightweight mass can also be found in Hekkenberg (2012). Some of the so-called "local" masses: accommodation (W_1), machinery with equipment and outfitting (W_2), piping outside of engine room (W_3) – can be calculated by separate equations, given in Table 8.14. Added to steel weight mass, calculated by a function presented in Table 8.12, they deliver the total lightweight. Formulas are valid for: $L/B = 4 \div 20$, $L = 40 \div 180$ m, $B = 5 \div 25$ m and $T = 1.5 \div 4.5$ m. Expected deviations from the actual masses are between $5 \div 10\%$.

Table 8.13 Mathematical model and corresponding coefficients. $W_{lightweight} = c_1(LBT)^2 + c_2(LBT) + c_3$

Dry bulk vessels						
	Transverse framing			Longitudinal framing		
***T* [m]**	c_1	c_2	c_3	c_1	c_2	c_3
1.5	5.34E-06	2.96E-01	4.98E+01	2.59E-05	2.33E-01	6.93E+01
2	1.48E-06	2.35E-01	4.86E+01	1.49E-05	1.84E-01	7.08E+01
2.5	2.88E-06	1.82E-01	6.29E+01	1.07E-05	1.44E-01	8.09E+01
3	2.00E-07	1.66E-01	5.85E+01	6.34E-06	1.31E-01	7.85E+01
3.5	2.01E-06	1.33E-01	8.27E+01	3.10E-06	1.30E-01	6.74E+01
4	3.60E-06	1.10E-01	1.06E+02	2.15E-06	1.21E-01	6.90E+01
4.5	2.73E-06	1.05E-01	1.15E+02	1.22E-06	1.15E-01	7.33E+01
Container vessels						
	Transverse framing			Longitudinal framing		
***T* [m]**	c_1	c_2	c_3	c_1	c_2	c_3
1.5	2.81E-05	2.68E-01	6.30E+01	3.30E-05	2.33E-01	7.16E+01
2	1.05E-05	2.30E-01	4.93E+01	1.73E-05	1.92E-01	6.40E+01
2.5	7.25E-06	1.81E-01	6.19E+01	1.19E-05	1.49E-01	7.64E+01
3	5.32E-06	1.53E-01	6.95E+01	8.05E-06	1.32E-01	7.51E+01
3.5	3.19E-06	1.40E-01	7.05E+01	4.74E-06	1.26E-01	7.06E+01
4	3.83E-06	1.19E-01	9.24E+01	2.26E-06	1.28E-01	5.81E-01
4.5	4.50E-06	9.94E-02	1.19E+02	1.48E-06	1.18E-01	6.79E+01

Source Hekkenberg (2012)

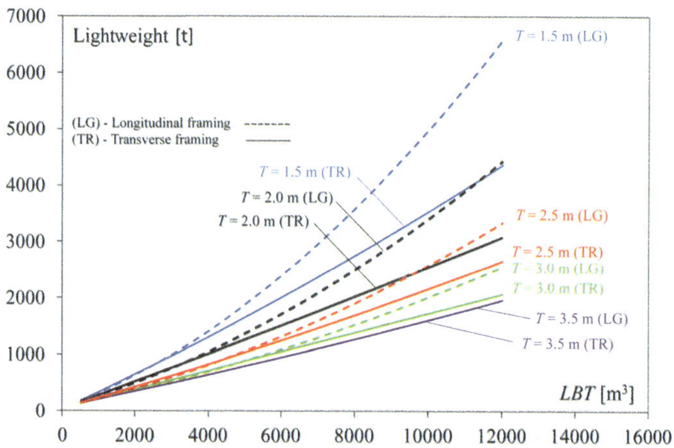

Fig. 8.19 Lightweight of dry bulk vessels

Table 8.14 Equations for estimation of local masses

Local mass	Equation
W_1	$0.173 \cdot 2.5 \cdot \max\left[\frac{L}{4} \cdot (B-2), 100\right]$
W_2	$28.04 + 4.605 \cdot T + 2.097 \cdot 10^{-2} LB + 2.24 \cdot 10^{-3} LBT - 4.258 \cdot 10^{5} \frac{1}{L^3}$
W_3	$-2.723 + 6.232 \cdot 10^{-2} L + 5.048 \cdot 10^{-2} B + 9.968 \cdot 10^{-2} T + 1.343 \cdot 10^{-4} LBT$

Local masses calculated by the above formulas, for a range of *L/B* ratios and design draught of $T = 2.5$ m, are presented in Fig. 8.20.

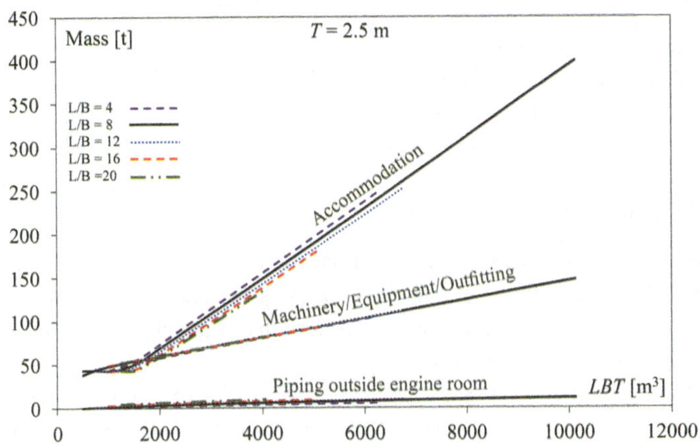

Fig. 8.20 Mass of other weight items – $T = 2.5$ m

8.7 Design Recommendations for Conventional Selfpropelled Container Vessels[9]

Restrictions in *water draught* connected to waterway depth, in *air draught* connected to the height of bridges, and *restrictions in beam and length* connected to the size of lock chambers, make numerous and serious challenges to the designer. Therefore, good IW container vessels differ significantly, not only from a seagoing ship, but even from one waterway to another. An optimal Danube container vessel would certainly not be the same as the optimal vessel for the Rhine or some other waterway, see Hofman (2006).

[9]The material presented in this section originates from COVEDA (2003) and Hofman (2006).

8.7.1 Assessment of Main Dimensions

The number of transported containers (which directly influences transport efficiency) depends on vessel length, beam and draught. Nevertheless, the inverse is true too, i.e. the design of an optimal container ship—her main dimensions—is directly influenced by the number of containers that should be transported onboard.

Number of Containers Abreast

Assuming the widths of the gangways of 0.65 m and of a TEU container with clearance between them $b_c = 2.54$ m, follows $B \approx 2.54 \cdot n_B + 1.3$, where n_B is the number of TEU containers abreast. A reasonable number of carried containers on IWW can vary from three to six abreast (rarely two, see Fig. 4.17). So, a proper beam of Danube container vessels should change in steps (see Sect. 3.2) in the following manner:

$B \approx 9$ m, for 3 containers abreast,
$B \approx 11.4$ m, for 4 containers abreast,
$B \approx 14$ m, for 5 containers abreast,
$B \approx 16.5$ m, for 6 containers abreast.

Number of Containers Along the Ship

The number of containers in the longitudinal direction is directly connected to the vessel length L by the relation $L \approx l_C \cdot n_L + l_f + l_a$, where l_C is the container length with the average clearance, and l_f and l_a are lengths of the ship ends, fore and aft of the hold, respectively, (see Fig. 8.21).

The average Rhine cargo vessels with a breath of 11.4 m (see Hofman 2006) have $l_f \approx 10$ m and $l_a \approx 19$ m (or $0.9 \cdot B$ and $1.2 \cdot B$, respectively) and the hold length of $6.23 \cdot n_L$, where n_L is the number of TEU along the ship. Therefore, for the Rhine vessels, follows an approximate relation: $L \approx 6.23 \cdot n_L + 29$. It is reasonable to assume that the same relation is valid for the Danube vessels too. Consequently, the relation between vessel length and the total number of containers ($n = n_H \cdot n_B \cdot n_L$) is:

$$L \approx 6.23 \cdot n/(n_B \cdot n_H) + 0.9 \cdot (2.54 \cdot n_B + 1.3) + 19,$$

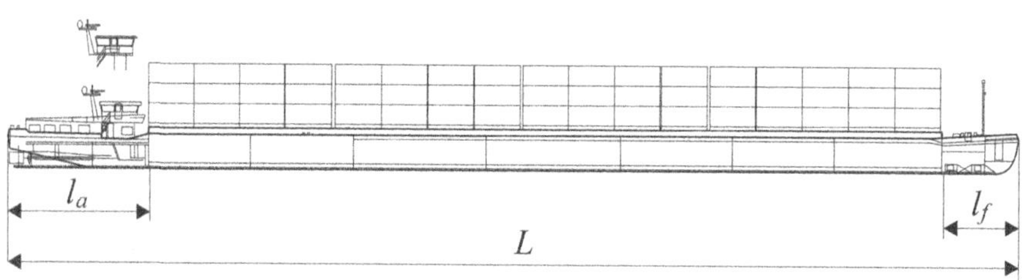

Fig. 8.21 Vessel dimensions in the longitudinal plane

where the ratio $n/(n_B \cdot n_H)$ has to be an integer number. This relation, for $n_H = 2, 3, 4$, and properly chosen vessel beam of 9 m, 11.4 m, 14 m and 16.5 m (corresponding to $n_B = 3, 4, 5, 6$, respectively) is presented in Fig. 8.22. The vessel length is given for the wide range of length to beam ratios $L/B = 7 \div 12$, which practically covers all existing IW container vessels.

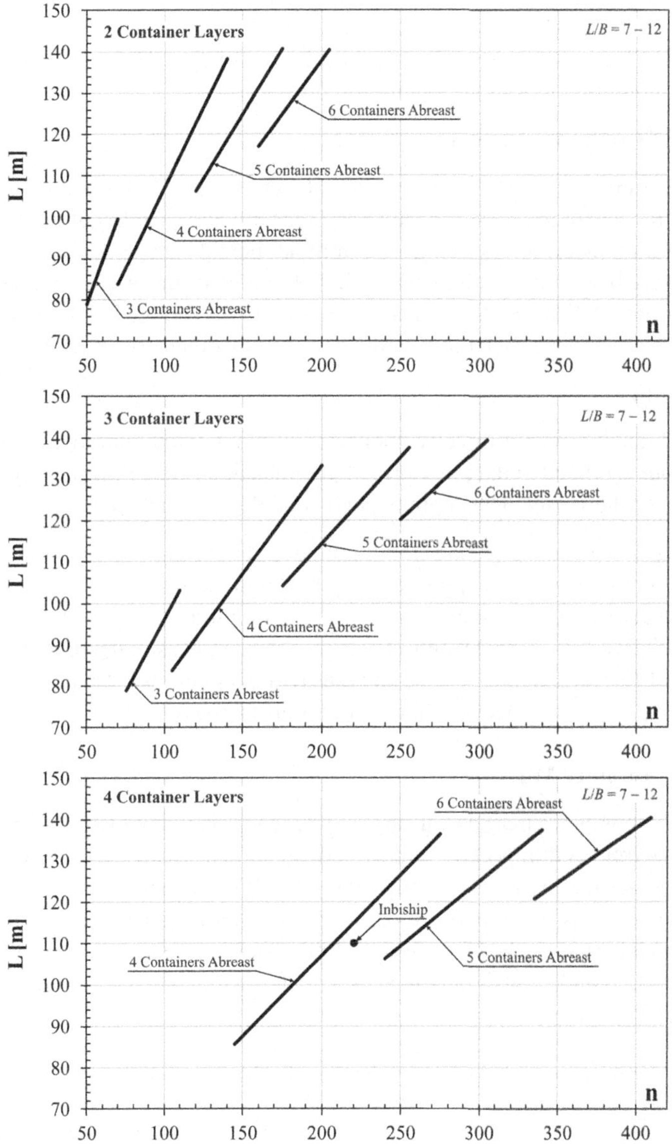

Fig. 8.22 Relation of vessel length to number of TEU containers

Figure 8.22 essentially gives the relationship between the number of TEU containers and vessel's main dimensions.[10]

The choice of vessel length and the beam is straight forward, except for the regions where the lines in the diagram overlap. In these regions, the designer has to decide between two options that have different *L*/*B* ratios. This decision depends on numerous stability, resistance, propulsion and strength considerations. For instance, if the vessel length is limited to 135 m (as on the Rhine), the maximal number of TEU containers onboard is approximately 204, 306 and 408 for 2-, 3- and 4-layer vessel (for 6 layers abreast in all cases, i.e. vessel of $B \approx 16.5$ m). However, note that the draught restrictions are not imposed here yet.

Number of Container Layers

The allowed number of container layers that can be carried onboard is $n_H < (h_a + T - h_d) / h_c$, where h_a, h_d and h_c are vessel air draught, heights of double-bottom and a container layer, respectively. The minimal height (above the high water level when passing bellow the bridges) is 5.25, 7.00 and 9.10 m for 2, 3 and 4 container layers respectively (see Sects. 2.4 and 4.1.1, Table 4.1—bridge heights). Sometimes this could be overcome by ballasting the ship, but this increases transport costs.

Vessel Depth

The vessel depth *H* is not directly connected to the maximal number of containers onboard, but is related to the other principal dimensions and influences significantly the quality of the vessel. The depth depends mainly on two factors:

- Freeboard requirements and longitudinal strength limitations. The minimal freeboard could be taken approximately as 0.6 m, giving $H \approx T + 0.6$.
- The typical strength limitation is $L/H \leq 35$, stemming from the GL Rules (GL, 2011). The vessels on the Danube, having a considerably smaller draft, have higher values of *L/H*, typically above 40 and thus require additional strength calculations (see Sect. 10.3.2 and discussion on structural strength).

Average Container Mass

The average container mass is varying and depends on many factors. For example, it varies in export and import, or in upstream and downstream navigation etc. Moreover, it changes randomly from a trip to a trip and in the first place depends on the percentage of empty containers onboard. The long-term statistics for standard TEU containers (see Statistisches Bundesamt 2001) show average values for the combination of loaded and empty containers as $9 \div 10$ t, while the average mass of loaded

---

[10]The INBISHIP concept, for instance, doesn't follow the abovementioned "rule" (see Sect. 10.1.2), i.e. it illustrates that the number of containers could be increased (or *L* reduced) if some special attention is paid for better utilization of cargo space. The INBISHIP concept has $B = 11.4$ m and 4 container layers abrest; thus, her shift from the usual vessels of the same beam and number of layers has been accomplished mainly by the application of newly designed diesel-electric propulsion that allowed better allocation of the machinery. See also SendoLiner, Fig. 4.19 and Sect. 10.2.2.

ones is 13 ÷ 14 t. For further considerations assumed is an average mass of 13 t (here also referred as required container mass).

Note that uneven vertical load distribution is almost always necessary (particularly when the number of vertical layers is larger than the number of containers abreast, i.e. when $n_H > n_B$). Otherwise, transversal stability may be endangered (see Sect. 8.3).

Vessel Draught

The *proper relation between cargo space and cargo weight* is desirable. Namely, the average *available container mass* (m_c) is limited and is in direct correlation to restricted draught T and the number of container layers (n_H), while the other parameters are much less important. In particular, the importance of the ship breadth (thus n_B) is insignificant, while the length (thus n_L) is of small influence. Figure 8.23 shows average values $m_C = f(n_H, T, L)$, i.e. the influence of n_B is ignored, see Hofman (2006). A particular combination of T and n_H indicates a *well-balanced vessel* (endorsing, for instance, $m_c \approx 13$ t).

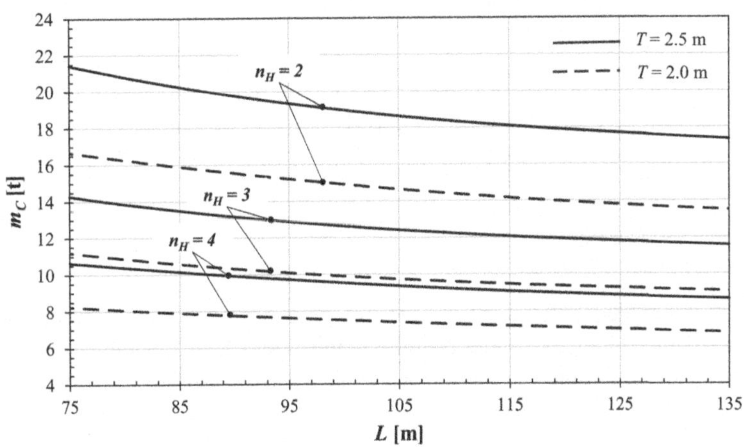

Fig. 8.23 Average available container mass for 2-, 3- and 4-container layers

The available container mass can be increased, without decreasing the number of containers or increasing draught, by reducing the lightship weight (see Sect. 8.6.2). It is also possible to increase the vessel breadth, without increasing the number of containers abreast, i.e. for 4 containers abreast, the vessel beam could be 12 ÷ 13 m, rather than 11.4 m, but this is not regarded as a practical solution.

The inverse problem are the draughts that would give the required average container mass of 13 t, see Fig. 8.24. So, draught should be between 3.1 ÷ 3.5 m for the 4-layer vessels which is usual on the Rhine. 3-layer vessels with $T = 2.4$ ÷ 2.7 m, and 2-layer vessels with $T = 1.65$ ÷ 1.85 m are more appropriate for the Danube. Note that $n_H = 4$ can be achieved on the Danube (if some bridge heights are ignored) but then m_C should not excide 10 t. The above-mentioned conclusions are valid for $n_B = 4$ ($B = 11.45$ m) and other assumptions given in Hofman (2006).

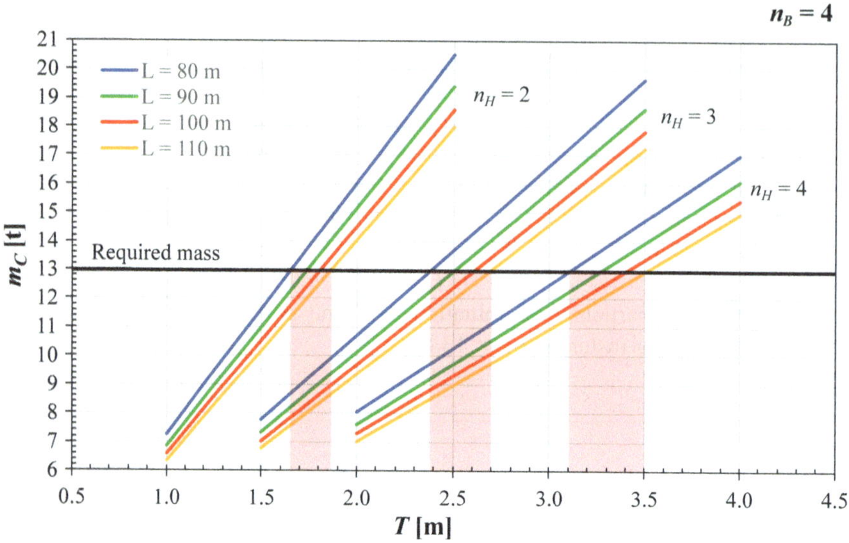

Fig. 8.24 Draughts for two, three and four-layer container vessels

It may be concluded that due to *water and air draught restrictions*, waterborne transport of containers on the Danube is inherently less efficient than on the Rhine. On the Danube, a vessel of the same size can transport one to two container layers less than on the Rhine!

8.7.2 Procedure for Estimating Main Dimensions

The starting point for preliminary determination of vessel main dimensions is (a) the required number of containers n that should be transported onboard, and (b) the transportation route. These input parameters follow from transportation and financial analyses and should be supplied by the shipowner. From that starting point, the dimensions are determined in the following steps, also presented in a flowchart in Fig. 8.25.

Choice of Number of Container Layers

The number should be $n_H = 2, 3 \text{ or } 4$. The greater n_H means the better utilization of the vessel and better financial effects.

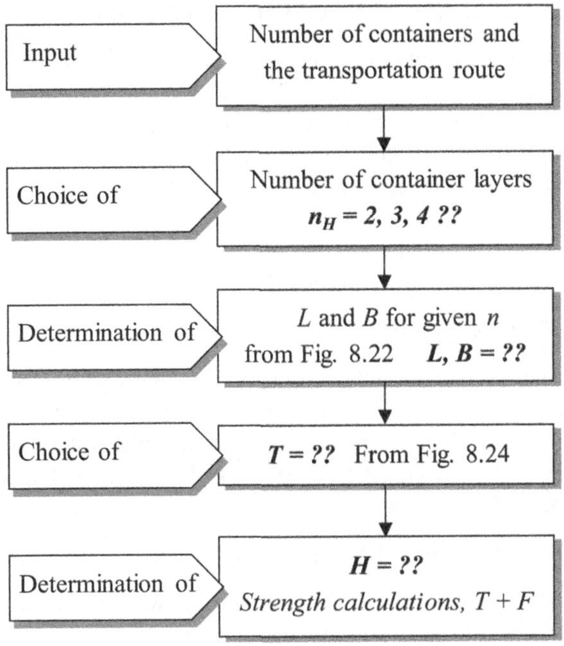

Fig. 8.25 An algorithm for estimating vessel principal dimensions

4-layers

- Implies high draughts (up to 3.5 m) to obtain the proper container mass.
- A vessel must have at least 4 container rows abreast to obtain the proper stability.
- This vessel is unable to sail at high, but also at low water levels along the whole Danube, due to bridge height and water depth restrictions (see Sect. 2.4).

3-layers

- Results in considerably smaller transport efficiency (ratio of container cargo volume and vessel displacement), but much higher chances to sail fully loaded during the year.
- Well balanced vessel for drafts appropriate for the Danube waterway.

2-layers

- Less than 2 layers is often regarded as inefficient, i.e. 2 is on the efficiency edge.
- Should be employed on the Upper Danube and during low water levels at the Lower Danube.

Conclusion: The natural and good compromise for the Danube is a vessel designed for 3 container layers as the main loading case, but carrying 4 and 2 container layers should be a feasible option too. On the Lower Danube, attention should be paid to the *restrictions of low water draught* and high air draught (which is not the case on other Danube stretches) so one might come to a wrong assumption that stacking a larger number of containers is possible.

Determination of Vessel Length and Beam

After the number of container layers is assessed, the length and beam follow from diagrams given in Fig. 8.22. That choice is straight forward, depending only on the known number of containers onboard. In some (small) regions of the diagrams the results overlap. In such regions, the choice has to be made between the vessels of very different ratios $L/B \approx 7$ and $L/B \approx 11$. Long vessels should be advantageous from the resistance point of view,[11] while the beamy vessel would be beneficial in stability and hull-weight considerations; both ratios satisfy stability requirements. So, a compromise should be made between resistance and weight considerations.

Choice of Draft

The assessment of the draft is in direct correlation to the chosen number of container layers. The required draft for the proper 4-layer vessel would be over 3.5 m, which is not available at most parts of the Danube waterway. The drafts of 3-layer vessels should be (according to Fig. 8.24) $2.4 \div 2.7$ m. The draft of 2-layer vessels should be $1.65 \div 1.85$ m.

Determination of Vessel Depth

After length, beam and draft are known, the vessel depth follows from minimal freeboard and/or strength requirement.

8.7.3 Other Important Aspects

Analysing other characteristics, not mentioned in Sects. 8.7.1 and 8.7.2, some unexpected but important results for the design of IW container vessels were derived (see COVEDA 2003, also Hofman 2006). These are:

- The profit in the first place depends on the number of containers onboard, and all other factors are by far less important.
- For IW container vessels, the reduction in hull weight is important because of their limited draught. A rough analysis (for three container layers) by choosing a beamy instead of a long vessel, indicates that the hull weight reduction is approximately $10 \div 20\%$. This allows an increase of available container mass by approximately $5 \div 10\%$ (Hofman 2006).
- The influence of a propeller diameter on overall efficiency is relatively insignificant, thus somewhat smaller diameters may be advantageous.[12] It follows that the propeller should be designed according to the minimal draught requirements.

---

[11]In shallow water, for high, near-critical speeds (see Sect. 7.2.1), the wave resistance depends mainly on parameters L/h and L/B, and both of these parameters are reduced if the beamy vessel is chosen. However, the usual speeds of the fastest selfpropelled vessels correspond to much lower Fn_h values, certainly below $Fn_h = 0.65$, thus the deep-water logic – the longer, the better – is valid here.

[12]According to analysis and assumptions adopted in COVEDA (2003) and Hofman (2006).

Moreover, splitting the total power to more propellers, but of a smaller diameter, can enable navigation in shallow water. Note that once chosen, the propeller diameter cannot be changed.

8.8 Worked Examples

8.8.1 IW Selfpropelled Cargo Vessel

IW selfpropelled ship Deligrad (Fig. 8.26) built in 2004 for JRB in the Shipyard "Belgrade", will be taken as a case example in this section. Deligrad is very similar to the Ukrainian ships of Kapitan Antipov and Kapitan Širkov class (see Sect. 4.3.1). Deligrad's main particulars and waterway data used here are given in Table 8.15.

Fig. 8.26 IW Ship Deligrad pushing a barge. *Source* JRB

Table 8.15 IW vessel characteristics, waterway restrictions and other required data

Ship particulars		
L_{OA}	[m]	95.06
L_{WL}	[m]	93.76
B_{OA}	[m]	11.05
T	[m]	2.7
∇	[m^3]	2490
DWT	[t]	1892
P_B	[kW]	2x638
D[1]	[m]	1.6

Waterway restrictions		
h	[m]	5
b	[m]	400

Other data		
g	[m/s^2]	9.81
ρ	[t/m^3]	1

[1] two propellers in nozzles

Power Evaluation

Input data required for the estimation of the delivered power of IW selfpropelled cargo vessel according to the Complex MM (see Sect. 8.4.1) are given in Table 8.15. Having in mind the waterway restrictions, i.e. depth and breadth of the waterway, the transitional speed between the sub-critical and critical regimes can be determined from:

$$V_{TR} = h \cdot (0.0000023 \cdot b^2 + 0.0013493 \cdot b + 2.3236268) \\ + 0.000009 \cdot b^2 + 0.004332 \cdot b - 3.600712 = 12.05 \text{ km/h}$$

Accordingly, if the speed is less than 12 km/h, the influence of waterway restrictions on delivered power can be neglected and the required delivered power will be approximately the same as in deep water.

Independent variables, required for the MM, which do not depend on the vessel's speed are the following:

$$X_1 = 6.9178 \quad X_2 = 4.0926 \quad X_3 = 400$$

Independent variables X_4 and X_5 should be determined for each considered speed, as explained in Sect. 8.4.1. Table 8.16 shows the calculation procedure.

Table 8.16 Evaluation of delivered power of IW selfpropelled vessel

No.	V	[km/h]	6	8	10	12	14	16	18
1	Fnh	[-]	0.24	0.32	0.40	0.48	0.56	0.63	0.71
2	h_{EQ} or h	[m]	2.5760	3.3775	4.1789	4.9804	5.0000	5.0000	5.0000
3	R_{hEQ} or R_h	[m]	2.3732	3.1218	3.8647	4.6020	4.6199	4.6199	4.6199
4	X_4	[m]	2.5760	3.3775	4.1789	4.9804	5.0000	5.0000	5.0000
5	X_5	[-]	0.3454	0.4016	0.4511	0.4961	0.5777	0.6602	0.7427
6	C_D	[-]	0.9628	1.3098	1.7174	2.2528	3.4704	5.5547	11.9633
7	P_D	[kW]	39	71	117	183	330	603	1461

1. Depth Froude number, based on actual waterway depth, $Fn_h = V/\sqrt{g \cdot h}$. These values are not required for the application of MM, but provide a better understanding of sailing regime of the vessel (see Sect. 7.2.1).
2. Waterway depth (see Fig. 8.13):

   • if $V < V_{TR}$ then the equivalent depth calculated from:

$$h_{EQ} = (V - 0.000009 \cdot b^2 - 0.004332 \cdot b + 3.600712 \\ /(0.0000023 \cdot b^2 + 0.0013493 \cdot b + 2.3236268)$$

   should be used for further calculations,

   • if $V \geq V_{TR}$ then the actual waterway depth should be used.

3. The hydraulic radius (see Fig. 8.13):
   • if $V < V_{TR}$ then the equivalent hydraulic radius calculated form $R_{hEQ} = (b \cdot h_{EQ} - B \cdot T)/(2 \cdot h_{EQ} + 2 \cdot T + b + B)$ should be used for further calculations,

4. Independent variable X_4 is determined from:
   • if $V < V_{TR}$ then $X_4 = h_{EQ}$
   • if $V \geq V_{TR}$ then $X_4 = h$

5. Independent variable X_5 is determined as follows:
   • if $V < V_{TR}$ then $X_5 = V/\sqrt{g \cdot R_{hEQ}}$
   • if $V \geq V_{TR}$ then $X_5 = V/\sqrt{g \cdot R_h}$

6. The coefficient of delivered power is estimated according to the MM given in Table 8.6.
7. Delivered power is determined from: $P_D = C_D \cdot \rho \cdot g \cdot \nabla \cdot V$

Note that results presented in Table 8.16 are given in a graphical format in Fig. 7.9 (a curve for $h = 5$ m). The other curves are shown in Fig. 7.9 corresponding to other waterway depths and are evaluated similarly.

Modified Energy Efficiency Design Index (*EEDI)**

The calculation procedure for estimation of energy efficiency for IW ship Deligrad is given in Table 8.17. *EEDI** (previously discussed in section) is evaluated for water depth $h = 5$ m and is based on the results presented in Table 8.16.

Table 8.17 Evaluation of the energy efficiency according to the *EEDI** concept (see Sect. 6.2.3)

No.	V	[km/h]	6	8	10	12	14	16	18
1	*Fnh*	[-]	0.24	0.32	0.40	0.48	0.56	0.63	0.71
2	P_D	[kW]	39	71	117	183	330	603	1461
3	P_{Bref}	[kW]	40	73	119	187	336	615	1491
4	*EEDI**	[g CO_2/t·km]	2.26	3.07	4.03	5.29	8.14	13.03	28.07
5	a	[-]	-	-	39.765	37.460	54.313	66.592	-
6	c	[-]	-	-	-0.30932	-0.26385	-0.25601	-0.21687	-
7	$EEDI^*_{req}$	[g CO_2/t·km]	-	-	3.854	5.116	7.870	12.965	-

1. Depth Froude number, $Fn_h = V/\sqrt{g \cdot h}$.
2. The delivered power required for each considered speed (estimated by the Complex Method – row 7 of Table 8.16).
3. Required engine power, $P_{Bref} = P_D/\eta_S$, where η_S is assumed to be 0.98 (i.e. 2% for shaft and gearbox losses).
4. The attained *EEDI** is calculated from:

$$EEDI^* = \frac{P_{Bref} \cdot SFC \cdot CF}{m_{DWT} \cdot V}$$

where $SFC = 200$ g/kWh and $CF = 3.206$ g CO_2/g fuel (see Sect. 6.2.3).

5. The coefficient a (according to Table 6.1, for $h = 5$ m) can be calculated from:

$$a = 93.712 \cdot F_{nh}^{-3} - 516.38 \cdot F_{nh}^{-2} + 886.54 \cdot F_{nh}^{-1} - 414.86$$

   Note that this equation is applicable for $0.4 \leq F_{nh} \leq 0.65$.

6. The coefficient c (according to Table 6.1, for $h = 5$ m) can be calculated from:

$$c = -0.4181 \cdot F_{nh}^{-3} + 2.5716 \cdot F_{nh}^{-2} - 5.2767 \cdot F_{nh}^{-1} + 3.3485$$

   The applicability range is the same as for the coefficient a.

7. The required *EEDI** is calculated from:

$$EEDI_{Req}^{*} = a \cdot m_{DWT}^{c}$$

The *Attained* and *Required EEDI** are shown in Fig. 8.27. Note that the required *EEDI** curve (evaluated in Table 8.17) lies on the required *EEDI** surface, for shallow water navigation for $m_{DWT} = 1892t$, see Fig. 6.2.

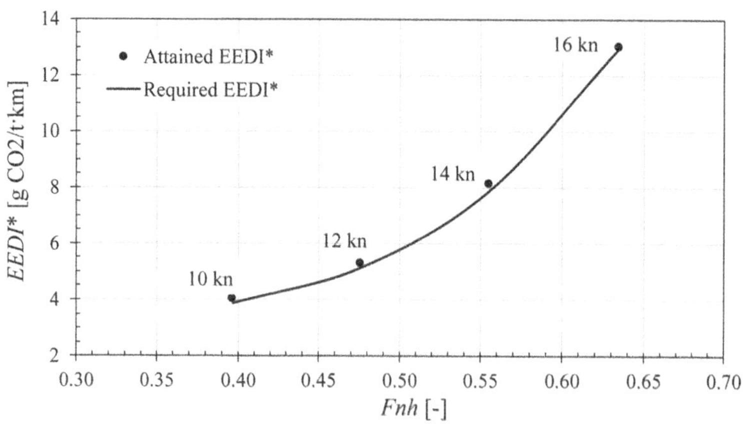

Fig. 8.27 Attained and Required *EEDI** for IW ship Deligrad ($T = 2.7$ m, $h = 5$ m)

These results can be presented in another, i.e. common format, with m_{DWT} instead of Fn_h on the abscissa (Fig. 8.28). The calculation procedure for required *EEDI** curves is given above (rows 5–7 of Table 8.17), except that now m_{DWT} is varied, see Table 8.18.

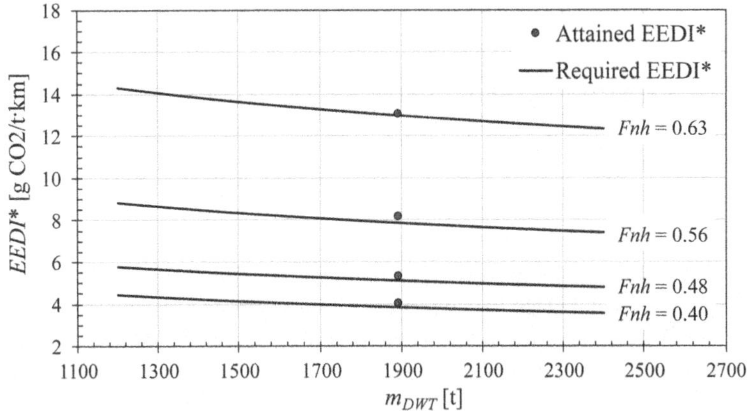

Fig. 8.28 Attained and Required *EEDI** for IW ship Deligrad ($T = 2.7$ m, $h = 5$ m)

Table 8.18 Required *EEDI** curves for $h = 5$ m

No.	m_{DWT}	[t]	1200	1400	1600	1800	2000	2200	2400
1	V [km/h] Fnh	10 0.40	4.44	4.23	4.06	3.91	3.79	3.68	3.58
2	V [km/h] Fnh	12 0.48	5.77	5.54	5.35	5.18	5.04	4.92	4.81
3	V [km/h] Fnh	14 0.56	8.84	8.50	8.22	7.97	7.76	7.57	7.41
4	V [km/h] Fnh	16 0.63	14.31	13.84	13.44	13.11	12.81	12.55	12.31

The results indicate the following:

- The energy efficiency of a seventeen years old IW ship Deligrad is close to the energy efficiency of an average IW selfpropelled vessel with similar cargo-carrying capacity used for the development of the Modified EEDI concept.
- The required *EEDI** surface (Fig. 6.2) may be considered as a practical benchmark criterion for the design of new IW selfpropelled vessels, or the assessment of energy efficiency of existing vessels.

German Proposal for Energy Efficiency Evaluation ($EEDI_{IWV}$)

The German proposal for energy efficiency evaluation is explained in Sect. 6.2.4. Note, however, that due to the applicability range, this method is valid for one draught only, which in the case of IW ship Deligrad is:

$$T = 1.5 \cdot D = 2.4m$$

Accordingly, a new estimation of delivered power in shallow water ($h = 5$ m) is required. Thus, for $T = 2.4\,m$, follows $L_{WL} = 93.33\,m$, $\nabla = 2206\,m^3$ and $m_{DWT} = 1607\,t$. Corresponding values of $P_D = \mathrm{f}(V)$ are given in Table 8.19, and Fig. 8.29.

Table 8.19 The delivered power versus speed for $T = 2.4m$ in shallow water $h = 5$ m

No.	V	[km/h]	6	8	10	12	14	16	18
1	*Fnh*	[-]	0.24	0.32	0.40	0.48	0.56	0.63	0.71
2	P_D	[kW]	29	56	96	154	267	473	1016

According to this method, the estimation of energy efficiency of selfpropelled IW dry cargo vessel sailing in shallow water ($h < 7.5$ m), should be performed for a *speed over ground that corresponds to the delivered power* which should be evaluated from the following equation (see Tables 6.2 and for coefficients α_6, β_4, γ_4, δ_2 and ε_1 Table 6.4 for Zone 3):

$$P_D = \left(\alpha_6 + \beta_4 \cdot EXP(-\gamma_4 \cdot B) - \delta_2 \cdot EXP\left(\frac{h}{-\varepsilon_1}\right)\right) \cdot m_{DWT} = 492\ \text{kW}$$

Consequently, the estimated speed through the water that can be achieved with $P_D = 492$ kW is $V = 16.1$ km/h, see Fig. 8.29.

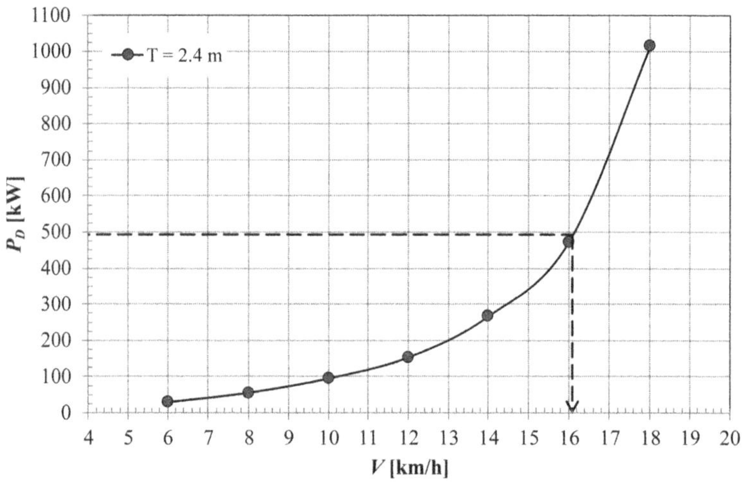

Fig. 8.29 The delivered power with respect to the speed through the water

Follows evaluation of energy efficiency design index from:

$$EEDI_{IWV} = CF \cdot SCF \cdot P_D/(V_S \cdot m_{DWT})$$

Speed over ground (V_S) depends on the current speed, thus the current speeds of 2, 4 and 6 km/h are assumed. Under the condition that propulsive efficiency for all assumed speeds in upstream navigation remains the same, the attained $EEDI_{IWV}$ may be evaluated, see Table 8.20.

Table 8.20 The attained $EEDI_{IWV}$ with respect to the current speed

V_C	[km/h]	2	4	6
V_S	[km/h]	14.1	12.1	10.1
$EEDI_{IWV}$	[g CO_2/t·km]	15.31	17.84	21.37

Note that here $SCF = 220$ g/kWh, while for the $EEDI^*$ concept $SFC = 200$ g/kWh was applied.

Required energy efficiency design index should be evaluated according to the equation given in Table 6.2 (equation's coefficients $\alpha_7, \beta_5, \gamma_5, \delta_3, \varepsilon_2, \zeta_1, \eta_1$ and θ_1 are given in Table 6.4 for Zone 3):

$$EEDI_{Req} = \left(\alpha_7 + \beta_5 \cdot V_C + \gamma_5 \cdot V_C^2\right) + \left(\delta_3 + \varepsilon_2 \cdot V_C - \zeta_1 \cdot V_C^2 + \eta_1 \cdot V_C^3\right) \cdot EXP\left(\frac{m_{DWT}}{-\theta_1}\right)$$

Evaluated values of $EEDI_{Req}$, depending on assumed current speeds and m_{DWT}, are given in Table 8.21.

Table 8.21 The values of the required energy efficiency design index ($EEDI_{Req}$)

Vc [km/h]	Mass of deadweight (m_{DWT} [t])							
	200	600	1000	1400	1800	2200	2600	3000
2	32.43	28.77	26.54	25.20	24.38	23.88	23.58	23.40
4	40.42	35.33	32.24	30.37	29.23	28.54	28.12	27.87
6	59.30	49.38	43.36	39.71	37.49	36.15	35.34	34.84

Comparison of estimated values of the attained $EEDI_{IWV}$ (points) and the required $EEDI_{Req}$ (lines) for three assumed current speeds is shown in Fig. 8.30.

Results indicate that the energy efficiency design index evaluated for IW vessel Deligrad according to the German Proposal, comply with the recommendations for all assumed current speeds.

*EEDI** Versus $EEDI_{IWV}$

Summarizing, the energy efficiency design indexes for IW ship Deligrad were assessed according to two different concepts, i.e. according to the so-called Modified method ($EEDI^*$) and German Proposal ($EEDI_{IWV}$). Both can be used for the validation of energy efficiency of IW vessels as suggested in Sects. 6.2.3 and 6.2.4, although none of them is mandatory nor is accepted by the industry. The conclusions derived, however, are different (see Figs. 8.28 and 8.30). This may be explained with the following:

- $EEDI^*$ and $EEDI_{IWV}$ are different methods, thus both attained and required *EEDI* are differently evaluated for each method.
- $EEDI^*$ and $EEDI_{IWV}$ are based on different values of SFC, i.e. 200 g/kWh and 220 g/kWh, respectively.
- Masses of deadweight for $EEDI^*$ and $EEDI_{IWV}$ are different, i.e. 1892 t (corresponding to the full draught of $T = 2.7$ m) and 1607 t (corresponding to $T = 2.4$ m, stemming from the method's restriction $T = 1.5 \cdot D$).

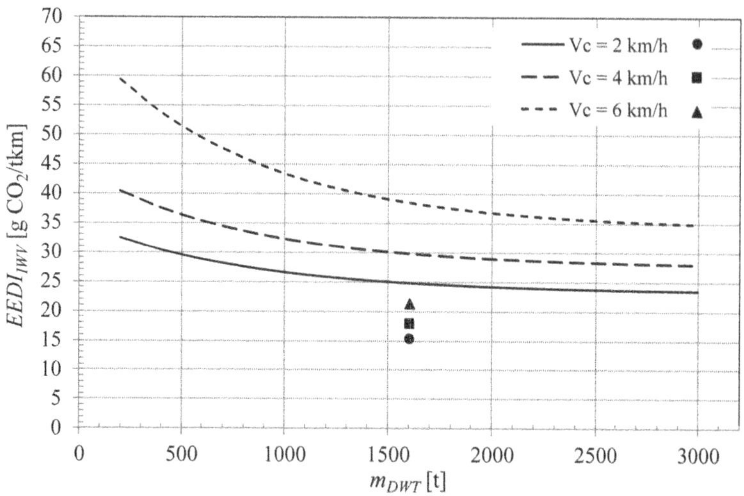

Fig. 8.30 Attained and Required *EEDI* according to the German Proposal for IW ship Deligrad ($T = 2.4$ m, $h = 5$ m)

Squat

Three methods for the evaluation of squat are explained in Sect. 8.5. Obtained results for IW ship Deligrad, corresponding to shallow but unrestricted waterway navigation, are given in Tables 8.22 and 8.23, and in Fig. 8.31.

Table 8.22 The bow and the stern squat for h = 3.5 and 5 m evaluated according to Romisch (1989)

h [m]	V	[km/h]	6	8	10	12	14	16	18
3.5	C_V	[-]	0.066	0.117	0.184	0.275	0.444	0.833	1.748
	S_B	[m]	0.035	0.062	0.096	0.144	0.233	0.437	0.916
	S_S	[m]	0.032	0.056	0.088	0.131	0.211	0.397	0.833
5.0	C_V	[-]	0.044	0.075	0.117	0.169	0.234	0.328	0.498
	S_B	[m]	0.027	0.047	0.074	0.106	0.147	0.206	0.312
	S_S	[m]	0.025	0.043	0.067	0.096	0.133	0.187	0.283

Table 8.23 The dynamic sinkage for h = 3.5 and 5 m evaluated according to Raven (2016)

h [m]	V	[km/h]	6	8	10	12	14	16	18
3.5	Fn_{hd}	[-]	0.100	0.134	0.167	0.201	0.234	0.268	0.301
	Fn_h	[-]	0.284	0.379	0.474	0.569	0.664	0.758	0.853
	S	[m]	0.031	0.057	0.094	0.145	0.220	0.333	0.536
5.0	Fn_{hd}	[-]	0.100	0.134	0.167	0.201	0.234	0.268	0.301
	Fn_h	[-]	0.238	0.317	0.397	0.476	0.555	0.635	0.714
	S	[m]	0.020	0.036	0.059	0.089	0.129	0.184	0.260

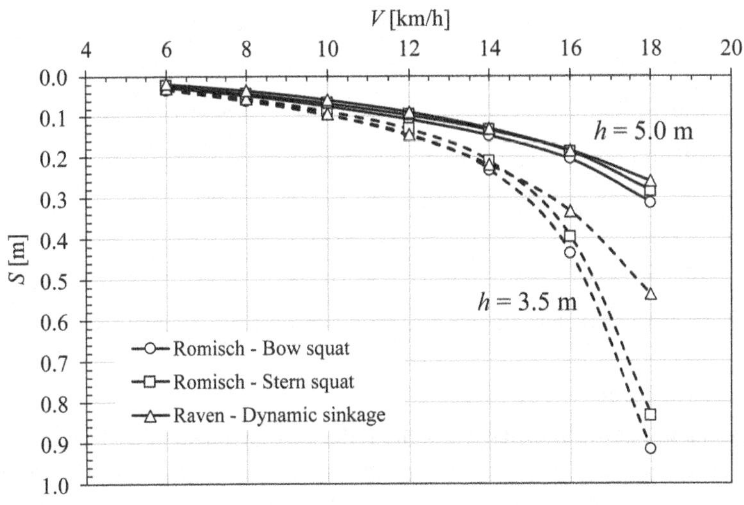

Fig. 8.31 Squat/dynamic sinkage of the IW ship Deligrad

Obtained results indicate that, in the case of a waterway depth of 3.5 m, the speed must be reduced to provide a safe under keel clearance.

The influence of a canal width (W) on the maximal squat of the IW ship Deligrad is evaluated according to Barrass (Briggs et al. 2009) method and results are given in Table 8.24 and Fig. 8.32. Two canal widths are considered—30 and 50 m, while the canal side grade is assumed to be 1:1.

Table 8.24 The influence of canal width on maximal squat of IW ship Deligrad, ($h = 3.5$ m)

W [m]	V	[km/h]	6	8	10	12	14	16	18
30	S	[m]	0.049	0.088	0.137	0.198	0.269	0.351	0.444
50	S	[m]	0.035	0.062	0.096	0.138	0.188	0.246	0.311

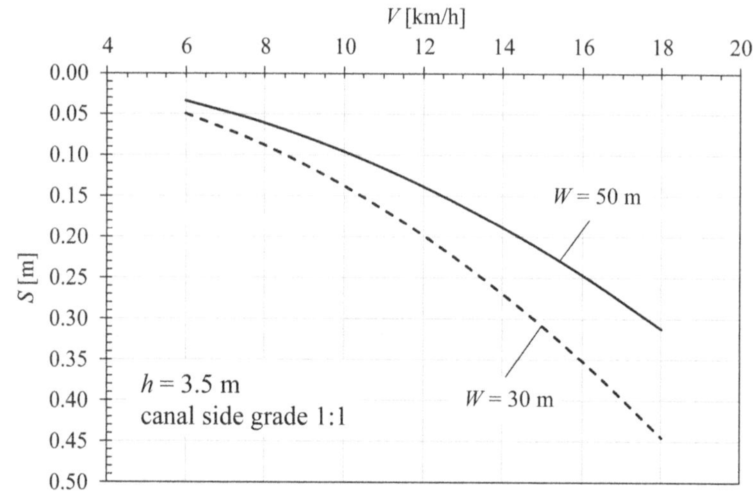

Fig. 8.32 Maximal squat of the IW ship "Deligrad" with respect to the canal width

Weight Estimation

IW ship "Deligrad", designed to carry dry bulk and containers, is used as an example for weight calculations. The weight estimations, derived according to Sects. 8.6.1 and 8.6.2, for steel weight and lightweight are given in Tables 8.25 and 8.26, respectively. Note that actual ship framing system is transverse; a case of longitudinal framing is included just for comparison.

Table 8.25 Steel weight

	Steel weight [t]		
Vessel type	Transverse framing	Longitudinal framing	
Dry Bulk	457* (452**)	439* (450**)	Hekkenberg (2012, 2014) - Table 8.11 & 8.12
Container	487* (488**)	454* (469**)	

*Table 8.11, **Table 8.12.

Table 8.26 Lightweight

	Lightweight [t]		
Vessel type	Transverse framing	Longitudinal framing	
Dry Bulk	574*(618**)	546*(616**)	Hekkenberg (2012) - (Table 8.13, 8.12 & 8.14)
Container	599*(654**)	563*(634**)	
General	654		Heuser (1986)
General/Container	605		Hofman (2006)

*Table 8.13, **Table 8.12 & 8.14.

Estimated values of steel weight and lightweight are slightly higher if the vessel is regarded as a container ship rather than a dry bulk carrier, at least in the particular case, see Hekkenberg (2012, 2014). Moreover, transversely framed weights more than a longitudinally framed vessel, which is a well-known fact.

Deligrad's lightweight is 598 t. This corresponds very well to the estimations (see Table 8.26) made by:

- Hofman (1.1% higher estimate), and
- Hekkenberg, equations for transversely framed dry bulk vessels which predict 4% less and 3.3% higher values (Tables 8.12, 8.13 and 8.14 respectively).

According to Table 8.14, the total difference between the lightweight and steel weight is 165.8 t, consisting of the following masses: accommodation (93 t), machinery/equipment/outfitting (68.4 t), and piping outside engine room (4.4 t).

Heuser's equation overestimates the lightweight of the particular vessel by 9.3%.

8.8.2 Pushed Convoys

The pushed convoy consisting of a pushboat and six Danube barges is analysed in this section. Delivered power and economic indicators for two different convoy configurations (P + 2 + 2 + 2 and P + 3 + 3, see Fig. 8.14) and three water depths (h = 3.5, 5.0 and 7.5 m) are evaluated, but only a case for *convoy configuration* $P + 2 + 2 + 2$ *and* $h = 5$ *m is explained in detail.* These results are shown in tabular and graphical format (Tables 8.29, 8.30, 8.31, 8.33, 8.34 and 8.35 and Figs. 8.33, 8.34, 8.35 and 8.36). The characteristics of examined pushboat and barges are given in Table 8.27.

Table 8.27 The characteristics of the considered pushboat and barges

Pushboat			Danube Barge		
L_{OAp}	[m]	35	L_{OAb}	[m]	76.50
L_{WLp}	[m]	33.9	L_{WLb}	[m]	74.32
B_p	[m]	14	B_b	[m]	11.33
T_p	[m]	1.75	T_b	[m]	2.0
∇_p	[m^3]	525	∇_b	[m^3]	1566
			m_{DWTb}	[t]	1236

Note that the characteristics of a pushboat cannot be altered as are embedded in the MM itself (see Sect. 8.4.2). Concerning the barges, the convoy consists of the Danube barges: steel weight of each barge is 330 t, length and volumetric displacement of the barge are determined according to equations given in Table 8.9 for a draught of 2 m.

Power Evaluation

Economic indicators, discussedCharacteristics of the pushed convoy P + 2 + 2 + 2 and other data required for the estimation of delivered power are given in Table 8.28.

Table 8.28 Characteristics of P + 2 + 2 + 2 convoy, waterway restrictions and other required data

Pushed convoy (P+2+2+2)			Waterway restrictions		
L	[m]	256.86	h	[m]	5
B	[m]	22.66			
T	[m]	2.0	**Other data**		
∇	[m^3]	9921	g	[m/s^2]	9.81
m_{DWT}	[t]	7416	ρ	[t/m^3]	1

Delivered power will be evaluated according to the MM presented in Sect. 8.4.2. Independent (input) variables that do not depend on the speed are the following:

$$X_1 = L/B = 11.33537 \quad X_2 = B/T = 11.33000$$
$$X_3 = L/h = 51.372 \quad X_4 = L/\nabla^{1/3} = 11.95393$$

The independent variable X_5 is a Froude number which should be determined for each assumed speed. The calculation procedure is given in Table 8.29.

Table 8.29 Evaluation of delivered power for pushed convoy P + 2 + 2 + 2 and $h = 5$ m

No.	V	[km/h]	4	6	8	10	12	14	15
1	X_5	[-]	0.02213	0.03320	0.04427	0.05534	0.06640	0.07747	0.08301
2	C_D	[-]	0.60172	0.97697	1.59836	2.42549	3.53116	5.49109	7.33556
3	P_D	[kW]	65	158	346	656	1146	2078	2975

1. Independent variable X_5 based on the convoy overall length and speed through water, $X_5 = V/\sqrt{g \cdot L}$.
2. The coefficient of delivered power is estimated according to the MM given in Table 8.10.
3. The delivered power, $P_D = C_D \cdot \rho \cdot g \cdot \nabla \cdot V/1000$

Evaluated delivered power (with respect to speed through the water) for other waterway depths, are shown in Table 8.30. Note that the value of a variable X_3 is now 73.38840 and 34.248 for $h = 3.5$ m and $h = 7.5$ m, respectively. Final results for all considered waterway depths are given in Fig. 8.33.

The delivered power for the pushed convoy P + 3 + 3 ($L = 182.54$ m, $B = 33.99$ m) is given in Table 8.31 and Fig. 8.34.

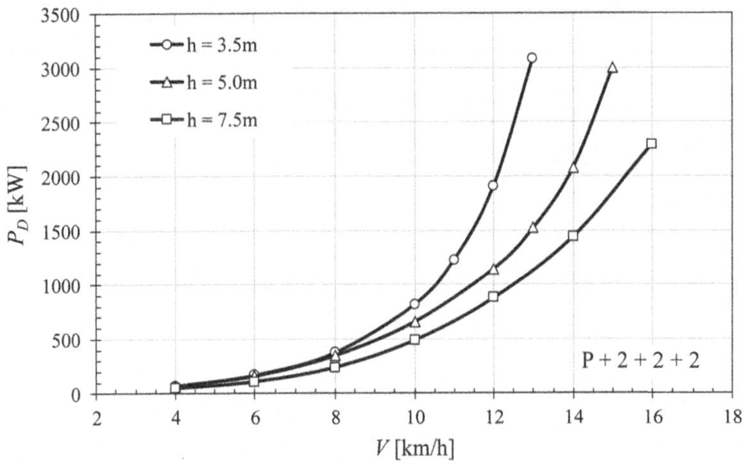

Fig. 8.33 The delivered power of the pushed convoy (P + 2 + 2 + 2) with respect to waterway depth and the speed through the water

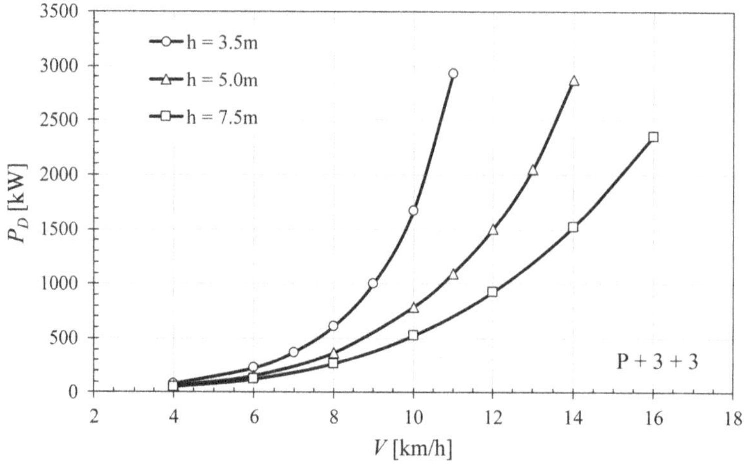

Fig. 8.34 The delivered power of the pushed convoy (P + 3 + 3) with respect to waterway depth and speed through the water

Table 8.30 Evaluated delivered power for pushed convoy P + 2 + 2 + 2 for h = 3.5 and 7.5 m

h = 3.5m	V	[km/h]	4	6	8	10	11	12	13
	P_D	[kW]	69	168	374	816	1229	1905	3081
h = 7.5m	V	[km/h]	4	6	8	10	12	14	16
	P_D	[kW]	50	110	241	489	879	1443	2294

Table 8.31 Evaluated delivered power of the pushed convoy P+3+3 for h = 3.5, 5.0 and 7.5 m

h = 3.5m	V	[km/h]	4	6	7	8	9	10	11
	P_D	[kW]	77	224	371	610	1002	1672	2935
h = 5.0m	V	[km/h]	4	6	8	10	12	13	14
	P_D	[kW]	60	152	367	786	1505	2057	2885
h = 7.5m	V	[km/h]	4	6	8	10	12	14	16
	P_D	[kW]	54	124	268	525	928	1522	2357

Economic Indicators

Economic indicators, discussed in Sect. 6.3, are applied for the cost-performance assessments for the pushed convoys P + 2 + 2 + 2 and P + 3 + 3 navigating in shallow water. Three waterway depths are analysed $h = 3.5$, 5.0 and 7.5 m. The river current is assumed to be $Vc = 3$ km/h and the convoy is supposed to sail upstream in all examined cases, regardless of configuration. Other required data are assumed to be the same as in Hofman et al. (2013) and are given in Table 8.32.

The calculation procedure for evaluation of the Transport efficiency index (*TEI*), Economic efficiency coefficient (*EEC*), Profit rate indicator (Π_t) and Profit coefficient (p) (see Sects. 6.2.2, 6.3.1, 6.3.2 and 6.3.3, respectively) for pushed convoy P + 2 + 2 + 2 and $h = 5$ m is shown in Table 8.33.

Table 8.32 Required data for assessment of cost-performances

k_f	[€/(t·km)]	0.01
k_c	[€/t]	847.4
SCF	[g/(kWh)]	219
m_{ref}	[t]	10000
V_{ref}	[km/h]	10

Table 8.33 Economic indicators for pushed convoy P + 2 + 2 + 2 and $h = 5$ m

No.	V	[km/h]	4	6	8	10	12	14	15
1	V_S	[km/h]	1	3	5	7	9	11	12
2	*TEI*	[t·km/kW·h]	113.97	140.39	107.26	79.17	58.26	39.25	29.91
3	*EEC*	[-]	6.14	7.56	5.78	4.27	3.14	2.12	1.61
4	*Πt*	[€/h]	62.08	193.07	306.64	397.43	454.84	430.06	337.81
5	*p*	[-]	0.062	0.193	0.307	0.397	0.455	0.430	0.338

1. Speed over ground, $V_S = V - V_C$.
2. $TEI = (m_{DWT} \cdot V_S)/P_D$, where $m_{DWT} = 7416$ t (from Table 8.28) and $P_D = f(V)$ (from Table 8.29).
3. $EEC = \frac{k_f}{k_c \cdot SFC} \cdot TEI$, where k_f, k_c and SFC are from Table 8.32.
4. $\Pi_t = k_f \cdot m_{DWT} \cdot V_S \cdot \left(1 - \frac{1}{EEC}\right)$, where k_f and m_{DWT} are from Tables 8.32 and 8.28, respectively.
5. $p = \frac{\Pi_t}{k_f \cdot m_{ref} \cdot V_{ref}}$, where k_f, m_{ref} and V_{ref} are from Table 8.32.

Estimated values of the Profit coefficient with respect to convoy speed for the other waterway depths are shown in Table 8.34. Obtained results for all considered waterway depths are shown in Fig. 8.35. Note that the curve for $h = 7.5$ m is also shown in Fig. 6.10, but with respect to the speed through water (V).

The Profit coefficients for the pushed convoy P + 3 + 3 ($L = 182.54$ m, $B = 33.99$ m) are given in Table 8.35 and Fig. 8.36.

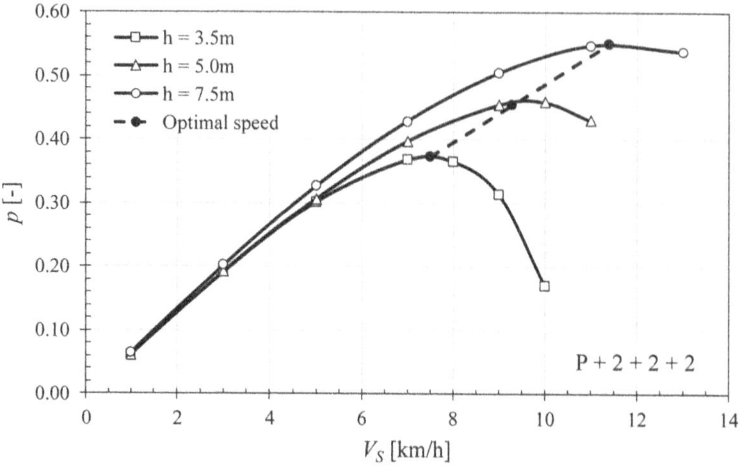

Fig. 8.35 The Profit coefficient of the pushed convoy (P + 2 + 2 + 2) with respect to waterway depth and the speed over ground

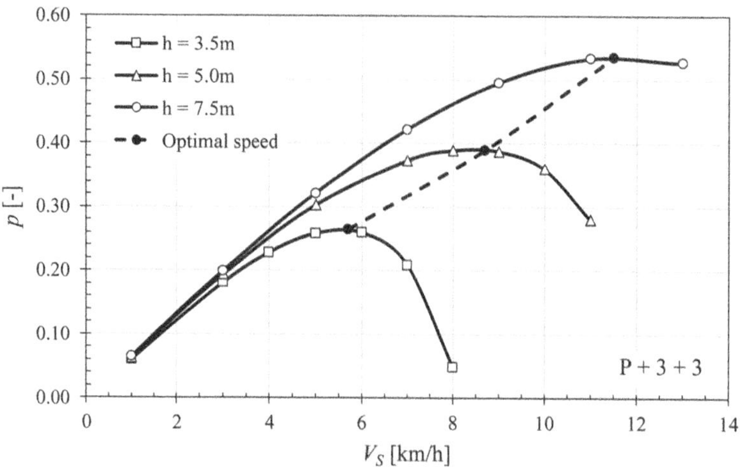

Fig. 8.36 The Profit coefficient of the pushed convoy (P + 3 + 3) with respect to waterway depth and the speed over ground

Table 8.34 Profit coefficient for pushed convoy P + 2 + 2 + 2 and h = 3.5 and 7.5 m

h = 3.5m	V	[km/h]	4	6	8	10	11	12	13
	V_S	[km/h]	1	3	5	7	8	9	10
	p	[-]	0.061	0.191	0.301	0.368	0.365	0.314	0.170
h = 7.5m	V	[km/h]	4	6	8	10	12	14	16
	V_S	[km/h]	1	3	5	7	9	11	13
	p	[-]	0.065	0.202	0.326	0.428	0.504	0.548	0.538

Table 8.35 Estimated Profit coefficient corresponding to the pushed convoy P+3+3 with respect to water depth

h = 3.5m	V	[km/h]	4	6	7	8	9	10	11
	V_S	[km/h]	1	3	4	5	6	7	8
	p	[-]	0.060	0.181	0.228	0.258	0.259	0.209	0.049
h = 5.0m	V	[km/h]	4	6	8	10	12	13	14
	V_S	[km/h]	1	3	5	7	9	10	11
	p	[-]	0.063	0.194	0.303	0.373	0.388	0.360	0.280
h = 7.5m	V	[km/h]	4	6	8	10	12	14	16
	V_S	[km/h]	1	3	5	7	9	11	13
	p	[-]	0.064	0.199	0.321	0.422	0.495	0.533	0.527

The profit coefficient (p) is a useful indicator for various cost analyses. For instance, for a given mass of cargo and pushboat power, it may assist in the assessment of the optimal speed of various pushed convoy configurations. A numerical example is given in the following.

Assuming the input data given in Tables 8.27 and 8.32 and upstream navigation ($Vc = 3$ km/h), a comparison of various pushed convoy configurations is shown in Fig. 8.37. The influence of river speed (Vc) in upstream navigation is shown in Fig. 8.38. A higher value of the Profit coefficient (p), indicates a more cost-effective solution. Note that in both cases the convoy speed is relative to the ground.

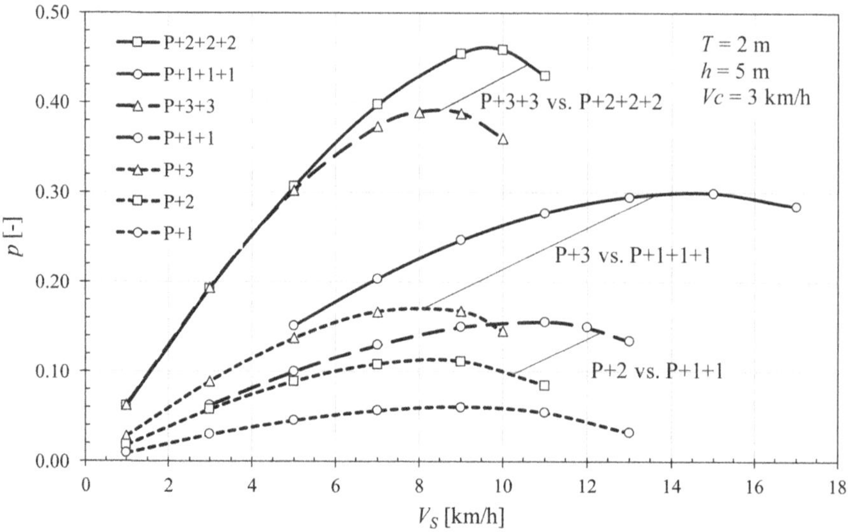

Fig. 8.37 Comparison of various pushed convoy configurations

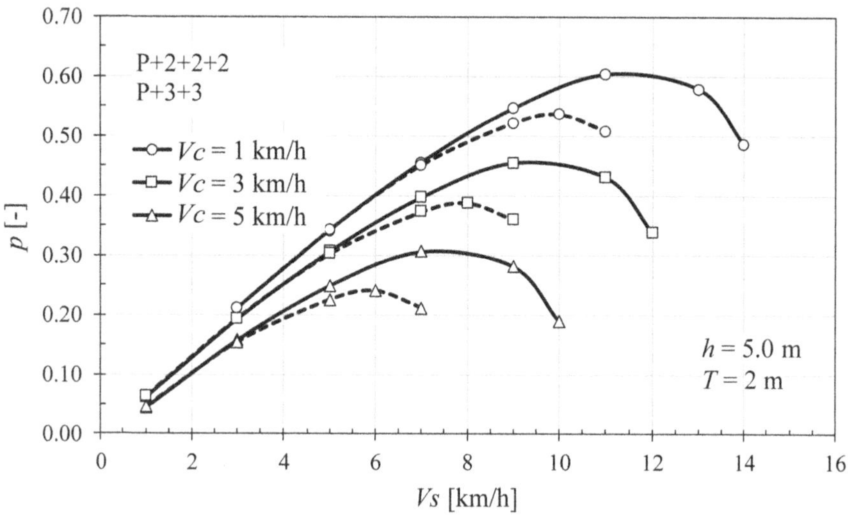

Fig. 8.38 Influence of river speed for two convoy configurations

8.9 Concluding Remarks

This chapter consists of a number of basic expressions and design guidelines specific for IW vessels navigating in shallow waterways and can be useful in the early design phases. The presented topics can be grouped as follows:

- Design guidelines (concerning mainly the hydrostatics) for selfpropelled ship and pushboat hull forms (Sects. 8.2.1 and 8.2.2, respectively), comments on stability and safety (Sect. 8.3), and recommendations for choosing the main dimensions of container vessels (Sect. 8.7). The lightship weight estimation for selfpropelled cargo vessels (Sect. 8.6) is placed here, though strictly speaking, it does not belong to hydrostatics.
- Essential expressions (concerning the hydrodynamics) for evaluation of required power of IW selfpropelled ships and pushboat convoys (Sects. 8.4.1 and 8.4.2, respectively) and assessment of squat (Sect. 8.5). Squat assessment probably belongs to an operational rather than ship design cluster of expressions but is placed here as it is needed for the determination of maximal allowed speed which is limited by the waterway depth.
- Worked examples for selfpropelled vessel and pushed convoys. Given are detailed procedures for assessment of:
- power, squat, energy efficiency indexes (modified *EEDI** and German proposal) and lightship weight for the selfpropelled ship Deligrad (Sect. 8.8.1) and
- power and economic indicators for two pushed convoys configurations (P + 2 + 2 + 2 and P + 3 + 3) at three water depths (Sect. 8.8.2).

Note that the Complex MMs for power evaluations of IW selfpropelled vessels and pushed convoys (given in Sects. 8.4.1 and 8.4.2, respectively) are released here for the first time.

References

Bačkalov I (2015) Impact of contemporary stability regulations on stability of shallow-draught inland container vessels. Safety Science 72

Bačkalov I, Kalajdžić M, Momčilović N, Rudaković S (2016) A study of an unconventional container vessel concept for the Danube. In: Proceedings of PRADS 2016, Copenhagen, Denmark 2016

Bilen B (1997) Innovations in pushboat convoy technology (in Serbian). Institute of Technical Sciences of Serbian Academy of Sciences and Arts, Belgrade, ISBN 86-80321-01-X

Bilen B, Bilen-Katić B (1997) Modified pusher tug design for the Danube river. International Shipbuilding Progress 44(438)

Briggs M, Vantorre M, Uliczka K, Debaillon P (2009) Handbook of coastal end ocean engineering. Chapter 26: Prediction of squat for underkeel clearance. World Scientific, Imperial College Press

Carlton JC (2012) Marine propellers and propulsion. Butterworth-Heinemann, ISBN 9780080971230

COVEDA (2003) Container Vessels for the Danube waterway. DPC Project

GL (2011) GL rules for classification and construction of Inland Navigation Vessels. Germanisher Lloyd

Hagesteijn G, Meij van K, Thill C (2015) Distributed propulsion, a novel concept for inland vessels. In: Proceedings of the ASME 2015 34th international conference on Ocean, Offshore and Arctic Engineering

Hekkenberg RG (2012) Inland ships for efficient transport chains. Doctoral Thesis, Technische Universiteit Delft

Hekkenberg RG, Hopman H (2014) New estimation methods for the steel weight of European inland dry bulk ships. J Ship Prod Des 30(3)

Heuser H (1986) Verdrängungsschiffe auf flachem Wasser. Teil II. Schiffstechnik Band 33, Heft 1

Heuser H (1990) Binnenschiff der Zukunft, VBD Report 1260

Hofman M (2006) Inland container vessel: optimal characteristics for specified waterway. RINA International Conference on Coastal Ships & Inland Waterways, London

Hofman M, Bačkalov I (2010) Risk-based analyses of inland vessel stability. In: Proceedings of the 11th international stability workshop, Wageningen

Hofman M, Krstović M, Kalajdžić M (2013) WP3 performance indicators and transport efficiency of pushed convoys. Innovative Danube Vessel

Kalajdžić M, Hofman M (2009) Power assessment of push boat—barge formations and computer code CONVOY (in Serbian). Report No. 002/2009, University of Belgrade, Faculty of Mechanical Engineering

Latorre R (1994) Design of a river towboat for operating in 2.5–3.5 m water depth. WEGEMT 21st Graduate School, Gerhard Mercator University, Duisburg

Marchal JLJ, ShenYD, Kicheva D (1996) An empirical formula to estimate the resistance of a convoy in a restricted waterway. J Ship Res 40(2)

Molland AF, Turnock SR, Hudson DA (2017) Ship resistance and propulsion—practical estimation of ship propulsion power. Cambridge University Press, ISBN 9781316494196

Mulder C, Williams P (2006) Improving the performance of shallow draft tugs in Northern Canada. SNAME Trans 114

Pompée PJ (2015) About modelling inland vessels resistance and propulsion and interaction vessel-waterway—key parameters driving restricted/shallow water effects. PIANC Smart Rivers, Buenos Aires 2015

Radojčić D (2009) Environmentally friendly inland waterway ship design for the Danube river. World Wide Fund for Nature International, Danube-Carpathian Programme (WWF-DCP)

Raven H (2016) A new correction procedure for shallow water effects in ship speed trials. In: Proceedings of PRADS, Copenhagen

Romisch K (1989) Empfehlungen zur Bemessung von Hafeneinfahrten. Wasserbauliche Mitteilungen der Technischen Universitat Dresden, Heft, p 1

Simić A (2012) Energy efficiency of inland waterway selfropelled cargo vessels (in Serbian). PhD dissertation, University of Belgrade, Faculty of Mechanical Engineering

Statistisches Bundesamt (2001) Wirtschaft und Statistik, 7/2001, Wiesbaden

Zibell HG, Müller E (1996) Binnenschiffe fur extrem flaches Wasser – Ergebnisse des VEBIS-Projektes. Handb. D. Werften, vol XXIII

Zöllner J (2003) Development of propulsion within inland navigation. Ship's Propulsion—News and Further Developments (in German). Duisburg

Part III
Innovative Technologies and Design Solutions for Contemporary Inland Waterway Vessels

Chapter 9
Technologies for Improvements of Energy Efficiency and Environmental Performance

9.1 Introduction

Improvements of the vessel's energy efficiency and environmental performance can be sublimated into the *greening* of the vessel. Namely, according to Dahlke-Wallat et al. (2020a) greening consists of three aspects:

- increase of vessel's efficiency,
- reducing the amount of emitted CO_2 per t km, and
- reducing the emitted air pollutants.

The objective of this chapter is the analysis of IW vessel greening technologies[1] that are *available today or in the nearby future.* These technologies are divided into seven logical groups, but are treated here in five sections (Sects. 9.2–9.6), see Fig. 9.1. The reason for this is that some technologies (or greening techniques) affect each other, i.e. are interconnected, while the other may be treated independently. Specifically:

- New engine types require new fuels, both are leading to *fuel-efficient and clean engines* (Sect. 9.2)
- The hull form should be optimized together with the propulsor, while the propulsor goes together with the transmission; all three are leading to a *hydrodynamically efficient ship* (Sects. 9.3 and 9.4)
- Hull lightening leads to displacement reduction, thus to resistance reduction too, but the shipowner would rather transport a *larger payload* and keep the same displacement (Sect. 9.5)
- Automation leads to *efficient ship navigation* (i.e. *speed reduction*) and *smaller crew* (Sect. 9.6).

[1] The potential for IWT greening is not only in the greening of the vessel; it also means that the vessel's operation and the logistics planning should be greener, as well as the waterway infrastructure.

D. Radojčić et al., *Design of Contemporary Inland Waterway Vessels*,
https://doi.org/10.1007/978-3-030-77325-0_9

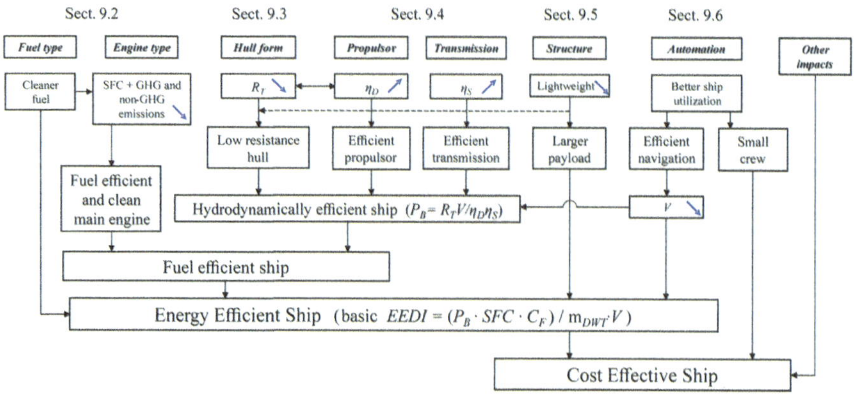

Fig. 9.1 Greening technologies that lead to a cost-effective ship

The abovementioned greening technologies treat the *vessel's efficiency* only, which just partly contribute to waterborne transport efficiency (i.e. the transhipment is ignored, see Chap. 3 and Fig. 3.1). Furthermore, the cost-efficiency of a vessel is influenced by several issues out of the book's scope. Consequently, it is reasonable to assess here only the *cost effectiveness*, instead of cost efficiency (which is supposed to be a measurable quantity). For that, the IW vessel has to be *energy efficient* (see Chap. 6), which further requires fuel efficiency. That essentially means the ship should a) have a *fuel efficient and clean engine*, and b) be *hydrodynamically efficient*. Both can be further segmented as depicted in Fig. 9.1 and call for the application of greening technologies.

The fragmented approach enables better insight into energy-saving methods. Note that several technologies treated here are state-of-the-art technologies, while others are innovative—presentation of the first are often necessary for understanding the second.

The *energy efficient ship*, as denoted in Fig. 9.1, is the one that releases the least CO_2 per t·km; this stems from the formulation of the *EEDI*. The benefit assessment of the employment of innovative technologies requires all 5 terms of the *EEDI* formula, while only the first term of the numerator is involved in the basic *EEDI* (see Chap. 6). The innovative energy efficiency technologies, however, are categorized[2] as follows (see IMO, 2013 and ABS, 2019):

- Category A—Technologies that *directly influence the speed-power curve*. Essentially this means that these technologies lead to the reduction of Propulsive Power (PP) at constant speed V_{Ref}. Category A technologies *cannot be separately assessed from overall vessel performance.*
- Category B—Mechanical technologies that reduce PP but do not generate electricity (5th term technologies).

[2]This categorization is valid for the seagoing ships, but may also be regarded as a framework for the innovative technologies for IW vessels which are treated in this chapter.

- Category C—Electrical technologies that generate electricity and provide power for ship propulsion (4th term technologies).

The innovative energy efficiency technologies presented in this chapter (Sects. 9.2–9.5) practically all belong to Category A. Besides those, amongst the innovative technologies of interest for IW vessels, is hull air lubrication, but it belongs to Category B (see IMO, 2013) and is discussed in Sect. 10.1.3. Other Category B and C technologies (e.g. wind and solar power, respectively) are presently not applicable for the *propulsion of IW cargo vessels.* Note, however, that the efficiency of hybrid propulsion, fuel cells and biofuels cannot be assessed within the current *EEDI* scheme for seagoing ships.

Consequently, the technologies treated in this chapter, except those in Sect. 9.6, affect the first term of the numerator only and the denominator of *EEDI* seagoing formula, i.e. the main engine emissions and transport work, respectively. Technologies treated in Sect. 9.6 inherently do not affect *EEDI*, but are treated here as they lead to speed reduction.

By far the largest energy savings may be obtained with *speed reduction.* This, however, is not always feasible (due to competitiveness, potential necessity to hire additional capacity etc.). However, if the speed reduction is a result of wise voyage planning (Sect. 9.6), substantial savings may be expected.

Clever design is probably the second area that can lead to noticeable savings (this is the subject of Chap. 10). On the other hand, the clever design incorporates the energy efficiency technologies which are treated in this chapter.

An overview of where and why the energy is consumed on the seagoing vessels, and potential energy saving methods is given in Lassesson and Andersson (2009). A detail underlined there is paraphrased here:

> For the energy efficient ship, it is of utmost importance to *optimize it for the conditions it will be working under*, i.e. for the particular operational profile. Moreover, all ship components should be optimized for the same conditions. If the working conditions change, the ship should be re-optimized for new working conditions.

This is essentially symbolized in Fig. 9.1, which indicates how interdisciplinary optimization of various technologies leads toward the energy efficient ship, thus cost effective too.

9.2 Innovations in Propulsion Plants and Fuels[3]

While the life of IW vessels is long, 50 or so years, the life of ship engines is typically 15–20 years, the older engine types have even longer service life. The low renewal rate of the IW vessels' engines is a cause for relatively high emissions of nitrogen oxides and particulate matter. The small number of IW ship engines (compared to those for road or construction vehicles) further leads to longer product cycles and thus to delayed uptake of new technologies. As a consequence, the position of IWT as the most environmentally friendly mode is questioned.

The following important remarks concern this section:

- The engines are essentially converting one source of energy into another, hence they are also named *energy converters*. Accordingly, various kinds of fuels are called *energy carriers*. This terminology will be used in this section only.
- Different types of drivetrains are mentioned throughout this section, but are treated in Sect. 9.4.1 which bears the title "Drivetrains of IW vessels—Basics". Here they are viewed with the focus on the energy converters (prime movers) while there the focus is on the propulsors. Therefore, *cross-referencing is suggested.*

9.2.1 Regulatory Framework—an Overview

Background

For all transport modes it is desired to progressively improve air quality by reducing the pollutant emissions and by gradually phasing out the most polluting engines. In the EU, Non-Road Mobile Machinery (NRMM), which includes engines on IW vessels too, are regulated by EU Directive 97/68 (EC, 1997), while in the US emission standards are managed by the EPA regulations for marine vehicles. Before 2003 engines of IW vessels were not regulated regarding exhaust gas emissions. The CCNR adopted its Stage I emission regulations in early 2000 and one year later the key elements for a revision of the regulations with the stricter limit values of the subsequent Stage II. Stage I applies to all diesel engines installed from the beginning of 2003 on ships licensed under the Rhine Shipping Regulations and Stage II from 1st July 2007. Similar limits became mandatory for all EU member states when Directive 97/68/EC Stage IIIA came into force.

Current Situation

Since the beginning of 2019 (2020 for engines with 300 kW or more output), stricter exhaust emission limits apply to new inland waterway engines commissioned in accordance with EU Regulation 2016/1628 (EC 2016). The EU Regulation sets the emission limits for so-called Stage V engines, replacing preceding stages and thereby

[3]The main sources of information in this section are GRENDEL (2020) and Dahlke-Wallat et al. (2020b).

the limits according to CCNR II. The Stage V calls for limit values for emissions of carbon monoxide (CO), hydrocarbons (HC) and nitrogen oxides (NO_X) for internal combustion engines installed in IW vessels. It as well sets the limits for emissions of particulate matter (PM). Table 9.1 shows the CCNR II and Stage V emission limits applicable to engines exclusively for use in IW vessels for direct or indirect propulsion (IWP) and auxiliary engines (IWA) exclusively for use in IW vessels. As marinized Non-Road Engines (NRE) up to 560 kW (class NRE-6 with variable (v) and constant (c) speed) and Euro VI truck engines may also be installed on inland vessels, the corresponding limits are listed as well. However, the limits are not fully comparable according to the differences in underlying test cycles.

Emission limits for NRMM engines are differentiated according to power ranges and applications. At least for large engines these limits are only achieved with Particulate Filters (PF) and Selective Catalytic Reduction (SCR) systems (see Sect. 9.2.3). The procedures for engine manufacturers for placing their engines on the EU market are also required. Proof of compliance with Stage V limits requires a dedicated type approval for engines, which is not only time-consuming but is also costly procedure in view of a small market. In addition, the type approval includes the specification of the fuel. Therefore, it is important to consider the (future) use of non-fossil drop-in

Table 9.1 Emission limits according to CCNR II and EU Regulations 2016/1628 (EC 2016) and 595/2009 (EC 2009a) for truck engines

Engine sub-category	Reference power (kW)	CO (g/kWh)	HC (g/kWh)	NO_x (g/kWh)	PM (g/kWh)	PN (#/kWh)
CCNR II	$130 \le P < 560$	3.50	1.00	6.00	0.20	
CCNR II	$P \ge 560$	3.50	1.00	$45n^{-0.2}-3$	0.20	-
IWP/IWA-v/c-1	$19 \le P < 75$	5.00	($HC + NO_x \le 4.70$)		0.30	-
IWP/IWA-v/c-2	$75 \le P < 130$	5.00	($HC + NO_x \le 5.40$)		0.14	-
IWP/IWA-v/c-3	$130 \le P < 300$	3.50	1.00	2.10	0.10	-
IWP/IWA-v/c-4	$P \ge 300$	3.50	0.19	1.80	0.015	1×10^{12}
NRE-v/c-6	$130 \le P \le 560$	3.50	0.19	0.40	0.015	1×10^{12}
Euro VI – WHSC		1.50	0.13	0.40	0.010	8×10^{11}

n: engine speed in revolutions per minute
PM: mass of ultrafine particulate pollutants
PN: number of particles larger than 23 nm
CO: carbon monoxide
HC: hydrocarbons
NO_x: nitrogen oxides

fuels[4] like Hydrotreated Vegetable Oil (HVO) and Power-To-Liquids (PTL), which can be used as blends or pure fuels.

9.2.2 Diesel Engines and Fuels—State-of-the-Art

The diesel cycle based on compression ignition was invented in the late nineteenth century. Due to its high efficiency, the robustness and compactness, as well as the low fuel costs, these engines replaced other prime movers within about 50 years. Today, apart from a few gas engines and battery-electric drives, almost all IW vessels are powered by diesel engines. Future diesel engines, or better various kinds of internal combustion engines, are expected to be greener. Greening requires cleaner and innovative fuel types, which are elaborated in the following sections. Since 2011, according to Directive 30 (EC 2009b), the sulphur content for gasoil used in IW shipping is limited to 10 mg/kg (0.001%). This is an important prerequisite for the use of exhaust gas aftertreatment. In comparison, the IMO reduced the maximum sulphur content for seagoing ships outside Emission Control Areas (ECA) to 0.5% from 1st January 2020, while inside the ECAs a limit of 0.1% applies. A diesel engine is also a starting point for all benchmarking in IW shipping.

Diesel engine's main advantage, compared to other prime movers, is its efficiency which in the last 100 or so years rose from some 30% to about 40 $\div$ 45%, and for low-speed engines even up to 50%. Their fuel consumption steadily drops (thus CO_2 emissions too) with a present rate of around 0.3 $\div$ 0.4% per year. Typical energy losses in the shipboard propulsion systems (assuming diesel engine efficiency is 40%) are depicted in Fig. 9.2. This means that 60% of the fuel energy is lost before it becomes rotational power.

A diesel engine essentially converts the chemical energy (stored in a fuel) into mechanical energy, thus is a classical *energy converter*. More about diesel engines can be found in any topical textbook, the *focus in the following is on the greening measures* of diesel engines that are applicable in IW shipping.

9.2.3 Exhaust Gas Aftertreatment

With the Stage V emission limits for CO, HC, NO_X and particle matters are greatly reduced. In addition to the reduction of the particle mass, the particle number for larger engines is also restricted. In order to meet this new emission standard, technical solutions become more complex. Internal combustion control concepts were sufficient to comply with the previous regulations. To fulfil the new limits, external aftertreatment is required. Sulphur oxides are not only harmful to the environment

[4]Drop-in fuels are a synthetic and widely interchangeable substitute for conventional petroleum-derived hydrocarbons as further described in Sect. 9.2.4.

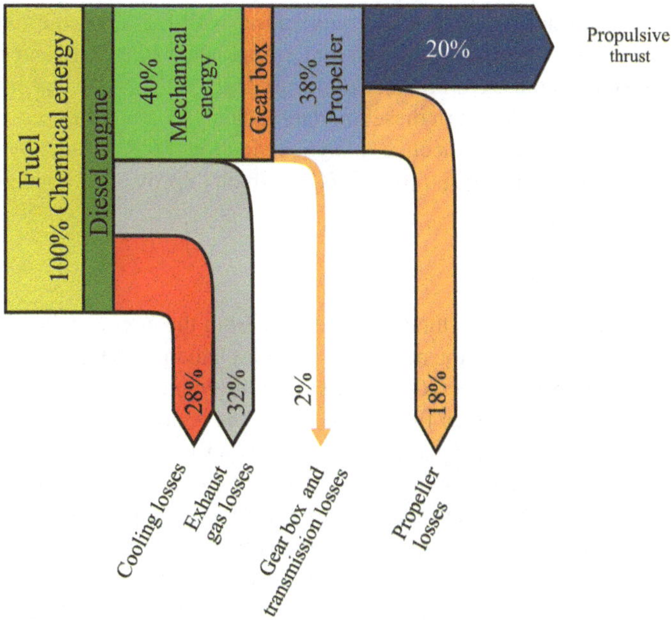

Fig. 9.2 Typical energy losses in shipboard propulsion system

but also toxic. Though there is no direct specification of gasoil quality (i.e. cetane number, density, viscosity etc.) the sulphur content is limited to 10 mg/kg for IW shipping since 2011. This enormously simplifies the use of exhaust aftertreatment systems.

In the following, four technologies are presented which contribute to the compliance with the limit values of Stage V.

Exhaust Gas Recirculation (EGR)

The nitrogen oxide concentration in the exhaust gas of a diesel engine can be reduced by the EGR. The concentration of NO_x depends on the production rate which grows exponentially with rising temperature. To reduce the peak temperature and the oxygen concentration a part of the exhaust gas is redirected to the intake and the combustion chambers. This effect can be enhanced by cooling the recircled exhaust gas. Lowering the combustion temperature to reduce the NO_x concentration is limited, because the production of soot particles is favoured at lower temperatures and the engine power and efficiency are affected. In addition, a residue of HC and CO remains in the exhaust gas.

Diesel Oxidation Catalyst (DOC)

To remove the residue of HC and CO from the exhaust gas a DOC is used. With the aftertreatment component DOC the HC and the CO are oxidized into water

(H_2O) and carbon dioxide (CO_2). In combination with the EGR the exhaust gas will be influenced for further aftertreatment components like SCR and DPF. The process conditions in the DOC are not suitable to degrade the NO_X concentration by reduction. Nevertheless, modern diesel oxidation catalysts are optimised to precondition the exhaust gas for the selective catalytic reduction (SCR) system by oxidizing nitrogen oxide (NO) to nitrogen dioxide (NO_2), which improves the conversion rates in the SCR.

Diesel Particle Filter (DPF)

A diesel particle filter reduces the particle matter from the engine exhaust gases. In addition, for Stage V engines the number of particles leaving the DPF is restricted by a number of 1.0e12 particles per kWh. The size of the counted particles is larger than 23 nm. Separating the particles from the exhaust gas is done by flowing through a porous material. The separated particles reduce the flow through the DPF and change the pressure difference between in- and out-flow of the DPF. At a maximum pressure difference the collected particles must be removed by changing the filter or filter regeneration. During the filter regeneration the collected particles, mainly soot, are burned to clean the filter. The ash residue increases with each filter regeneration. The increase of ash is much slower than the growth of soot and therefore the time is much larger until the filter is plugged with ash. If possible, the filter could be cleaned or it must be replaced. To start the filter regeneration a higher temperature is needed. This can be done by heating or a catalyst in combination with injecting fuel in the exhaust stream to initialise the burning of soot. The separation of soot is more than 90%. Future filters will have even smaller mesh sizes and higher separation rates.

Selective Catalytic Reduction (SCR)

To reduce the NO_X concentration in the exhaust gas a selective catalytic reduction technology can be used. A reagent is injected in the exhaust gas before entering the SCR which starts a chemical reaction to convert nitrogen oxides, with the aid of the reactant as catalyst, to diatomic nitrogen (N_2) and water (H_2O). Ammonia (NH_3) is suitable but toxic. Therefore, the handling will be problematic on board. To avoid these problems an urea-water solution (AdBlue) is used as reagent, which is not classified as a dangerous good. After injecting the urea in the exhaust gas the ammonia is built by dissociating the urea. Afterwards the NO_X is oxidised to N_2 and H_2O. The dosing of the urea is important. At low urea concentration the NO_X emission rises, because not all NO_x could be oxidised. If the urea dosing is too high the toxic ammonia leaves the SCR. The usage of urea raises the operational costs slightly, which may be compensated by more efficient engine control. The NO_x emissions can be reduced by 70% up to more than 90% depending on the system and configuration.

Combined Aftertreatment Systems

The exhaust gas aftertreatment components explained above must be combined to fulfil the Stage V emission standard with engines above 300 kW power. This is usually done in the following sequence: EGR, DOC, DPF and SCR as shown in Fig. 9.3.

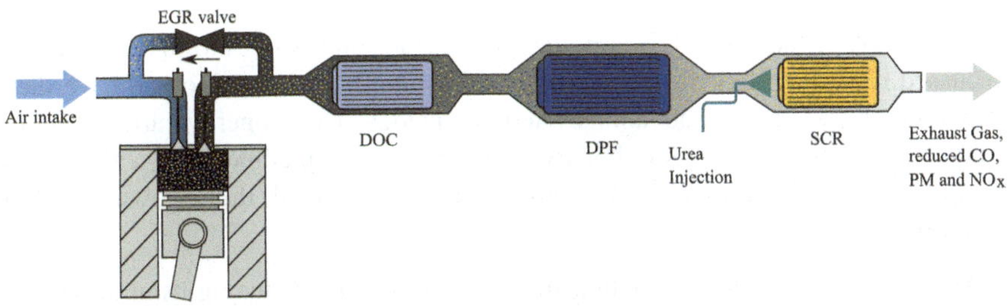

Fig. 9.3 Exhaust aftertreatment system with EGR, DOC, DPF and SCR

Current developments aim at a combination of DOC and DPF in one, more compact component. For retrofitting combinations of DOC, DPF and SCR are possible, depending on the emission standard of the installed main engine and other prerequisites.

For new engines aftertreatment systems are an integral part of the engine design and required to obtain the type approval. Similar environmental performance can be achieved by retrofitting aftertreatment systems to existing engines. The design of the corresponding aftertreatment systems depends on the following aspects:

- Exhaust gas temperature
- Allowable back pressure of the engine
- Operational profile (e.g. operational hours and utilisation)
- Available space in engine room or on the deck
- Mass flow rate of exhaust gas
- Engine maintenance condition.

Depending on the engine, operational profile and other prerequisites exhaust aftertreatment can reduce air pollutants massively. Fuel consumption and the coupled CO_2 emissions remain constant or can even be lowered by optimisation of engine control parameters.

For example, most CCNR II engines are optimised to reach emission limits without aftertreatment at the cost of reduced efficiency. The aftertreatment system allows to regain efficiency. For proper use-cases, i.e. suitable operational profile with sufficient exhaust temperatures, suitable engine in terms of back-pressure limits and sufficient space/heat discharge capacity, retrofitted aftertreatment systems are a very effective measure to improve the environmental performance of inland ships. However, there is a risk that when the engine needs to be replaced the aftertreatment system becomes worthless.

9.2.4 New Energy Carriers

In the context of the desired decarbonization various alternatives to fossil diesel are discussed regarding their future use in the IW shipping sector which all have their individual characteristics, advantages and downsides. These energy carriers can be structured in different ways either by their production method, physical state or chemistry. Even though a certain overlap cannot be avoided, the following groups are used here:

- The first are hydrocarbons with different chain lengths reducing from diesel and diesel-like fuels with ten to twenty carbon molecules to methane with only one.
- The second group are pure hydrogen and liquids used as hydrogen carriers. There is some overlap with the first group, when hydrocarbons or alcohols are used with reformers to separate the hydrogen.
- The third are storage technologies for electric energies, though batteries could rather be considered as the tank system than the energy carrier.

In the following, diesel-like fuels, methane, methanol and hydrogen are considered as energy carriers for inland vessels. Many further solutions are being researched and may play significant roles in the energy transition of the transport sector in the future. For example, sustainable production and marine applications of green ammonia as a fuel for internal combustion engines and fuel cells are studied in several current R&D projects. However, as of today ammonia has not reached sufficient Technological Readiness Level (TRL) yet.

The individual energy sources that can be used have different energy densities, which can be seen in Fig. 9.4 (1 kWh = 3.6 MJ). The figure shows the volumetric and gravimetric energy densities for the pure fuels and including an exemplary storage system. The efficiency of the energy converter is not taken into account here. While methanol and diesel can be stored at ambient pressure and temperature, other solutions require high-pressure tanks or very good thermal insulation. As the proportion of carbon in the energy carrier decreases, so does its volumetric energy density, which is a challenge for mobile applications including inland waterway vessels.

Drop-In Fuels

A diesel engine is the workhorse of almost every commercial ship nowadays and will be in the mid-term future as it offers high performance over a long period of time. To reduce dependence on fossil fuels and mitigate climate change by releasing hydrocarbons, sustainable alternatives are gaining attention. *Drop-in fuels are synthetic and interchangeable substitutes* for conventional fossil hydrocarbons (gasoline, jet fuel, and diesel), meaning it does not require significant adaptation of the engine or the fuel

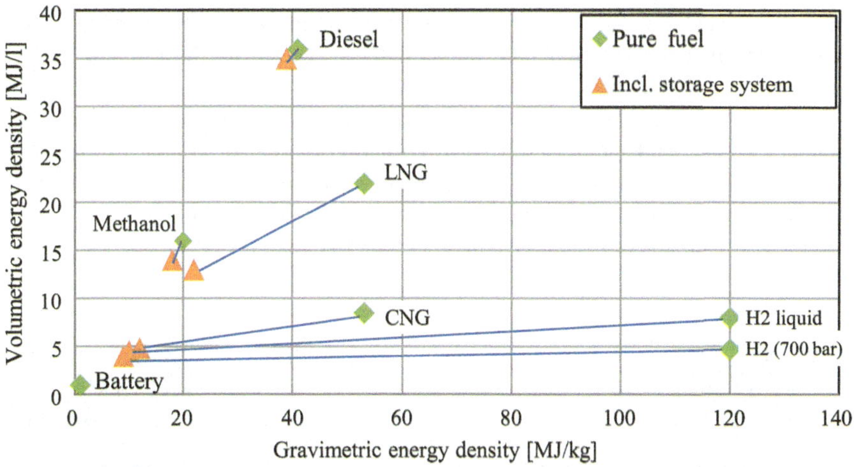

Fig. 9.4 Volumetric energy density over gravimetric energy density for different fuels and storage conditions

system. Usually, they are standardised as paraffinic fuels according to EN 15940 and can be used "as is" in currently available engines either in pure form and/or blended in any ratio with conventional fuels. However, lubricants and some engine control parameters might need to be changed in coordination with the engine manufacturer to improve efficiency and/or environmental performance. Therefore, the fuel has to be specified in the manufacturer's fuel directive and the type approval for each engine series according to the recent emission standards. Since the type approval process is elaborate and costly compared to the small market, standardization and the future usage and availability of blends or pure drop-in fuels have to be coordinated far in advance. The guidelines related to this topic published by the European Association of Internal Combustion Engine Manufacturers can be surveyed online (EUROMOT 2020) and are subject to continuous further development.

Among the synthetic fuels that are considered important for inland navigation are *Gas To Liquid* (*GTL*) and *Hydrotreated Vegetable Oil* (*HVO*). GTL is produced with the Fischer-Tropsch synthesis, a process generally called X-To-Liquid (XTL) that was developed by Franz Fischer and Hans Tropsch in 1925. The "X" is a variable and is replaced by an abbreviation of the original energy carrier, e.g. "G" for gas. Within this process various liquid synthetic fuels such as GTL, lubricating oils and other paraffinic products for the chemical industry can be obtained from natural gas, other gasified fossil fuels or biomass. If biomass is used as a starting material, also the term Biomass-To-Liquid (BTL) is commonly used, replacing the "X" with "B".

HVO is a mixture of straight-chain and branched paraffins, the simplest form of hydrocarbon molecules under the aspect of clean and complete combustion. Typical carbon numbers[5] are C15 … C18. In addition to paraffins, fossil diesel fuels contain also significant amounts of aromatics and naphthenes. Aromatics impair a clean

[5]The carbon number is the number of carbon atoms in each molecule of hydrocarbons.

combustion. HVO, on the contrary, does not contain aromatics, and its composition is similar to that of GTL and BTL diesel fuels, produced by the Fischer-Tropsch synthesis from natural gas and gasified biomass.

HVO should not be mistaken with Biodiesel (see also Fig. 9.5). Biodiesel is a chemically fatty acid methyl ester (FAME) and could cause trouble in long-term storage and being used as a fuel substitute in a conventional engine. Nevertheless, today this is the most important BTL fuel used as a 7% blend with fossil diesel (B7). Increasing the blends of FAME is a greater challenge than for HVO and not covered by usual test fuels.

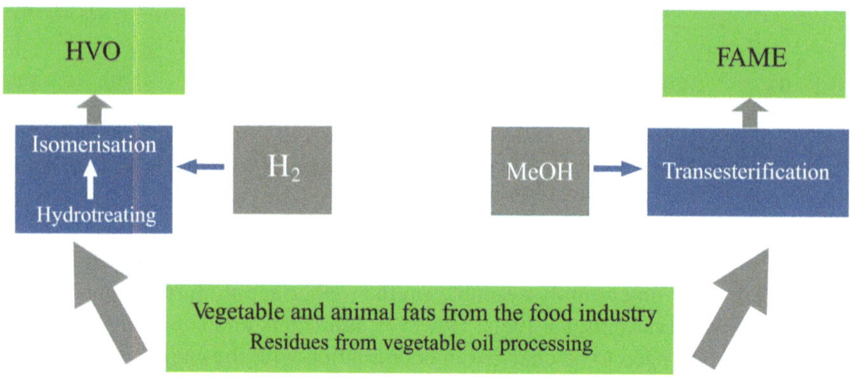

Fig. 9.5 Process of HVO production and difference to the production process for biodiesel (FAME) (based on Neste Corporation, 2016)

The feedstock for HVO consists of renewable sources. These can be residual plant and animal fractions from the food industry or residues from vegetable oil processing. The fuel HVO is considered to be climate neutral in the tank-to-wake cycle. This is in line with the Intergovernmental Panel on Climate Change (IPCC) assumptions (Lasco et al. 2006) and also confirmed by the 2019 energy transition outlook published by DNV-GL (2019). It is explained that carbon contained in biomass is eventually absorbed from the atmosphere by photosynthesis by the plants replacing the burned plants. Other factors such as potential additional emissions due to e.g. deforestation to make room for crops producing biofuel are accounted for under Agriculture, Forestry and Other Land-Use (AFOLU) instead of the transport sector (Waldron et al. 2006).

The synthetic fuel produced entirely from renewable energy sources is called PTL. Here the "P" stands for power. An electrolyser is operated with electricity generated from renewable sources to separate hydrogen. Then, again using a Fischer-Tropsch process, a synthetic, diesel-like fuel is produced from the hydrogen and added carbon. The output of today's PTL refineries is still very low; and therefore, an immediate switch to this fuel is unfeasible. However, as market interest in this fuel increases, it can be expected that production capacity will increase significantly. Besides the sustainable feedstocks the viability of these fuels is highly dependent on cheap renewable electricity.

In combination with the latest emission standards, the use of drop-in fuels can make a major contribution both to reducing climate-impacting emissions and lowering air pollutants.

Methane (LNG)

Methane (CH_4) is an odourless, colourless, non-toxic and non-corrosive gas with a flammability range of 5 ÷ 15% of fuel-air mixture. It is the main component of fossil natural gas with the rest mostly ethane, propane, butane and nitrogen. In its liquefied state the well-known abbreviation LNG for Liquefied Natural Gas is used. In the gaseous compressed state the abbreviation CNG is used. Besides usage as a fuel in combustion engines it can also be used as a hydrogen carrier for fuel cells in combination with a reformer or directly in Solid Oxide Fuel Cells, not covered here. LNG shall not be mistaken for LPG—Liquefied Petroleum Gas (mainly consisting of propane and butane).

Due to its lowest possible carbon content methane has a potential to reduce CO_2 emissions when used as fuel. Nevertheless, since methane is a very climate-impacting gas, methane slip must be kept under control when LNG is produced, distributed and used as fuel in order to maintain the advantage of low emissions from combustion; and to ensure reductions in greenhouse gas (GHG) emissions while using LNG. The 20-year Global Warming Potential (GWP) of methane is as high as 84 times the GWP of CO_2. Due to the chemical persistence of methane the GWP for a period of 100 years is lower. However, the emission of 1 kg of methane still would heat up the earth by the same amount as 28 kg of carbon dioxide.

LNG is produced by cooling down the natural gas to minus 162 °C, thus converting it to liquid state for ease of storage and transport. Methane could also be produced as a power-to-X fuel. Just like bio methane, it can be blended with natural gas and used directly as a renewable substitution. When the methane is produced from biomass or based on electric power it should not be called "natural gas". Therefore, bio-LNG is a misnomer and the term Liquefied Methane Gas (LMG) should be used instead.

In case LNG is spilt, it evaporates, forming visible "clouds". Portions of the cloud could be flammable or explosive under certain conditions. A fuel-air mixture of about 10% methane in air (the middle of the 5 ÷ 15% flammability limit mentioned above) and at atmospheric pressure might be ignited if it does encounter an ignition source (a flame or spark or a heat source with 540 °C or more). Otherwise, the vapour will generally dissipate into the atmosphere.

LNG contributes to a significant reduction of sulphur oxides emissions (SO_X), nitrogen oxides emissions (NO_X), particulate matters (PM) and carbon dioxide emissions (CO_2) from ship engine exhaust emissions in comparison to traditional fuels. In comparison to diesel CO_2 emissions are reduced by up to 25%. When the methane slip is included, the GWP or CO_2 equivalent is about 10% lower with the current technology mix. In the future further reductions of methane-slip can be expected. Emissions of particulate matter are reduced by nearly 100% and of NO_X by up to

90%. However, differences are substantially reduced by advanced emission legislation and exhaust gas aftertreatment. The reduction of SO_X known from the maritime sector only plays a minor role in IW shipping due to the use of low sulphur fuels.

Methanol

Methanol is the simplest member of the group of alcohols with the molecular formula CH_3OH making it rich in hydrogen with only a single carbon atom. It is a clear colourless liquid with a density of 0.79 kg/l. It is produced from fossil sources (natural gas), but can also be produced regeneratively. There are various ways to produce renewable methanol. One is to capture CO_2 from geothermal power generation which is then reacted together with renewable hydrogen (produced via electrolysis) into renewable methanol. Other methods are to convert biogas from fermentation or gasification of sustainable biomass into bio-methanol as well as producing it from solid waste feedstocks. It is also produced as a by-product of the kraft pulping process in paper production.

Methanol can be used in adapted combustion engines or as a hydrogen carrier for fuel cells. Reforming at 300 °C produces H_2-rich reformate gas. When used with a Proton Exchange Membrane Fuel Cell (PEM FC) see Sect. 9.2.5, a fine purification is necessary. Reforming reduces the system efficiency of a FC system.

Methanol is harmful to the health and environment (same water hazard class as diesel) but biodegradable. Due to the liquid property of methanol (it remains liquid up to a temperature of 60 °C at ambient pressure), handling is similar to that of diesel or petrol, i.e. it can be stored in simple tanks. In combination with the comparably high energy density, this is the strongest advantage of methanol. Due to the corrosiveness, materials for piping and tanks need to be selected carefully. Stainless steel or appropriate coatings are proven solutions.

Hydrogen

Hydrogen (H_2) is gaseous under normal conditions (0 °C and 1 bar) with a density of 0.0899 kg/m^3. Hydrogen is the most commonly known chemical element and can be transported as compressed gas or liquid. Generally, gases can be liquefied at high pressures or low temperatures. Hydrogen can not be liquefied at all above - −240 °C no matter how high the pressure is. Usually, liquid H_2 is stored at ambient pressure below −254 °C, about 90 °C below the temperature of LNG. Liquefaction of hydrogen is an energy demanding process, requiring about one third of the stored energy content. Pressurization up to 800 bar requires less energy and is more common today even though the volumetric energy density is lower (see Fig. 9.4).

Since hydrogen does not occur naturally on earth as a single molecule, but only as a chemical compound, it must always be separated to obtain pure hydrogen. Currently, there is much active research on how this process can be made as energy-efficient and climate-neutral as possible. For the time being the production of hydrogen is linked to significant GHG emissions. About 75% of the global hydrogen production is based on natural gas and almost all the rest from coal. A very small share of dedicated hydrogen production globally comes from water electrolysis today.

To differentiate between hydrogen generation technologies and their sustainability a colour scheme is used. Hydrogen is called grey when it is produced based on natural gas or coal, green when water electrolysis with renewable electricity is used. Between grey and green additional colours have been introduced. Blue hydrogen stems from steam reforming in combination with carbon capture and storage, pink hydrogen is based on nuclear energy, and there are even more less frequently used hydrogen production paths and colours.

When hydrogen is used in the PEM FC, attention must be paid to hydrogen purity, which is an advantage of hydrogen from electrolysis. In principle, any hydrogen contamination can impair the performance and service life of the membrane electrode assembly. The required purity is particularly difficult to achieve during the reforming process from natural gas or methanol. The hydrogen purity should be above 99.99 vol.%.

The internal combustion engine running on H_2 is also considered, see Sect. 9.2.5. For further reading on hydrogen as fuel for IW shipping the feasibility study written within the MariGreen project is recommended (MariGreen 2018).

Batteries

Batteries provide the possibility to store electrical energy and make it available on the move. Batteries are used in many applications with differing requirements. Thus, there is a wide range of battery types. Batteries can be characterized by the following factors: power density, capacity, cycle lifetime, capital costs, charging time, reliability and safety.

The following battery types are available:

- Lead-acid
- Li-ion
- Sodium sulphur
- Nickel-based
- Others: sodium-ion, magnesium-ion, zinc and aluminium.

Figure 9.6 shows an example of the functional principle of a battery based on a lithium-ion cell. Besides all varieties of different batteries, they share about the same construction. Each cell consists of two electrodes, the negative anode and the positive cathode, which are enclosed by an electrolyte. The electrolyte can be made of liquid, gel or solid materials. For both electrodes and electrolyte different chemical matters are used.

Discharging of a battery means to convert chemical energy to electrical energy. For charging a battery all reactions are reversed. However, this controlled reversal for most cases is not perfect, which limits the number of cycles and battery lifetime.

The different lithium-ion batteries are usually classified by their cathode chemistry. The five available types are:

- lithium cobalt oxide (LCO)
- lithium iron phosphate (LFP)
- lithium nickel cobalt aluminium oxide (NCA)

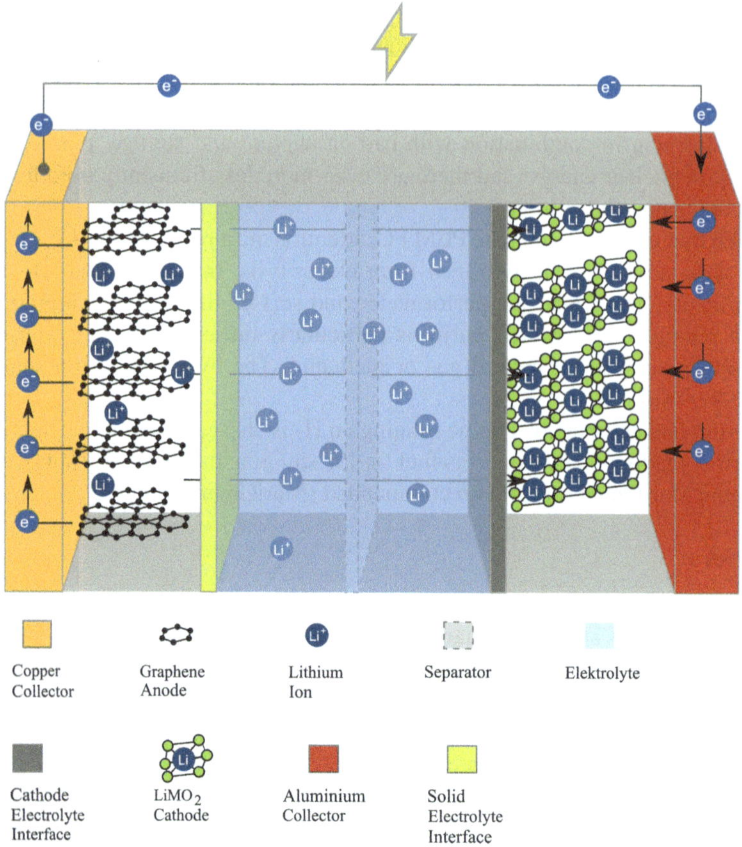

Fig. 9.6 Functional principle of a lithium-ion battery

- lithium nickel cobalt manganese oxide (NCM)
- lithium manganese oxide (LMO).

Further, the anode material as well as the electrolyte can change. An example is the Lithium Titanate Oxide anode (LTO).

Maritime battery systems typically consist of several thousand cells. It is, therefore, important that each cell works consistently with all other cells. The individual battery cells are interconnected to form battery modules, whereby the required voltage is reached. Due to the net-working of these units, large systems with a high capacity can be assembled. These systems can be installed in battery rooms or in containers. Empty battery containers can be exchanged with charged ones to avoid long stops for charging with corresponding infrastructure ashore. This solution is also discussed as a separate business model where the ship owner is not the owner of the batteries but pays a third party for this energy service (see De Alphenaar, Sect. 10.2.2).

Investment costs and weight of batteries are the most relevant barrier for most applications. The energy content (net heating value) of one kilogram of diesel is approximately 12 kWh. Lead-acid batteries have up to 0.05 kWh/kg and lithium-ion cells up to 0.26 kWh/kg. For complete heavy duty battery systems ready for use aboard ships 0.10 kWh/kg is realistic. Costs are in the range between 250 and 1000 €/kWh. For example a 2.5 MWh system sufficient for the propulsion of a 110 m cargo vessel for a few hours weighs about 25 t and costs about one million euros.

9.2.5 Advanced Energy Converters

This section describes the energy converters. This includes on the one hand the classic diesel engine, which, optimised and equipped with the latest exhaust aftertreatment systems and operated with a climate-friendly fuel, is a good option for many applications. On the other hand, new energy converters for inland navigation such as the fuel cell, purely electric propulsion concepts and the hydrogen combustion engine will be introduced.

Gas Engines and Fuel Systems

The technical solution for propulsion of IWW vessels depends on the type of vessel, the required speed and corresponding power and the sailing profile. Methane gas can be used in compressed or in liquefied state. The Liquefied Methane Gas (LMG) has a higher energy density and is the only form used in IW shipping today. Further information on the energy carriers can be found in Sect. 9.2.4.

To make gas propulsion[6] a suitable option, a vessel should meet the following criteria:

- High energy demand to maximise benefit from reduced energy costs
- LNG bunkering infrastructure within the operational area.[7]

The technology for using LNG on ships in combustion engines is commercially available. However, the comparatively high price of the cryogenic system components is a barrier, especially in IW navigation. Figure 9.7 shows the minimum necessary equipment for a LNG drivetrain.

---

[6]Gas-electric systems are similar, except that additional energy conversion is required, see Sect. 9.4.1, Fig. 9.22.

[7]On the Danube there is one LNG terminal and bunkering station in Ruse (Bulgaria). Concepts have been developed for Galati and Constanta (Romania).

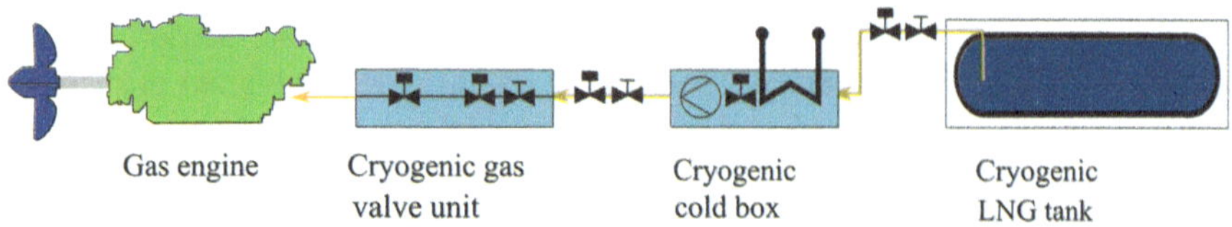

Fig. 9.7 General schema of the system components of a LNG powered drive train

Two different types of LNG tanks are available: membrane tanks and pressure tanks. While membrane tanks are used for LNG as a cargo, LNG as fuel is usually stored in pressure tanks (IMO Type C Tanks). They are mostly cylindrical and have either a vacuum or foam insulation as shown in Fig. 9.8. For the vacuum insulation the space between the inner and outer hull is filled with perlite, an insulation material, before the vacuum is drawn. With foam insulation the heat transfer is higher. Tanks and piping systems have safety valves to protect them from overpressure. All systems are redundant. This means that of each safety system there are at least two individual ones available in case one fails.

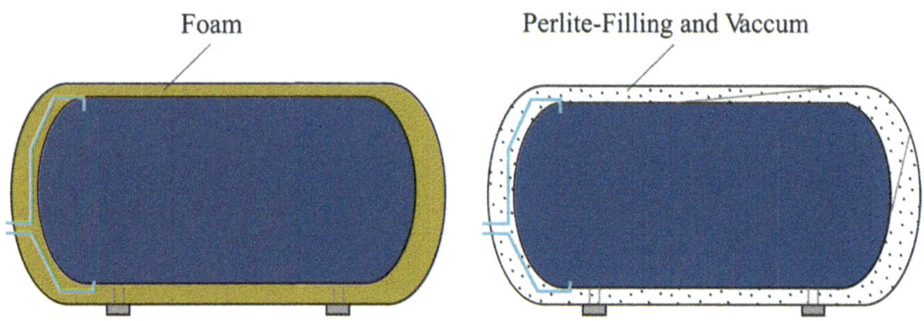

Fig. 9.8 Pressure tanks foam and vacuum insulated

In the cold box the LNG is evaporated. The energy (heat) for the evaporation process is often delivered by the cooling water of other engines on board. This part of the installation is also known as gas treatment system. The pipes are double walled. The space between the inner and the outer pipes is flooded with nitrogen. Each pipe has an automatic and a hand operated valve; each piping section also has a release valve. The automatic valves are closed at an emergency shutdown.

The Gas Valve Unit (GVU) controls the gas flow to the engine and can also perform an emergency stop. The two main engine types are *dual-fuel engines* running on diesel as well as gas and *pure gas engines* running on gas only. In the case of a dual-fuel engine, which is used in the majority of the current LNG fleet, an additional diesel tank is necessary.

Mono-fuel gas-engines work with the Otto principle and have a spark-ignition (see Fig. 9.9). They also have a different characteristic which is slightly more suitable for gas-electric applications in gensets than for direct drives.

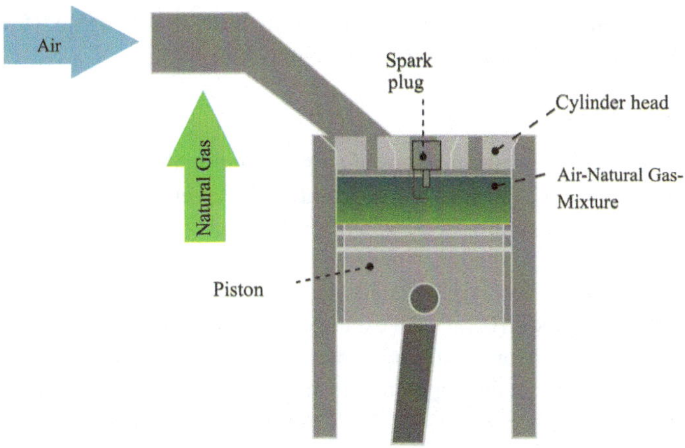

Fig. 9.9 Gas engine (Otto-cycle)

Dual-fuel engines use the Diesel-cycle with methane gas fed into the engine's intake system (see Fig. 9.10). The air-methane mixture is then drawn into the cylinder with a leaner air-to-fuel ratio compared to a spark-ignited engine. Near the end of the compression stroke, diesel fuel is injected and ignites the gas. A dual-fuel engine can operate on pure diesel fuel or a mixture of diesel and gas, delivering the same power

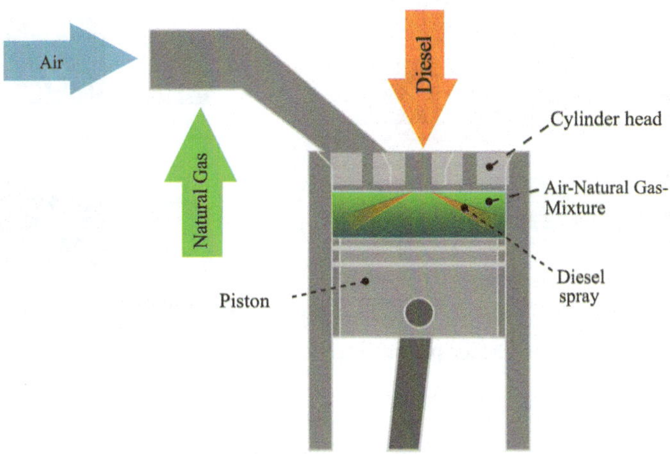

Fig. 9.10 Dual fuel engine (Diesel-cycle)

density, torque curve and transient response as the base diesel engine. However, the emission performance is lower than for the mono-fuel engines.

Electric Propulsion

Ship operation without local emissions of air pollutants and CO_2 can be realised with electric drives based on batteries and/or fuel cells. The latter are covered in more detail below. Batteries (discussed in Sect. 9.2.4) will be used more and more in different applications of inland navigation. Smaller passenger vessels or ferries with limited range of operation in densely populated areas are most suitable for early adoption (see E-ferry Ellen in Sect. 10.2.2). An exemplary setup of the required components is shown in Fig. 9.11. Even long haul cargo vessels can benefit from batteries and reduce emissions and exposure of the population to air pollutants temporarily. Batteries with moderate capacity can be used to switch off combustion engines of ships with hybrid propulsion (i.e. with more than one different energy sources for the electric motor, see Sect. 9.4.1) while operating in ports, passing sensitive areas or when berthed. Incentive schemes like the Dutch Green Award level Platinum (see Green Award 2021) with reduced fees for ships capable of several hours of emission-free operation contribute to business cases for these hybrid drivetrains.

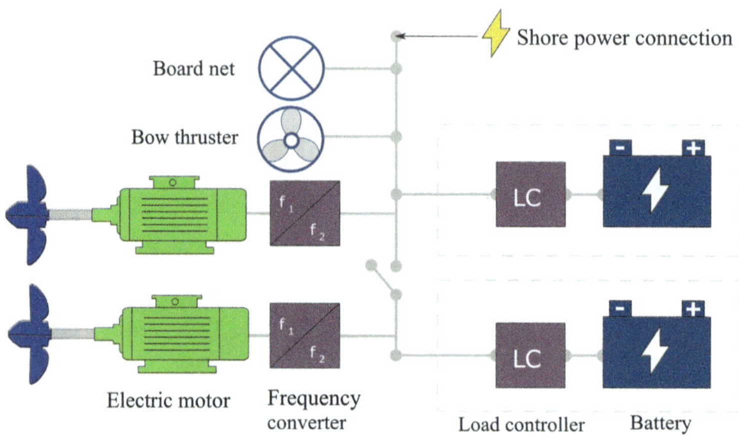

Fig. 9.11 General schema of the components of a battery electric drive system (synchronous electric motors)

The basic principle of *electric motors* is the conversion of current (electric power) into movement with the help of magnetic fields. Electric motors can be classified according to the type of current source. They can operate either with direct current (DC), alternating current (AC) or three-phase current. Alternating current and three-phase current driven motors can further be divided according to other specifications. Depending on the specific application, the optimum motor can be chosen.

The *asynchronous motor* is the most widely used industrial motor. It can be connected directly to the three-phase mains and is very robust and easy to build. The asynchronous motor takes its name from the fact that it does not rotate exactly with the mains frequency. It only has a torque if its speed deviates from the synchronous speed. In the operating range, the torque is proportional to this deviation. This type of electric motor is characterized by low investment costs and small dimensions. Its nominal rate of revolutions is usually too high to be used as a direct drive. A gearbox between electric motor and propeller shaft is necessary. The gearbox increases the investment costs, lowers the efficiency of the drive train and could be a point of failure. If the advantages of asynchronous motors and the disadvantages of the gearbox are balanced correctly, a cost and energy efficient drive train can be designed.

For *synchronous motors*, the speed of the motor is equal to the mains frequency divided by the number of pole pairs. The rotor of a synchronous motor is permanently magnetized and follows the rotating field of the stator with a phase difference. This type of electric motor is characterized by high energy efficiency, low nominal rate of revolutions and a good torque/speed characteristic. This motor can be used as a direct drive, without a gearbox between motor and propeller shaft. Its large outer dimensions are disadvantageous like the high investment costs. Using a synchronous electric motor for the propulsion system leads to an efficient drive train with a sensitive control.

Fuel Cell Systems

Fuel cell systems are the second option besides batteries capable of supplying an electric drive without harmful local emissions to the environment. The focus in this section is on *Polymer Exchange Membrane (PEM) fuel cell systems* as the technologically most mature solution for shipping applications. These usually run on hydrogen, which may be stored in and extracted from different energy carriers. Often hydrogen, methanol or natural gas are used as basis for the electrolytic process as shown in Fig. 9.12. Further options are so-called Liquid Organic Hydrogen Carriers (LOHC) or ammonia with a cracker to separate the hydrogen from nitrogen. The efficiency of the fuel cell can be as high as 60%. In real applications, however, temperature control or the utilization of hydrogen carriers lead to efficiency comparable to diesel engines.

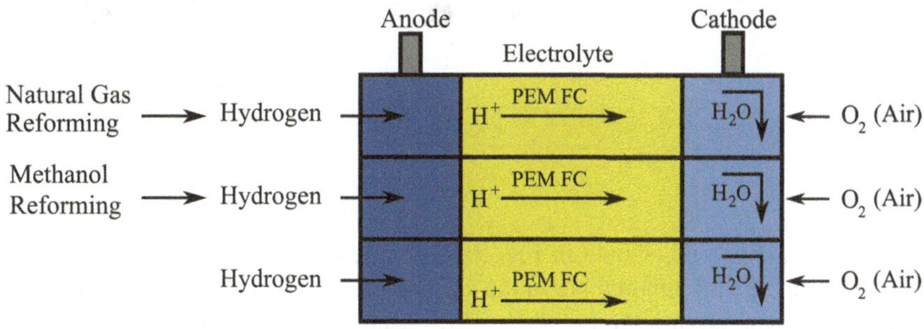

Fig. 9.12 Electrolytic process based on hydrogen, methanol or natural gas

All fuel cells consist of two electrodes—the anode and the cathode as shown in Fig. 9.13. These are separated by an electrolyte with an ion-permeable membrane. After the fuel has been supplied to the anode, it is divided into electrons and protons. The free electrons flow into an outer circuit between the anode and cathode to be used as an electric current. The protons spread through the electrolyte to the cathode. At the cathode, the oxygen from the air combines with the electrons from the outer circuit and protons from the electrolyte. This results in water and heat. The modularity allows the combination of cells in stacks and stacks in complete systems matching the required voltage and power up to MW range. From the fuel cell the electricity is provided as direct current (DC). If alternating current (AC) is required for further use, an additional inverter is required.

The fuel cell based propulsion system for a ship consists of several components. These include the fuel cell system itself, an electric motor, batteries and, depending on the energy carrier, a reformer or cracker. Fuel cell systems are less dynamic than combustion engines. This inertia is usually balanced by batteries, which also help in the start-up phase and for peak-shaving. The fuel cell system can be designed with moderate power and thus be operated at a largely constant load. Note, however, relatively high price of fuel cell systems of about 2.500 €/kW.

In the following, a *low temperature (LT PEM FC)* system with compressed hydrogen storage, and a *high temperature (HT PEM FC)* system with methanol are explained.

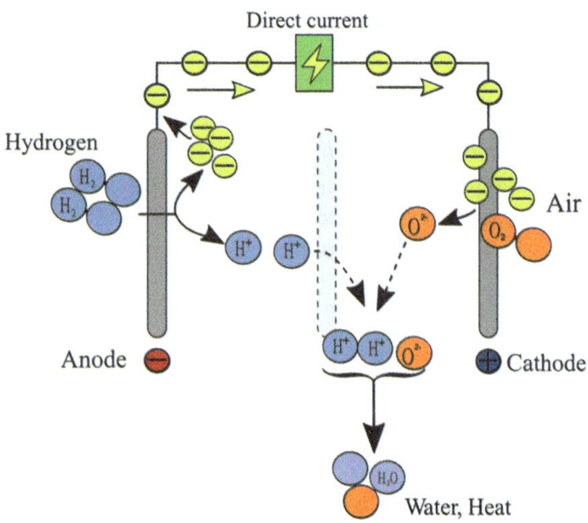

Fig. 9.13 Basic working principle of fuel cells

For the *LT PEM FC* shown in Fig. 9.14 the pressure reduction unit lowers the pressure from several hundred bar in the tank to a level suitable for the fuel cell. The hydrogen is then fed into the fuel cell, which usually has an operating temperature between 70 and 100 °C. The voltage of the electric current produced is transformed

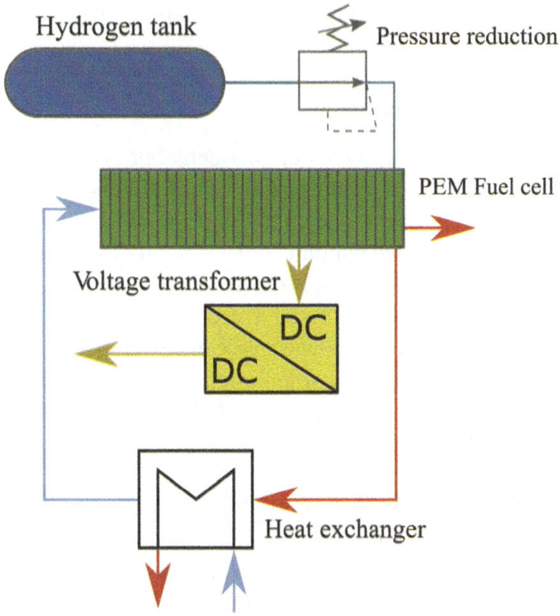

Fig. 9.14 General schema of the hydrogen LT PEM FC system

into the usual on-board voltage by the voltage transformer or inverter. The reaction heat is emitted in a separate heat exchanger system.

The *HT PEM FC* illustrated in Fig. 9.15 operates at temperatures between 100 and 200 °C. From the methanol tank the fuel is taken to the reformer unit to extract the hydrogen. The process needs heat which is produced by burning an amount of methanol in the heater. The pure hydrogen is then fed into the fuel cell. Some of the reaction heat in the fuel cell is fed back into the reformer. The remaining heat is emitted in a separate heat exchange system. The voltage of the electric current

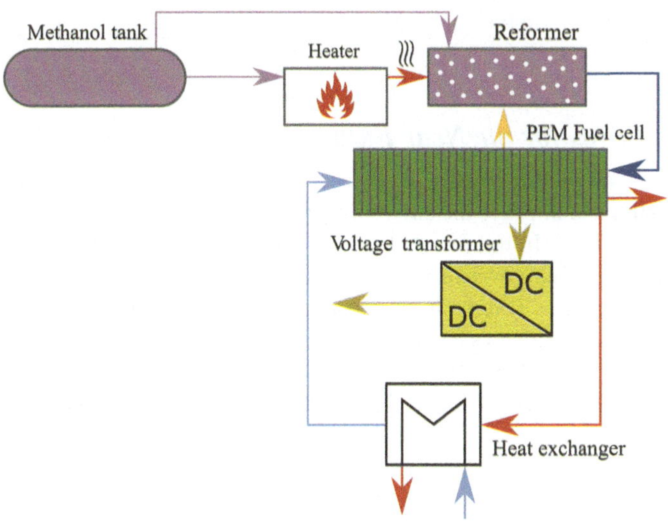

Fig. 9.15 General schema of the methanol HT PEM FC system

produced is transformed into the usual on-board voltage by the voltage transformer or inverter.

Hydrogen Combustion Engines

Hydrogen can be used for fuel cells but also for the classic Internal Combustion Engines (ICE). Recently, several manufacturers have started the development of commercially available engines (Stellmach 2020). In contrast to the PEM fuel cell, rare materials like platinum are needed for the production of the combustion engine. Being carbon-free, makes the hydrogen operation of the combustion engine, at least theoretically, CO_2, CO and hydrocarbon-free. In real operation, however, traces of hydrocarbons in the exhaust gas can be detected due to lubricating oil in the combustion chamber. The local emission of nitrogen oxides, though, must be taken into account (Eichsleder and Klell 2020). With current emission legislation it is likely that Hydrogen Combustion Engines require SCR to stay below the NOx emission limits.

The wide ignition limits of hydrogen allow good control over the entire operating range of the engine. In contrast to conventional fuels, hydrogen can theoretically be burned homogeneously up to an air ratio of lambda = 10. As with conventional fuels, the required ignition energy increases with the air ratio. To ignite a stoichiometric hydrogen-air mixture, only one tenth of the energy required to ignite a gasoline-air mixture is needed. In contrast, the self-ignition temperature of hydrogen is significantly higher than that of conventional liquid fuels. Although this can bring advantages in terms of knocking behaviour in the case of premixed combustion, it requires very high compression ratios or other measures to increase the charge temperature in the case of the self-igniting hydrogen engine.

The high laminar flame velocity of about 230 cm/s shows that extremely short, efficient burning times can be achieved with hydrogen. Even with lean mixtures, the laminar burning speed is significantly higher than that of conventional fuels. However, in the premixed combustion of stoichiometric mixtures the engine is more heavily loaded and induced by the rapid and thus higher pressure increase, which also leads to higher combustion noise.

9.2.6 Pathway for the Near Future

Multiple options to improve the environmental performance of inland shipping have been presented here. In the future decarbonization and the complete avoidance of air pollutants will require a mix of advanced solutions. In the coming years the combustion engines will remain the most important prime mover. In combination with modern aftertreatment systems air pollutants can be avoided to a large extent. A fleet fully equipped with Stage V IWP/IWA engines or equivalent retrofit aftertreatment would emit about 80% less NO_X and 95% less particles compared to the 2015 baseline. The use of NRE or Euro VI truck engines and a future emission regulation going beyond the Stage V limits for large IWP engines lead to further reductions.

The reduction of CO_2 emissions is even more challenging. They can be primarily reduced by decreasing the energy demand (these subjects are treated in Sects. 9.3–9.6). Secondary, alternative drop-in fuels with sustainable feedstock and upstream chain can play a major role to reduce the carbon footprint. These can be used as pure fuel or blended with increasing share with fossil diesel. If methane slip by production, transport and combustion is largely avoided in the future, CO_2-equivalent emissions of LNG drives can be about 20% below the diesel reference. Blending of liquefied bio-methane or synthetic methane gas would further increase the reduction potential. On a midterm basis, electric drives with modern combustion gensets, potentially already including a battery and/or fuel cell to avoid emissions in ports and urban areas and future proof power management can be considered as a precursor for the later implementation of affordable zero-emission power sources. Further technology-neutral developments and pilot applications are required. For the relatively small sector decisive technological leaps are unlikely to happen internally. Therefore, the developments in other sectors like long-distance road haulage should be monitored.

Given the long lifecycles of ships and drivetrains in the inland shipping sector the energy transition should be started as soon as possible. However, the legal framework, the market parameters and the cost structures will hardly bring the required momentum in the next years. Therefore, the model state aid scheme prepared in the GRENDEL project (GRENDEL 2020) and the CCNR financing energy transition studies (CCNR 2020a) are important steps to coordinate the Europe-wide fleet modernisation.

9.2.7 Final Notes

It should be noticed that the production of energy carriers was not discussed, thus all kinds of problems connected with the production are skipped. If that would not be the case, i.e. if production of primary energy source, its distribution etc. would be considered, which is often referred as *Well-To-Tank (WTT)* analysis, the overall picture concerning greening would be different. WTT analysis, however, is out of the book's scope. Using the same terminology, *Tank-To-Wheel (TTW)* analysis is related to subjects treated in this book. For the first, the energy suppliers are responsible, while for the second these are the vehicle manufacturers. Consequently, to get a better insight into which type of energy is best concerning efficiency, emissions, costs etc. a sum of the abovementioned impacts is needed, i.e. *Well-To-Wheel (WTW)* analysis which essentially clarifies what is the total cost of energy conversion. These terms stem from the road industry, while for marine vessels (IW too) TTW and WTW terms correspond to *Tank-To-Propeller (TTP)*[8] and *Well-To-Propeller (WTP)*, respectively.

[8]Term *Tank-to-Wake* is also used so that TTW acronym remains unchanged.

The same applies to the production of energy storage devices and energy converters. Today's low- or zero-emission technologies require the use of rare metals such as cobalt, nickel and lithium (batteries), rare earths (electric motors) or platinum (PEM FCs). Many R&D projects aim to improve sustainability for individual aspects and the entire complex life cycle chain.

9.3 Improvements in Hull Resistance

The hull resistance (see Sect. 7.2) may be decreased by reducing (a) the *viscous resistance*, and/or (b) *wave-making resistance*. This section is focused on the hull form, thus on the reduction of wave-making resistance and hull-propeller interaction (treated in Sect. 9.4.8). Viscous resistance, specifically frictional resistance, depends on surface smoothness (and fouling), i.e. on the coating type, as well as on hull cleaning intervals. Both are the question of maintenance which is an operational measure and is out of the book's scope. New ship-related technology which is oriented towards reduction of frictional resistance is air lubrication treated in Sect. 10.1.3.

As already emphasized in Chap. 7, the IW vessel's dimensions, hull form and speed should be adjusted according to particular waterway restrictions. Moreover, the owner's demands, concerning the type of a ship, speed and cargo-carrying capacity, are as important. Consequently, there are two possibilities on how to determine the hull form of a new vessel:

(a) to adopt the hull form of an existing IW vessel (which is not necessarily optimized for the required conditions)
(b) to develop a new hull form (taking into account all requirements and constraints).

The second option is more demanding and generally justified if not just one, but a series of vessels are to be built. It is also an unavoidable option if existing hull forms are not good enough for the specific fairway conditions and operational profiles.

9.3.1 Principal Dimensions

To enhance economic efficiency, i.e. to increase freight capacity as much as possible, the main dimensions of IW vessels are usually *close to upper limits* imposed by the waterway restrictions, i.e. draught depends on waterway depth, and length and/or breadth on lock chamber dimensions.

To reduce the power requirements, a general rule for the seagoing ships is to increase L/B and reduce B/T ratios with other parameters (Δ and C_B in the first place) unchanged. An increase of L/B tends to reduce the wave resistance, while the reduction of B/T tends to reduce the wetted surface, thus the frictional resistance.

However, for IW vessels that navigate in restricted fairway conditions, the above-mentioned power reduction rule cannot be unconditionally applied. Therefore, to determine the main dimensions of the ship, which is done in the early design stage, a comprehensive analysis that takes into account the fairway restrictions, is required.

9.3.2 Ship Form

Limited draught usually results in higher C_B, C_M, *B/T* and *L/T* ratios, compared to seagoing ships. The prismatic coefficient C_P ($C_P = C_{B/}C_M$), which indicates the longitudinal distribution of displacement, is also rather high. Other secondary hull form characteristics, as well as the form of the bow and stern, influence resistance and propulsion to a great extent (see Sects. 7.2 and 7.3 respectively). Essentially, the forebody and afterbody optimizations are crucial for good design, as the parallel middle body of IW ships is more or less predetermined (see Sect. 8.2.1).

It should be emphasized, however, that a low-resistance and propulsive-efficient hull form can be obtained only after:

- Comprehensive computational analysis, i.e. Computational Fluid Dynamics (CFD), and/or
- Detailed model-scale experiments are carried out in specialized towing tanks.

As a result of hull form optimization, contemporary inland vessels can have lower resistance than those of a few decades ago, in some cases of even up to 50% (Zöllner, 2003). Unfortunately, despite the huge potential of improvements of IW vessels hydrodynamics, a methodical approach has not been done yet.

Computational Fluid Dynamics (CFD) techniques may be used as an efficient engineering "tool" for hull form optimization to minimize ship resistance, or to improve the efficiency of the propulsors. CFD techniques, however, are generally developed within scientific and research institutions and proper application requires experience. Namely, the designers typically do not do CFD evaluations, these are usually done by the CFD specialists. This results in a "once removed" relationship similar to experimental facilities (i.e. designers vs. CFD specialists, corresponds to designers vs. towing tank, cavitation tunnel etc.). This may change in the future when CFD tools will become more reliable and convenient for everyday engineering use. To further optimize the hull form, CFD may supplement both the other numerical methods and design recommendations, and model experiments. It should be noted that CFD analyses usually give a very good *qualitative* illustration of waterflow around the hull—but the *quantitative* value of results is sometimes questionable.

Possibilities for the CFD calculations are demonstrated, for instance, in van Meij and Raven (2014) paper, showing how small changes of hull-form, and specially focusing on the tunnel design, can be effective in the hydrodynamic hull-form design. This kind of optimizations has to be done on a case-by-case basis.

Specialized shallow-water research institutes, often called towing tanks, differ from the usual towing tanks (which are also not so common) as they pose facilities specialized for shallow water experiments. The most important, of course, is an adjustable water depth towing tank. Equally important is the staff, i.e. experts for shallow water hydrodynamics, ship design etc., with vast experience and access to a library that contains public and restricted circulation reports of previous research. In Europe, two such institutions should be mentioned:

- DST—Development Center for Ship Technology and Transport Systems, Duisburg, Germany and
- MARIN—Maritime Research Institute Netherlands, Wageningen, The Netherlands.

In this context, the results of related projects, for instance, the VEBIS Project (Zibell and Müller 1996) and "Inland ship of the future" (Heuser 1990) are very useful. In both projects, optimal units/hull forms for variable transport tasks and regimes of operation are investigated. Derived recommendations for shallow water IW ships, can be applied to the Danube vessels too, see Figs. 9.16 and 9.17 (small and large selfpropelled vessels) and 9.18 (pushboat). These guidelines should be combined with those given in Sect. 8.2 that are valid for the Rhine waterway (which underpins deeper draughts than for the Danube).

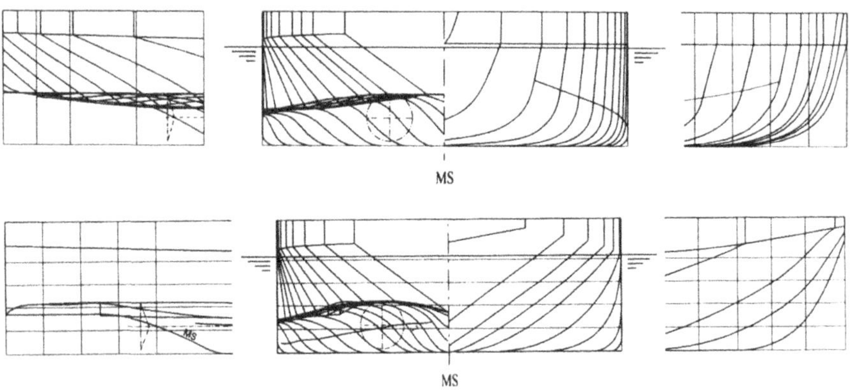

Fig. 9.16 Twin-screw ship (VEBIS Study: Type I and from it developed Type IV)—$L = 82$ m, $B = 9.5$ m, $T = 2.5$ m, TEU 77, propellers in nozzles, conventional rudders

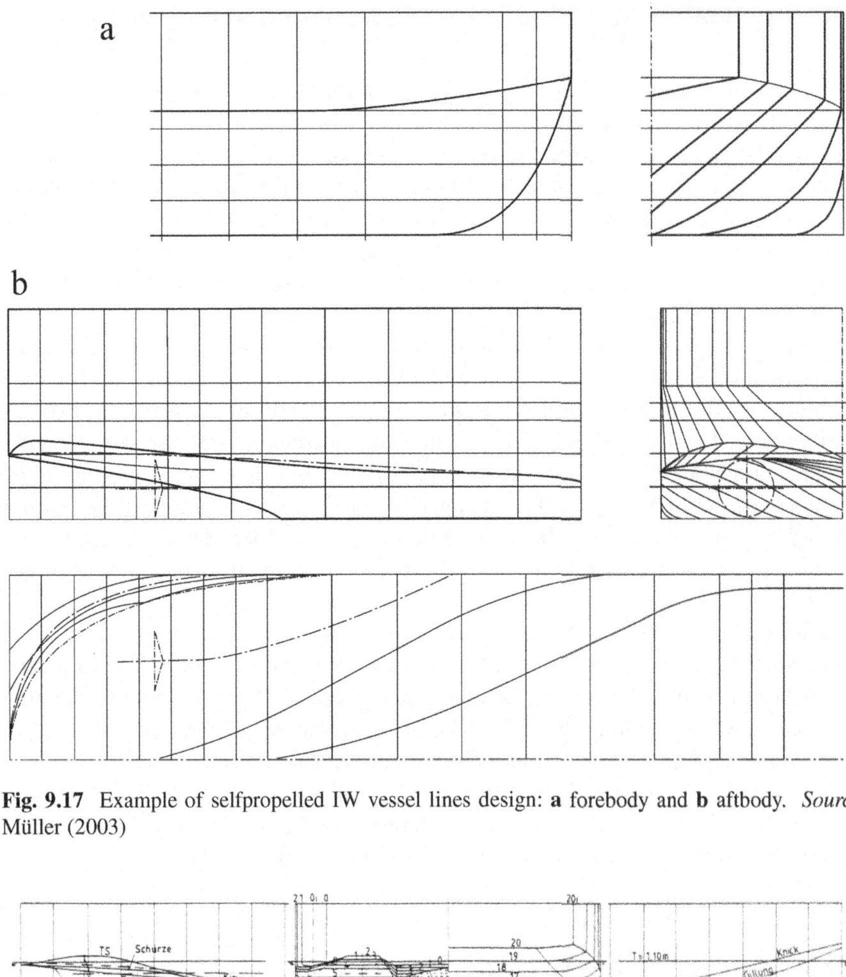

Fig. 9.17 Example of selfpropelled IW vessel lines design: **a** forebody and **b** aftbody. *Source* Müller (2003)

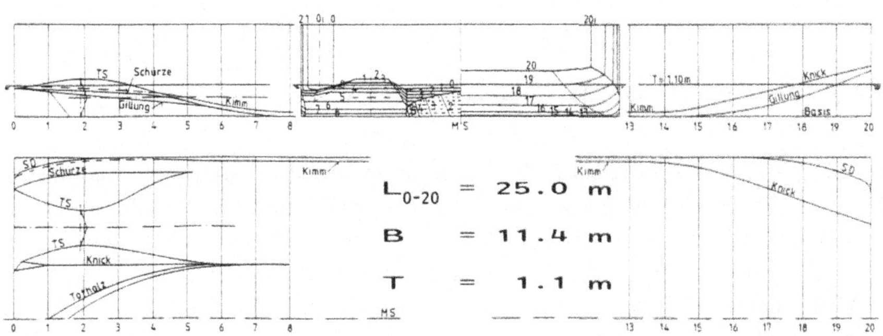

Fig. 9.18 VEBIS Pushboat with a propeller of 1.2 m in a nozzle

9.4 Innovations in Propulsion, Transmission and Steering

Propulsion is a subject of ship hydrodynamics which treats various aspects (hull forms, propulsors, conditions etc.) of all ship types, whereas the idea behind this book is to surface the peculiarities of IW vessels only. Therefore, a discussion that follows is qualitative rather than quantitative. For quantitative analyses, literature that addresses general ship hydrodynamics should be referred, e.g. Carlton (2018), Molland et al. (2017) etc. It should be borne in mind that the prime distinctiveness of IW vessels which influences propulsion are the fairway restrictions, in the first place limited water depth.

Propulsors that can be used on IW vessels, just for this analysis, will be divided into two groups—*Screw Propellers* and *Other Propulsors*—rather than divided according to the principle how they work (as is usually done in the textbooks). Furthermore, having in mind their potential application, propulsors will be analysed here together with the applied concept of power transmission, as they are often categorized according to that. For instance, *pod propulsors* (electrical) and *rudder propellers* (mechanical) use the same type of propellers but are considered to be different propulsor types, although the difference stems only from the mode of power transmission.

The way how the vessel has to be steered should also be considered. Some propulsor types inherently enable steering (rudder-propellers or azimuthing thrusters, for instance), while others need an additional steering device—a rudder. Steering and manoeuvring capabilities of IW vessels are discussed separately in Sect. 7.7, but steering devices—rudders (not including the steering mechanism) are examined in this section. Again, only the uniqueness of rudders used on the IW vessels is discussed, whereas a broader knowledge about steering devices, in general, can be found, for instance, in Molland and Turnock (2007), and Brix (1993).

9.4.1 *Drivetrains of IW Vessels—Basics*

The *drivetrain* or *powertrain* of IW vessels comprises all components, from the one that generates power (the prime mover) to those that deliver thrust to the water (i.e. the propulsor). The prime mover of IW vessels is almost always a diesel engine[9] (for other kinds of prime movers, or energy transformers see Sect. 9.2).

Many diesel engines in service are *marinized general-purpose engines*, which are also used in stationary generating-sets having 1500 or 1800 rpm for 50 or 60 Hz supply frequency. As a consequence, contemporary gearboxes have higher gear ratios than those few decades ago. These drivetrains are lighter than their predecessors, but their engines are still significantly heavier than in construction machinery or trucks.

[9]The term *engine* will be used for the prime mover, while the device that drives a propulsor is the *motor*.

Turbochargers and intercooling also help to reduce the size of the modern engines while maintaining the same power output.

Note that many IW vessels have a considerable *power reserve*. This is due to the relatively low cost of the engines, the wish for longer service intervals as a result of operation at reduced power, and the possibility for coupled convoy operation when additional power is required. Extra power is also advantageous for ensuring the stopping ability in some unexpected situations. As the innovative drivetrains are expensive, it is expected that the power reserve will be reduced, i.e. the *right-sizing* applied (see Sect. 11.2). This means that the installed power should be optimized according to the operational profiles.

Diesel Direct Propulsion

The most common drivetrain type for the IW vessels (both, selfpropelled and push-boats) is the so-called *diesel-direct* propulsion, see Fig. 9.19. In this case, the transfer of torque and thrust is straightforward and is as follows:

- The torque (power) is transmitted to the gearbox (which can be connected to the engine via the bellhousing or be a separate unit), propeller shaft, boss, propeller blades and finally to the water.
- As a result, the thrust is transmitted backwards from the water to propeller blades, boss, shaft, thrust block and to the vessel. The thrust generated by the nozzle is transmitted directly from the nozzle to the hull structure.

Note that for some other propulsion arrangements presented in the following sections, transfer of torque and thrust may be different, even intriguing (see for instance Figs. 9.40b and 9.46).

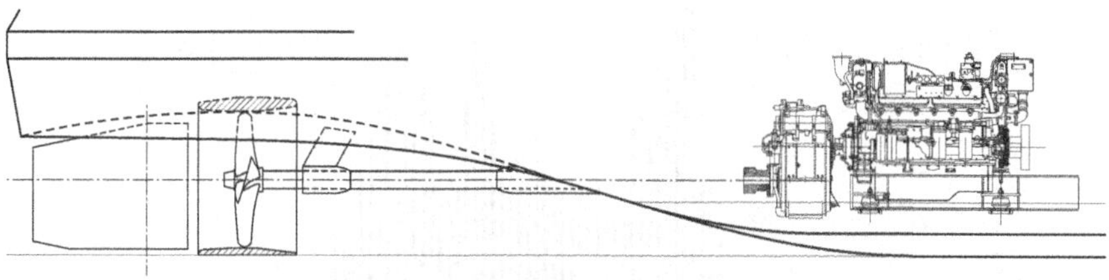

Fig. 9.19 The most common powertrain type—diesel-direct propulsion

Direct transmission means that in the sailing mode the engine (in this case diesel engine) and a propulsor are directly coupled and that both are running (i.e. idling is disregarded). Nevertheless, the IW vessels have *at least two* typical sailing or operational modes—*upstream* and *downstream* mode[10]—each having different power requirements, meaning that the engine will have to operate continuously in (at least)

[10]Others may be empty versus full loaded, shallow versus deep water navigation, combination of previous etc.

two regimes. Operation in the downstream mode will thus be inefficient as the engine will run underloaded what results in fuel consumption and emissions increase.

For example, the long-term (6 months) performance for a Rhine pushboat Veerhaven X (sistership is shown in Fig. 4.14) in a convoy with 4–6 barges (upstream loaded, downstream empty) on the Rotterdam-Duisburg-Rotterdam route was measured, see Godjevac and Drijver (2014). The pushboat's operational profile is shown in Fig. 9.20, clearly indicating three operational modes: upstream, downstream, and manoeuvring, which take, respectively, 59, 31, and 10% time of the total operational profile. Follows that in upstream, downstream and manoeuvring modes, the engines are mostly running at regimes that require around 2500 ÷ 3000, 1300 ÷ 1500 and 250 ÷ 300 kW, respectively, out of installed 4080 kW. It may be, therefore, concluded that the diesel engines are running for a long period of time underloaded in less efficient mode, generating higher emissions (the optimal regime is usually around 2/3 of the maximal output).

For other propulsion configurations that are given in the following, especially in the downstream mode, there is no need for all prime movers to be permanently engaged. Note that the operational profile, although typical for IW vessels, is valid for conditions given above; other routes, vessels etc. will result in a different operational profile.

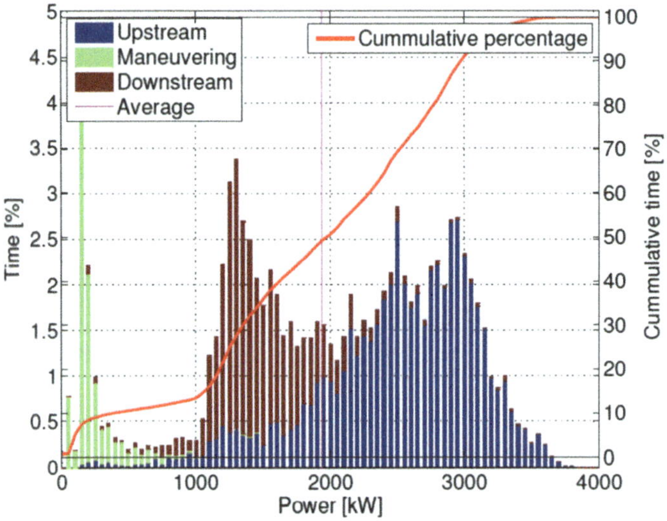

Fig. 9.20 Operational profile during 6 months. *Source* Godjevac and Drijver (2014)

Diesel-Electric and Diesel-Hydraulic Propulsion

The concept of electrical propulsion systems is not new. The first ships with diesel-electric propulsion were in operation as early as 1904 and on the Danube from 1939, see Sect. 10.2.2 and Fig. 10.21. The electrically driven vessels have made a lot of progress since then; it is estimated that around 200–300 diesel-electric vessels are currently in operation on the European IWW. Diesel-electric propulsion combines high efficiency in matching use-cases, low noise levels and environmental sustainability due to potentially lower emissions. Diesel-electric drives with future-proof power management can serve as a bridging technology towards future zero-emission technologies.

In the case of *diesel-electric* (same for *gas-electric*) propulsion, the diesel engine drives a generator[11] that feeds the electric motor that delivers power to the propeller. Specifically, primary fuel energy is transferred to mechanical (diesel engine), then to electrical (generator), back to mechanical (electromotor), then via propeller conveyed to a thrust which overpowers resistance and moves the ship through the water. Therefore, the prime mover is *indirectly* connected to the propeller.

The additional energy conversion causes additional losses so that the energy efficiency of diesel-electric drives is not necessarily better than that of direct drives. Moreover, an arrangement with one generator set per propeller is generally not advisable. According to the European Standard laying down Technical Requirements for Inland Navigation vessels (ES-TRIN, see CESNI 2019), a redundant supply from at least two independent energy sources is prescribed for electrical propulsion systems. One of the gen-sets or power sources must be able to ensure a safe ship operation for at least 30 min in case of a failure of the second source. Thus, a division into several generator sets is necessary anyway. Figure 9.21 shows the setup of a simple diesel-electric drivetrain with the main components.

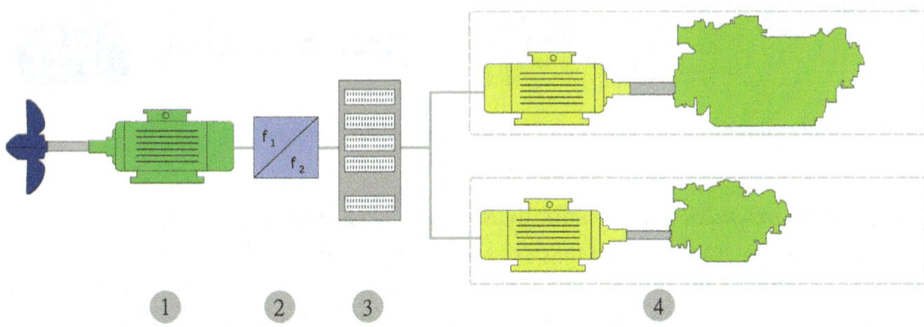

Fig. 9.21 General setup of a diesel-electric drivetrain

The electric motor (1) drives the propeller. Its advantage is a nearly constant efficiency at all load cases. Depending on the selected motor, a gearbox can be omitted. The frequency converter (2) supplies the electric motor with a frequency and voltage amplitude variable AC voltage. The converter can be supplied by any AC or DC on-board energy grid. The rotational speed of the electric motor is controlled by varying the output frequency. The main switchboard (3) distributes the energy from all sources to all loads. The loads are frequency converters at the propulsion

[11] In shipping and marine engineering, the term gen-set is common; it implies a unit which consists of a fuel powered engine that drives an electric generator.

systems, hotel load, pump systems and so on. It could be designed as a single AC or DC rail, which can be split, e.g. into a starboard and portside system. Each gen-set (4) can consist of any combustion engine and an electric generator. The generator can provide AC or DC power, depending on the selected main switchboard and frequency converters.

Since the propulsion systems of inland vessels often operate in the partial load range for long periods in many operating profiles, splitting the installed power to several generator sets is one of the biggest advantages of diesel-electric drives. Separate power sources can be operated selectively. This allows to run each engine at an efficient operating point or to switch it off, leading to extended service intervals. Note that the percentage share of total energy loss per component varies with vessel speed (e.g. at low speed auxiliary demand is high relative to installed power and vice versa, see PROMINENT 2018).

For *gas-electric* systems depicted in Fig. 9.22, the complexity is increased (see also Sect. 9.2.5—Gas propulsion concepts and gas-direct propulsion depicted in Fig. 9.7).

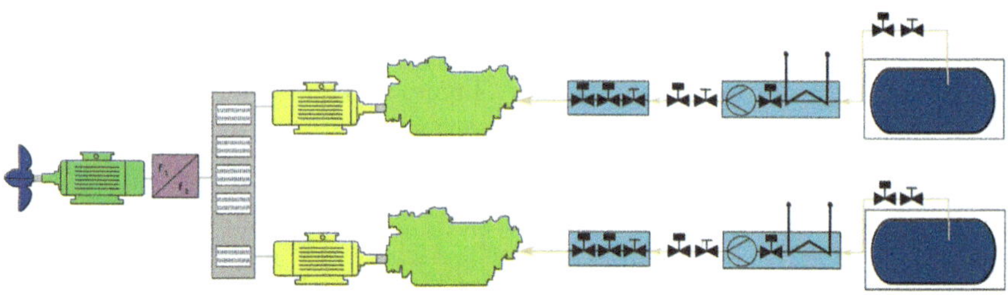

Fig. 9.22 General schema of the components of a gas-electric system

Hydraulic propulsion is essentially similar, i.e. a hydraulic pump substitutes a generator and a hydraulic motor takes the role of the el. motor. Therefore, the function of the generator + el. motor, and the hydraulic pump + hydraulic motor, are to convey power from the prime mover to the propulsor, thus replacing the function of the conventional gearbox and a shaft. This enables permanent running of the engine at the optimal regime (close to design point) and decreases emissions; the main disadvantages are higher investment costs and lower transmission efficiency (compared to diesel direct propulsion).

Drivetrains with an electrical power take in, or the so-called *father-son layout* with two internal combustion engines of differing size, acting together or separately via a common gearbox, are rare. However, good emission performance and quick amortization were reported for some ships equipped with it (see Bach 2015).

In general, it can be said that more components have to be accommodated in the engine room when installing electric drives. However, these are smaller and lighter than those of a conventional diesel system. Since no mechanical connections are necessary with the electric version, the components can also be distributed more flexibly in the engine room (and even along the ship). To summarize, the overall volume of the system components is similar to that of a diesel direct drive. Due to the smaller and more flexible components, the space available in the engine room can be better utilized, so that there is potential for savings in engine room size. This may impact ship design and result in potential indirect benefits (e.g. by creating more space for the payload), see for instance Fig. 4.19. A side effect is that the absence of a conventional heavy prime mover may lead to pronounced propeller-induced vibrations and noise.

Recapitulating, to benefit from a conversion to diesel-electric propulsion, a vessel should meet one or more of the following criteria:

- High electrical power demand with a high degree of varying loads
- A high degree of partial load for propulsion
- High comfort demands
- High demands towards redundancy
- High demands towards manoeuvrability.

Hybrid Propulsion

Hybrid propulsion means that two different engine types drive a propeller, e.g. diesel-direct and diesel-electric, or diesel-direct and battery-electric. The hybrid vessels, therefore, use two or more power sources for propulsion. Moreover, hybrid propulsion usually implies that several diesel engines are installed, as the main engines and generators are usually combined (which may be the case for diesel-electric propulsion too). Hence only a portion of the total available power may be used for propulsion, battery feeding, onboard electricity needs etc. The motors which drive the propulsor, whether diesel, electric or hydraulic, may be placed within the hull or on the deck, close or relatively far from the propulsor itself, or even out of the hull (e.g. in the pod of electrical pod propulsor).

As might be expected, the hybrid propulsion better complies with stringent environmental requirements. It can also be integrated with alternative energy sources such as solar or fuel cells (see Fig. 9.23). A sophisticated control system enables (a) run of as many diesel-powered generators as required to cover the power demand[12], (b) diesel engines operation at optimal efficiency (independent of the required propeller shaft speed) and (c) optimal distribution of generated power between that needed for propelling the ship and other ship demands. Amongst disadvantages, however, are high investment costs and higher weight compared to conventional diesel-mechanical systems.

[12]Note that maximal power is needed only occasionally, mostly in upstream navigation (see Fig. 9.20).

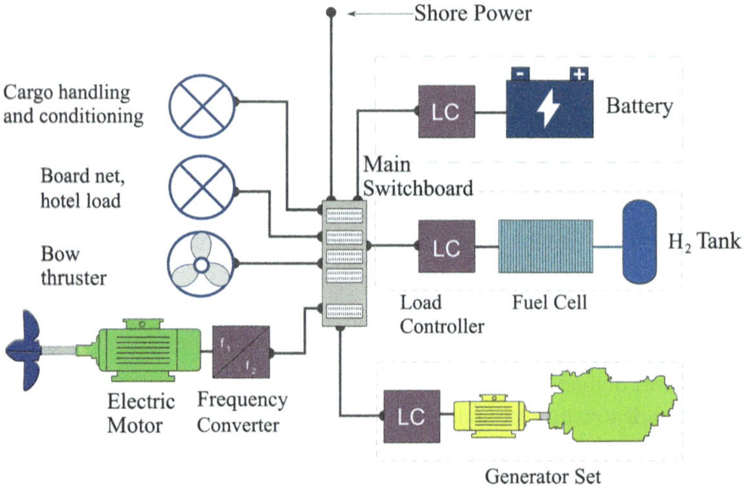

Fig. 9.23 Exemplary schema of an advanced hybrid system

To summarize, a variety of motor types, their placements and modes of power transmission exist. They are often functionally integrated with, or even within, the propulsor in a compact unit (drive). For a better insight into the powertrain components, its functional disintegration is necessary.

Transmission of Power

Transmission of power from the engine (prime-mover, usually a diesel engine) can be as follows:

(a) *mechanical—horizontal* (traditional and usual case, a rudder is necessary),
(b) *mechanical—vertical* (azimuthing thruster or rudder-propeller, usually rotates 360°, thus conventional rudder is not necessary),
(c) *electrical* (Diesel-electric propulsion—el. motor in the ship hull or in the body of an electric pod propulsor), or
(d) *hydraulic* (Diesel-hydraulic propulsion—hydrostatic pod propulsor).

Related transmission losses[13] from the engine to the propeller are, respectively:

(a) around $3 \div 4\%$ (with the gearbox losses),
(b) up to 10% (gearbox + 2 pairs of bevel gears),
(c) $10 \div 15\%$ (energy conversion losses, mechanical-electrical-mechanical), and
(d) $15 \div 20\%$ and more (energy conversion losses, mechanical-hydraulic-mechanical).

[13]Transmissions losses ad (c) and (d) vary with vessel speed, i.e. power output (see PROMINENT, 2018).

It should be noted that transmission losses in some cases are high, which is often forgotten (note that in the last 50 years of propeller development its efficiency has increased by few percents only; in that context, it is pity to lose energy on transmission).

Consequently, the overall advantage of indirect compared to direct transmission should be greater than the energy conversion losses. This subject is further discussed in Chap. 10. Comparing only the hydrodynamic efficiencies of the two most common cases, i.e. horizontal and vertical propulsion of a pushboat, according to Schot and Rotteveel (2020) it seems that conventional horizontal propulsion is better in shallow water and leaves more room to prevent ventilation at extremely small draughts, while vertical (rudder-propeller) is slightly better in deep water.

9.4.2 Steering Devices

Steering or control devices are converting energy into control forces (Brix 1993). Shipboard control devices (as may be external too, e.g. tugs) can be *passive* or *active*. The first (and these are rudders in the first place) are absorbing energy from the propeller slipstream or from the passing water, while for the second, energy should be supplied to enable steering (e.g. thrusters, propellers etc.). The first are controlled by the angle of attack, and the second by their output (i.e. number of revolutions, pitch etc.). In general, the passive devices are becoming more effective with the increase of ship speed, while the active ones are more effective at low and zero speeds. Consequently, at low speeds only the active devices should be used, for medium both, and for high speeds only the passive ones.

Passive Control Devices—Rudders

Due to the limited draught of IW vessels, imposed by shallow water, it is not easy to fulfil the manoeuvring requirements, particularly for pushed and coupled convoys. Consequently, various kinds of control devices—rudders—are in use, depending on the vessel type, propulsion system, draught etc. Specifics of rudders used on IW vessels are given in the following.

A rudder is essentially a vertical, hydrofoil that travels through the water. When deflected, the rudder does not just redirect the water; it also produces the difference in pressure (due to Bernoulli's principle) between the inner and the outer side, which generates a transverse force that creates a steering moment about the centre of gravity of the ship. The magnitude of the suction effect on the outer side usually exceeds the increase of pressure on the inner side of the rudder. These effects essentially depend on the rudder's (a) deflection angle, (b) position relative to the propeller (i.e. fluid velocity) and (c) rudder geometry—profile and area. Rudder lateral area A_R (or sum of areas of all rudders) is of particular importance and is usually expressed as % of ship's lateral area—$L_{WL} \cdot T$ (A_R is usually between $3 \div 5\%$ of ship's lateral area).

Specifically, the rudder should be located *close and in line with the propeller* due to a higher flow speed caused by the propeller slipstream.[14] The higher the flow speed around the rudder, the larger is the transverse force, i.e. the steering effect. However, the overflow at the tips of the rudder (between high-pressure and low-pressure area) reduces its efficiency. This effect is less pronounced if the rudder's length compared to its height is smaller, i.e. if the rudder's aspect ratio is higher. Due to restricted draught, the *aspect ratio of rudders used on IW vessels is rather low*—usually close to one or even less (see Figs. 9.19 and 9.24). Accordingly, the maximal transverse force is obtained at *much higher rudder deflection angles* (compared to those of the seagoing ships which is typically up to 35 deg.) which leads to enlarged rudder resistance. Its efficiency can be increased by adding *endplates* on the top and bottom sides, see Fig. 9.26.

Fig. 9.24 Stern propulsion/steering configuration of IW selfpropelled vessel

Another way to improve rudder efficiency is the application of *flapped rudders*, which are very efficient at smaller rudder angles. These rudders have a movable flap installed at the trailing edge of the main rudder blade, see Fig. 9.25. The ratio of flap deflection and main rudder angle deflection is constant (and is greater than one), thanks to the simple mechanical link system. Thus, the flap generates additional transverse force and improves the rudder's efficiency.

Having in mind the restricted draught, and for good manoeuvrability required rudder area, single rudder per propeller is often not enough. Therefore, *rudder systems* consisting of two, or even three rudders per propeller are used, see Fig. 9.26. *Advanced rudder systems* do not have the same deflection angle of each rudder, see Fig. 9.27, i.e. each rudder differently deflects the water flow in order to optimize the manoeuvring performances of the ship.

[14] If not in a propeller slipstream the rudder area should be at least 30% larger.

Fig. 9.25 Flapped rudders on a pushboat. *Source* Becker Marine Systems 2020

Fig. 9.26 An IW vessel equipped with two rudders per propeller. *Source* Liu et al. (2014)

The manoeuvrability of some twin-rudder seagoing ships was significantly improved with a special rudder deflection setting. Namely, according to Liu and Hekkenberg (2017) if both rudders (named clamshell rudders) are set outwards at 75 deg., stopping time and the stopping distance may be reduced by some 50% compared to a conventional reverse engine stopping. There is not much evidence of the application of clamshell rudders on IW ships, although they are commonly equipped with multiple rudders (see for instance clamshell rudders on large Rhine IW container ship, see Fig. 9.27).

Different rudder design is coming from Nautican, see Fig. 9.28, showing *Nautican Triple Rudders* that consist of 3 *high aspect ratio foils* per nozzle, which create high lift with less resistance. It is reported that this low-torque requirement rudder system is better than that of flapped rudders as creates a side force of more than 70% of the bollard pull.

Rudder's *hydrodynamic characteristics*, thus ship manoeuvrability too, very much depend on rudder sectional shapes. Typical rudder profiles are shown in Fig. 9.29.

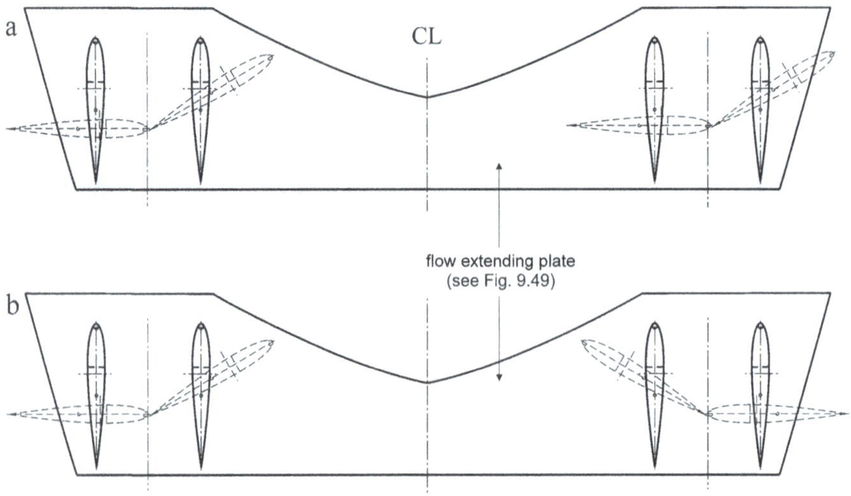

Fig. 9.27 Advanced rudder system according to Székffy patent, **a** full left manoeuver, **b** rudders at a clamshell angle—stopping manoeuver. *Source* Zöllner (2005)

Fig. 9.28 Nautican triple rudders. *Source* Nautican rudders (2021)

The first two are usually used on seagoing ships, the next three on the IW vessels and the last one (flap rudder) on both. The *Fishtail rudders* (also known as Schilling rudders) shift the center of pressure aftwards, and on IW vessels are usually used with endplates. In addition to the abovementioned rudders with constant cross-section along the height (the usual case), *twisted rudders* have specially adapted leading or trailing edge to reduce the swirl of propeller jet stream and thus increase the propeller efficiency. More about the hydrodynamic characteristics of these rudders can be found in Liu (2020) and in Liu and Hekkenberg (2015).

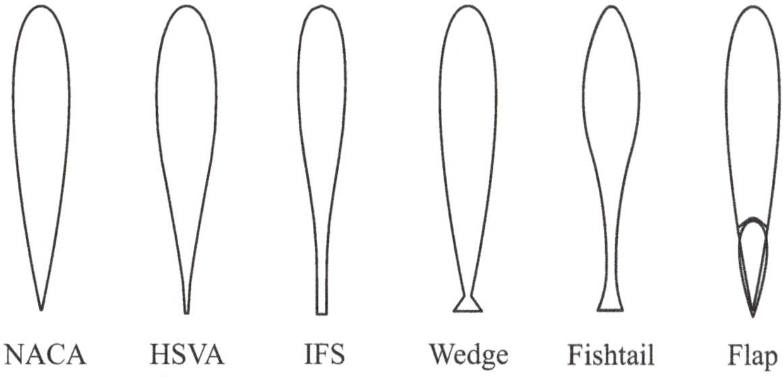

Fig. 9.29 Common rudder profiles

To improve manoeuvring performances during astern operations, some IW push-boats have so-called *Flanking rudders*. These rudders are located in front of the propeller and hence disturb the propeller water intake in ahead operations. In order to increase the propulsive efficiency during ahead operations, an innovative concept of fully retractable flanking rudders is developed, see Fig. 9.30 and Sect. 8.2.2. This system should be lowered only during astern operations. Note, however, that the flanking rudders are nowadays frequently replaced by bow thrusters, which will be discussed later in Sect. 9.4.6.

Fig. 9.30 Flanking rudders arrangements on IW vessel. *Source* DAMEN (2019a)

Bow rudders are rarely used on the IW vessels and occasionally on the seagoing ships, mainly on the ferries. Dutch company Kooiman produces retractable bow rudders for barges to increase manoeuvrability and course stability of pushed convoys, see Fig. 9.31. Pneumatically operated and pneumatically retractable bow rudder (to prevent damage) should be installed only on the front barge of the convoy.

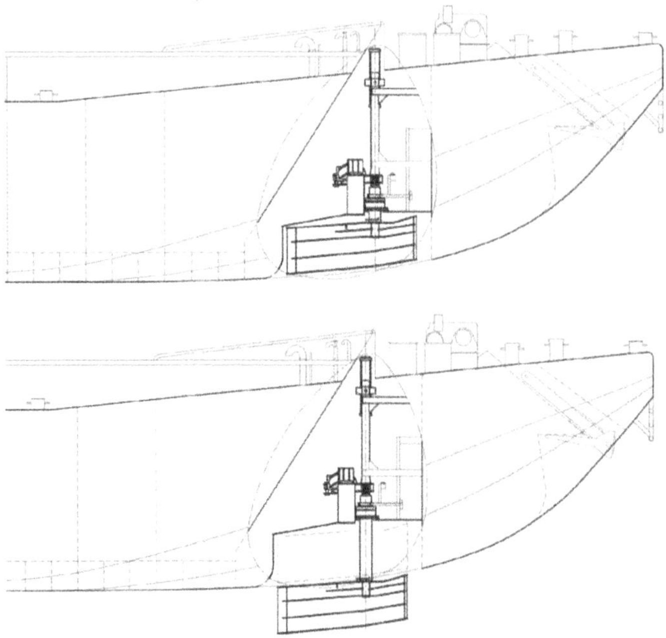

Fig. 9.31 Retractable bow rudder for push barges. *Source* Kooiman (2020)

Another Dutch company produces *rotor bow rudders* (see Van der Velden 2021) for pushed convoys and container vessels. Their system is based on the Magnus effect, i.e. a retractable rotating cylinder is placed into oncoming water current which results in the generation of side thrust. Apparently, the effect is larger than of a conventional bow rudder. Note, however, that energy has to be supplied for cylinder rotation (thus this is a semi-active or semi-passive device).

The steering nozzle creates a steering force by jet deflection, and thus can replace the combination of a conventional (fixed) nozzle and a rudder. It can produce a much larger side-force (its vector diagram is broader) than a conventional installation. Due to the kinematic reasons (steering nozzle's pivot axis must be located in the propeller plane, see Fig. 9.32), installations with steering nozzles are relatively more complicated and expensive. Moreover, they are also vulnerable to jamming and a bit less effective than fixed nozzle due to inherently larger clearance between propeller tips and nozzle's inner ring.

Summarizing, to accomplish required steering performances, it is necessary that (given in order of priority):

(1) The rudder is of sufficient area and positioned in the propeller slipstream.
(2) Rudder type and profiles (cross-sections) are appropriately chosen.
(3) The steering gear is adequately dimensioned to enable proper "rudder steering time".

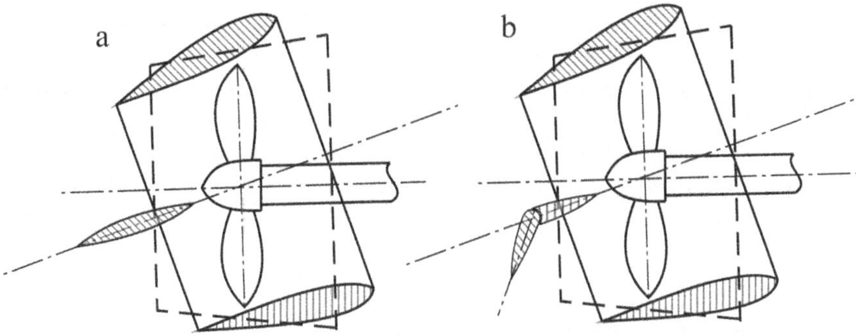

Fig. 9.32 Steering nozzle, **a** with fixed fin and, **b** with flapped fin. *Source* Brix (1993)

The most important rudder's feature, its lateral area, is not explicitly prescribed by rules and regulations—but implicitly, by definition of manoeuvrability requirements. Consequently, that feature, as well as the others, is usually copied from similar and successful vessels.

The following should also be noted:

(a) Rudder resistance is of secondary importance compared to its steering characteristics.
(b) Rudder-propeller-hull interaction may be important, amongst others, as it directly influences fuel consumption.
(c) Rudders require additional equipment (e.g. steering gear) which means that in its design and assessment the whole system should be considered (costs, weights, space etc.)—not just the rudder itself.

Active Control Devices

There is no need for rudders in cases when the power transmission line is vertical—often called *Z-drive* (motor is placed horizontally) or *L-drive* (motor is placed vertically). These propulsors are called *rudder-propellers* or *azimuthing thrusters* and provide complete directional thrusting capability by rotating around their vertical axis (usually 360 deg.). In general, enhanced steering capabilities of *vertical shaft-line-thrusters* have to be "paid" by "breaking" the shaft line, which results in reduced robustness and lower efficiency, compared to conventional transmission (not to mention higher investments costs of the drive itself).

The azimuthing thrusters may be installed *within the hull* (referred as *well installation*; these are Z and L drives mentioned above) or *outboard*, on the deck (denoted by Schottel, for instance, as *Navigator*). In the second case, the thruster can be retracted by tilting, or raised and lowered vertically to adjust the draught of the vessel.

Active control devices are also various kinds of bow or stern thrusters. For IW vessels longer than 110 m bow thrusters are mandatory. Some types of thrusters have water intake from the bottom (good for low draught). Besides steering (in some cases stopping) they may be used for propulsion too (e.g. various kinds of pump-jets and channel thrusters, discussed in Sect. 9.4.6).

For more information regarding azimuthing thrusters and steering devices in general, see Virendra et al. (2015), Brix (1993), and Molland and Turnock (2007). It is, however, expected that contemporary information about the rudder efficiencies and related topics will be available after the ongoing research project on IW rudders (DST, 2021) carried out in DST ends.

9.4.3 Screw Propellers—an Overview

The main propulsors which are nowadays used on IW vessels are based on a screw-propeller (or just propeller). These are the following:

- *Fixed-pitch propeller (FPP)* or monoblock propeller—there are several FP propeller types, which can have up to 7 ordinary or skewed blades (for reduced vibrations); simple and cheap.
- *Controllable pitch propeller (CPP)*—the thrust is controlled by changing the pitch of the blades, hence CPP can adapt to resistance variations (due to water depth, free-running or towing conditions etc.); advantageous for vessels whose loading varies.
- *Propeller in the nozzle (Ducted propeller)*—increases the thrust in bollard pull conditions and at low speed if propeller diameter is restricted (and thus propeller heavily loaded); most common case on IW cargo vessels.
- *Contra rotating propellers (CRP)*—two coaxial propellers positioned one behind the other, rotating in the opposite directions (thus eliminating mutual rotating wake); have the highest efficiency among all propulsors.
- *Tandem propellers*—two propellers on the same shaft rotating in the same direction; efficiency is between FPP and CRP.

Various combinations of the above are also possible (for instance CPP in a nozzle etc.).

Ducted Propeller (DP)

A pair that consists of a nozzle and a propeller, i.e. ducted or nozzled propeller, is considered as a new propulsor. However, each part, a nozzle and a propeller may be examined separately. The majority of IW cargo vessel types use it, thus DPs deserve few words on its own. Only the basics are presented here, for more see the already mentioned textbooks, Kuiper (1992), Zondervan et al. (2006) etc.

- DPs should be considered when the naked propeller is overloaded (restricted in diameter, operates at near bollard pull conditions or low speed etc., see Fig. 9.33), i.e. when $C_{Th} > 1.5 \div 2.0$. For equal power, DPs' bollard pull is up to 30% larger, while efficiency improvements may be up to $10 \div 20\%$ (depending on the ship speed, nozzle type etc.), resulting in that DP of considerably smaller diameter may

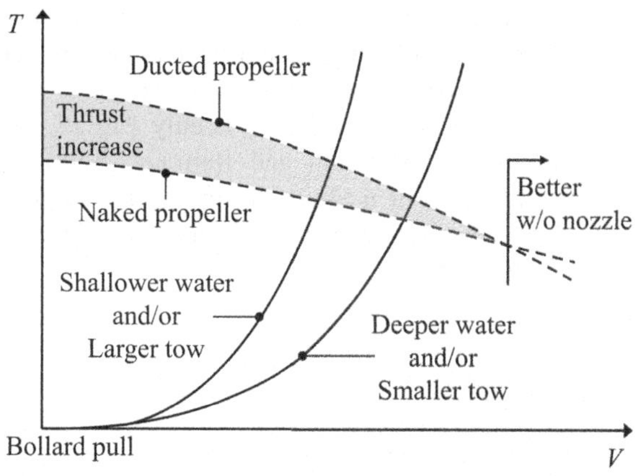

Fig. 9.33 Thrust-speed diagram for ducted and naked propeller

produce the same thrust as the naked propeller. The typical bollard thrust of DP is 100 N/kW (80 N/kW for naked propeller). Moreover, DP's power absorption is far less sensitive to ship speed than is the case with conventional naked propellers. DP astern performance, however, is generally inferior.

- The clearance between the propeller tips and inner nozzle ring should be $0.001D$, but is usually $0.01D$ (1% D); if it is larger the efficiency drops. DPs are susceptible to jamming of the propeller due to floating debris, ice etc. To install as large propeller as possible, the pushboat nozzles are usually a) cut/deformed at the underside, and b) have a relatively large angle of the nozzle ring countersunk in the tunnel (about 90 deg. or 25% of the circle). Both reduce DPs' efficiency but allow more power to be transferred to the water due to the larger propeller diameter.
- Nozzles are defined with their cross-section (profile) and length (expressed as ratio l/D, where l is nozzle's length and D is propeller diameter). Common nozzles are MARIN's Type 19A and 37 (which has better astern and a bit worse ahead performance than 19A); both have $l/D = 0.5$, see Kuiper (1992) for instance. For slow speeds and high thrust capabilities, longer nozzles are needed—the pushboats often have $l/D > 0.5$ ($0.6 \div 0.8$, rarely more); faster IW selfpropelled vessels have $l/D < 0.5$ (usually around 0.3). Faster vessels (e.g. passenger ships) should use a naked propeller. Common propeller series which are used with the nozzles are B, Ka, Kd, Kc, Meyne-VBD etc. see the abovementioned references.

Above DP basics is a starting point for familiarization with some other nozzle types and other propulsors (treated in the following sections). Standardized MARIN nozzles 19A and 37, mentioned above, are in use for decades; the new nozzle types with slightly different cross-sections are often marketed under different tradenames (e.g. Wartsila's HR nozzle) and are benchmarked against MARIN nozzles. Three general nozzle derivatives, amongst several, will be mentioned here as they may be of interest for IW vessels.

(1) *Simplified nozzles* are production-oriented as they have a simple form, thus are cheaper for production and later for repair, see for instance the Shushkin's nozzles (observe the fully developable conical and cylindrical surfaces), Fig. 9.34a. Their efficiency is only 1 ÷ 2% lower than comparable MARIN nozzles (Schneekluth and Bertram, 1998). *Unconventional wake-adapted* nozzle, also of a Shushkin type (for a single propeller on an 82 × 9.5 m, T = 2.5 m selfpropelled ship) is shown in Fig. 9.34b, see Zibell and Müller (1996).

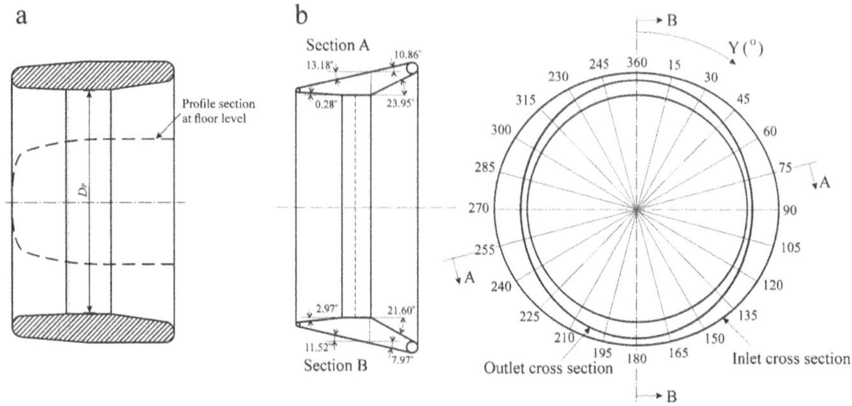

Fig. 9.34 **a** Shushkin nozzle for a pushboat ($l/D = 0.75$) (*Source* Schneekluth and Bertram 1998); **b** Wake-adapted nozzle for a single propeller, selfpropelled ship (*Source* Zibell and Müller 1996)

(2) *Slotted* and *flapped nozzles* belong to a family of unconventional nozzles; they are also called double nozzles or annular two-element nozzles. The slotted nozzle essentially incorporates an additional short nozzle installed *before* the propeller, while the flapped one has it *after* the propeller, see Fig. 9.35. Both are effective when the nozzle is short ($l/D = 0.3 \div 0.35$) and for a bit higher speeds. The expected increase of efficiency is around 5 ÷ 8% (compared to the conventional nozzle); the disadvantage, besides higher production cost, is red

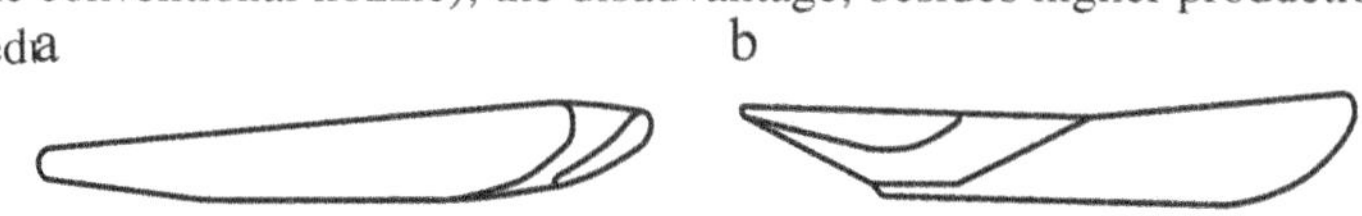

Fig. 9.35 **a** Slotted and, **b** flapped nozzle sections

(3) *Nozzle with circular internal grooves* which thicken the boundary layer and thus draw inward the inflowing objects preventing propeller jamming, see Fig. 9.36. Nozzle struts or forward control surfaces (i.e. ESDs, see Sect. 9.4.8) also prevent large floating objects to enter the nozzle. Note that the slotted nozzle (mentioned above) was initially developed to prevent debris and ice floes entering in the nozzle. Further development was a slotted nozzle with grooves.

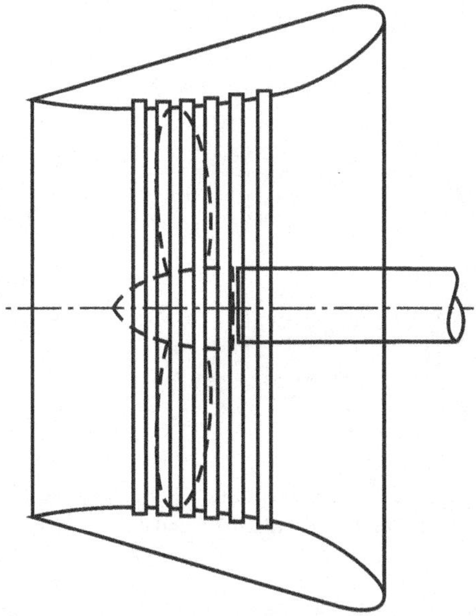

Fig. 9.36 Nozzle with circular grooves. *Source* Schneekluth and Bertram (1998)

9.4.4 Contemporary Propeller-Based Propulsors

Contemporary propeller-based propulsors for IW vessels would be the following:

- *Propellers in nozzles* (FPP and CPP), with usual (mechanical) horizontal or vertical power transmission—Figs. 9.19 and 9.37a, respectively.
- *Tandem and CRP* with mechanical transmission—Fig. 9.37b, c respectively.
- *Pod propulsors*[15] (electrical with FPP)—Fig. 9.38.
- *Combinations* of above (a) within a single azimuthing thruster, or (b) within a single vessel. An example for the first case is depicted in Fig. 9.39 showing the so-called integrated L-drive that has a single gear transmission in the pod (as opposed to Z-drive with 2 pairs of bevel gears, see Figs. 4.12 and 10.29). In this case el. motor is above the propeller in the hull, while the pod propulsor shown in Fig. 9.38 has the el. motor in the pod. Both are electrically driven rudder propellers with the propeller in a nozzle. An example for the second case is a combination of horizontal mechanical and azimuthing thruster one aside another (see Figs. 10.7 and 10.11), resulting in good manoeuvrability; or one behind the other (working as CRP) giving good efficiency.

[15]Pod propulsors (actually belonging to the family of azimuthing thrusters) incorporate an electric motor installed in the submerged pod. Probably the first units of the Azipod type used on the Danube were on the Austrian river icebreaker "Roethelstein" (see Fig. 10.22). Azipod is a trade name of the first pod propulsors on the market produced by Finish Kvaerner Masa + ABB. The well-known INBISHIP Project (see Sect. 10.1.2) was based on Azipods.

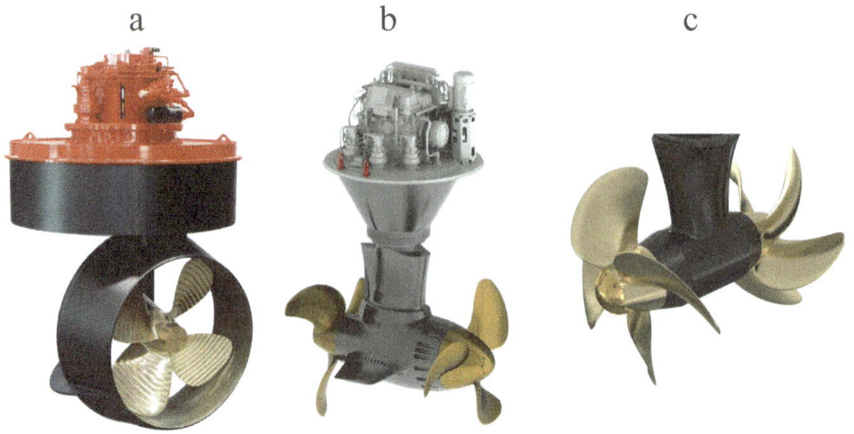

Fig. 9.37 Contemporary propeller-based propulsors—**a** Rudder-propeller in an integrated nozzle (VETH 2020), **b** Tandem propeller (SHOTTEL 2020) and **c** CRP (Veth Z-Drive)

Fig. 9.38 Electric podded propulsor—a compact electrical motor is located inside the pod and is directly coupled to the FP propeller in nozzle. *Source* Verhaar Omega (2020)

Fig. 9.39 Veth Integrated L-Drive with asynchronous and permanent-magnet el. motors, both within the hull. *Source* VETH (2020)

9.4.5 *Innovative Propeller-Based Concepts*

Tip-Driven or Rim-Driven Propeller (TDP or RDP)

If *power is transmitted to the propeller* via *blade tips* instead via boss (the second being the usual case), that is a tip-driven propeller. It does not comprise a classical shaft, which, amongst others, enables a good, unobstructed water inflow. Electrical Tip-Driven propeller with both stator and rotor integrated into the nozzle has been developed by few companies, e.g. Brunvoll, Rolls Royce, Voith, Schottel and others. These new propulsors seem quite promising for application on river vessels as well.

There are two types of tip-driven propellers (Radojčić 1997):

(a) When bearings supporting the propeller axle are connected to the hull through the strut arms (as shown in Fig. 9.40a)—*shaft-less* drive/propulsor.

(b) When the thin-section bearing is located in a nozzle. In this case, there is no need for an axle or a propeller hub, so the propeller may be of a novel (unusual) design, Fig. 9.40b; this is both a *shaft-less* and an *axle-less* or *hub-less* drive/propulsor.

Discussion: The tip driven propeller is essentially a kind of ducted propeller. Namely, the propeller has a ring attached to blade tips (this propulsor is called a ring-propeller) which is then countersunk in a nozzle. The core of the innovative technology is that: (1) the ring is also a rotor of electromotor, and (2) the nozzle contains a stator of an electromotor. The rest, i.e. the ring-propeller in combination with a nozzle, is already seen technology, at least from the viewpoint of hydrodynamics, see Oosterveld (1972). Accordingly, the shaft-less drive installation (case a) is conventional technology, while shaft-less and axle-less (case b) is a new propulsor

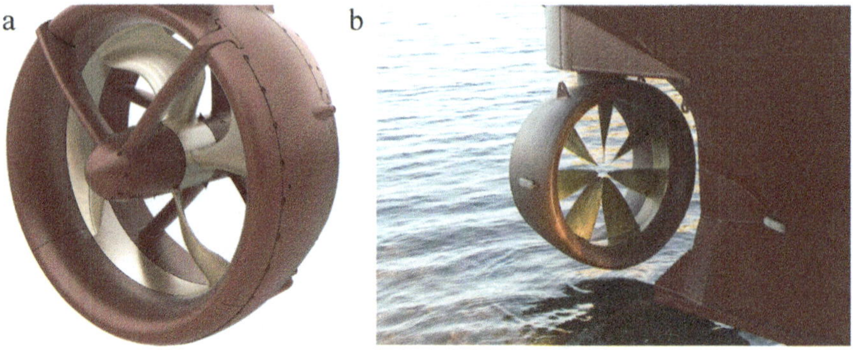

Fig. 9.40 Tip-driven (rim-driven) electric motor propellers: **a** Rolls Royce Azimuthing Permanent Magnet Thruster AZ-PM (*Source* Rolls Royce 2020), **b** BRUNVOLL azimuth RDT 1500 installed on MF Eiksund (*Source* BRUNVOLL 2014)

(see also Radojčić 1997). The efficiencies of the ring-propellers and ring-propellers in combination with a nozzle are generally lower than of conventional DP (due to additional drag of the rotating ring, thus the ring should be as narrow as possible) see Oosterveld (1972).

Surface-Piercing Propellers (SPP)

Surface-Piercing Propellers (SPP) have only the lower half of a disc immersed in the water and therefore are suitable for shallow-draught vessels. They are usually used for high-speed crafts. However, VOITH started the development of a slow-speed SPP-CPP. It exploits the fact that, besides the thrust, SPP generates a large side force too, which enables steering.[16] Hence a rudder is not required. Nevertheless, slow speed SPP are somehow clumsy (due to the large propeller disc, as only the bottom half produces thrust). Due to generated side force, they have to be used in pairs, Fig. 9.41.

Side-by-Side Propeller (SSP)

Recently a *Side-by-Side propeller* was presented, see Fig. 9.42. It is still in the test phase and according to the available information (see REINTRIEB 2019), SSP's efficiency with a diameter of 75% D is the same as that of a conventional propeller in a nozzle with diameter D.

[16]This principle was applied on MS "Turanor Planet Solar", the largest solar-powered boat in the world ($L = 31$ m, solar panels 537 m^2, 93 kW, launched in 2010).

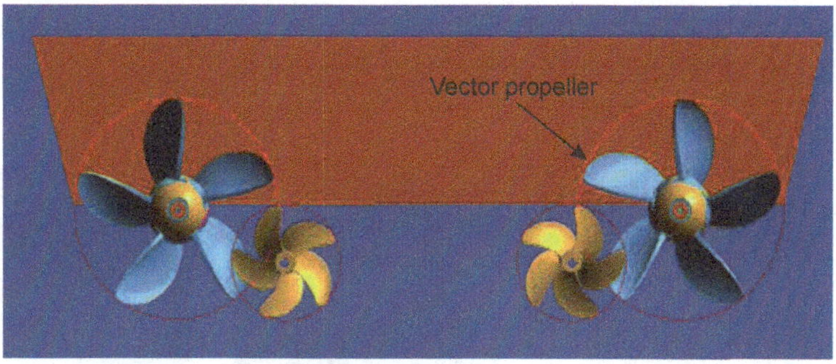

Fig. 9.41 Size comparison of Vector and conventional propeller. *Source* VOITH website which is not active any more

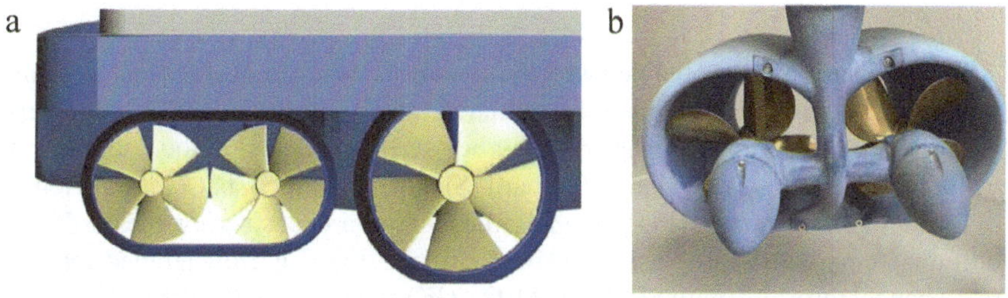

Fig. 9.42 Size comparison of Side-by-Side propeller with a conventional propeller in the nozzle, (**a**) and test model of Side-by-Side propeller, (**b**). *Source* REINTRIEB (2019)

9.4.6 Other Propulsors

Not counting the clumsy *side* or *stern paddle wheels* (which require low RPM but have good efficiency and are inherently adapted to shallow draught and thus are not beyond the consideration in some recent projects) only three propulsors, other than propeller-based, will be mentioned here:

- *Waterjet* or a *pump-jet*
- *Vertical propeller* (produced only by Voith and therefore often called Voith-Schneider propeller, sometimes Cycloidal propeller)
- *Whale or fish tale propulsors* (still in development phase).

Pump-jets

Pump-jets, with a vertical axis, are initially developed to be bow-thrusters, see Fig. 9.43. They consist of a mixed-flow pump placed in a special volute casing that can rotate about its vertical axis, enabling steering throughout 360 degrees. Water is drawn into the casing below the hull and is expelled through the outlet nozzle. Advantages are applicability to very shallow draught vessels, good manoeuvrability, robustness (even grounding is allowed) and reduced jamming. The disadvantage of the pump-jets is relatively high costs and low efficiency,

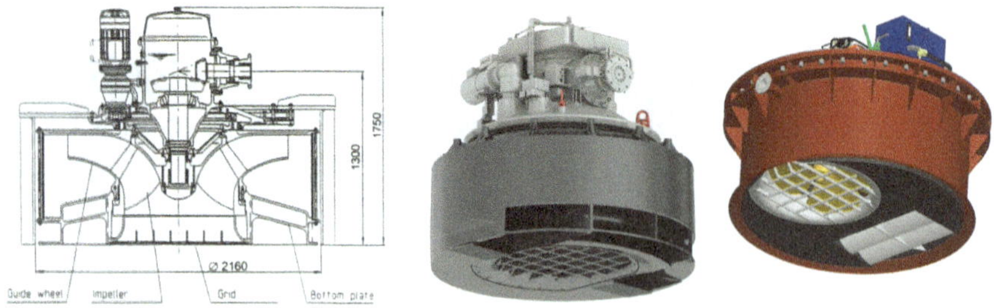

Fig. 9.43 Schottel Pump Jet (2020) and Veth's Compact-Jet

Similar results may be obtained with the so-called *360-degrees-steerable steering-grid* and *4-channel thrusters* which are primarily intended to be used as bow thrusters, while the abovementioned pump-jets may be used for propulsion too. Actually, 4-channel thrusters (Fig. 9.44) have starboard (SB), portside (PS), backwards oriented (BO) and forward-oriented (FO) channels; the first two are for steering, the third is for additional (emergency) propulsion and the fourth for stopping. Downgraded versions are 3-channel thrusters (with SB, PS and BO or FO channels) and 2-channel thrusters (with SB and PS channels). Thus, the last one is not a propulsor, its function is steering only, while a 3-channel thruster with FO tunnel has a stopping ability too (i.e. both belong to typical IW active control devices, see Sect. 9.4.2). Channel thrusters are well suited to low draught vessels and may be mounted in the hull or outside of the hull in a gondola.

Fig. 9.44 Bottom view of a model of a 4-channel thruster. *Source* Peeters et al. (2020)

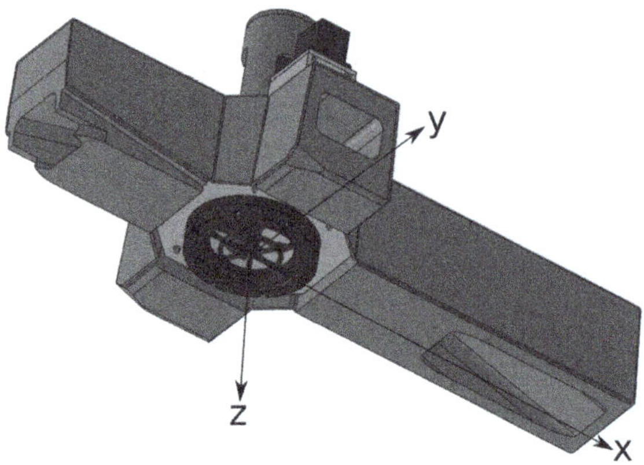

Vertical Propeller

A vertical propeller (see Fig. 9.45, Voith-Schneider propeller) might be used on vessels requiring exceptional manoeuvrability since they can produce controllable thrust throughout 360 degrees. Nevertheless, vertical propellers are relatively complicated and therefore expensive. They are not as efficient as ordinary propellers, since their vertical blades generate thrust only over a part of the revolution cycle with a cost of ever-present frictional resistance. Similar to Voith-Schneider is the Kirsten-Boeing propeller (both with a single rotor) and the new Bivortix propeller (with two coaxial and contrarotating rotors), see Baroni et al. (2016).

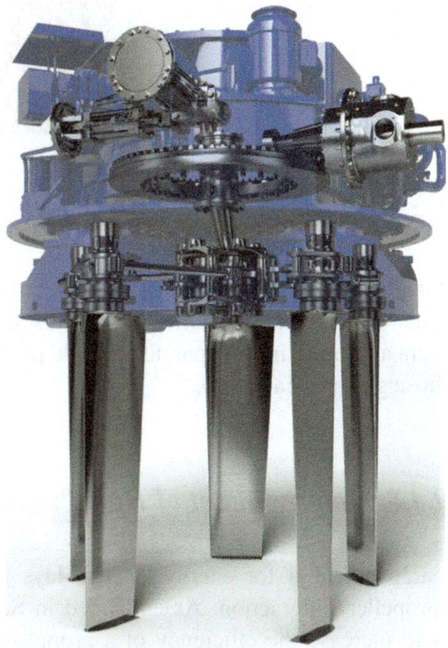

Fig. 9.45 Vertical Voith-Schneider propeller. *Source* VOITH (2020)

Whale Tail Propulsor

The principle of the vertical axis propeller can be applied to a cycloidal propulsor with a transverse horizontal shaft. This is the basic idea of the whale tail propulsor (Fig. 9.46) which is still in the development phase. Some research reported supreme propulsive efficiency, particularly when combined with IW vessels with restricted draught.

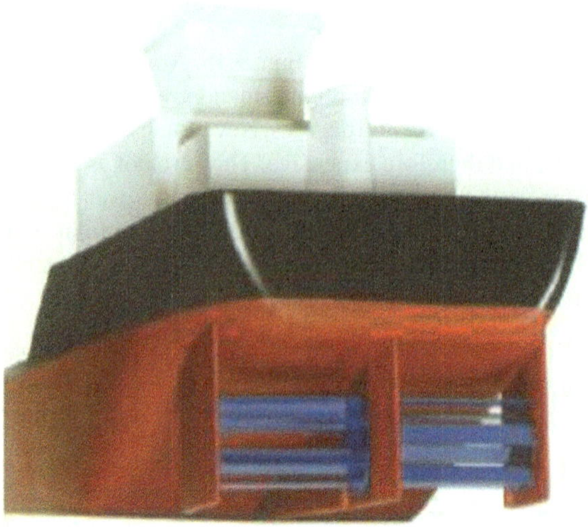

Fig. 9.46 Whale tail arrangement. *Source* CREATING (2005)

9.4.7 *Rating of Propulsors*

An attempt was made to compare all mentioned propulsors on the same basis, considering specific characteristics of IW vessels—see Table 9.2. Picking out just one of them and assessing it separately, would probably result in different conclusions from those given in Table 9.2.

9.4.8 *Improvement of Hull-Propulsor Interactions*

Large potential for energy savings lays in improvements of hull and propulsor (propeller) interaction. As explained in Sect. 9.1 (see also Fig. 9.1) the intention is to increase the efficiency of a propulsor (essentially of η_D) which depends on hull-propulsor interaction. In other words, the aftship body should be adapted to (a) particular propulsor, and (b) the waterway (i.e. water depth which varies from one river stretch to another). To achieve good hull-propeller interaction is often necessary to sacrifice resistance; i.e. resistance increase has to be overcome (paid) by the larger increase of propulsive efficiency. Typical problems arise when two different departments (persons) are in charge for resistance and propulsion, each optimizing own task within own sphere of competence.

Adjustable Retractable Tunnel

The aftship of IW ship should be designed in such a way that advantages of navigation in deep water are fully exploited (with a relatively undisturbed inflow of water to the propeller) while maintaining satisfactory efficiency in shallow water operation (in

Table 9.2 Propulsor applicability on a potential IWW vessel

No.	TYPE OF A PROPULSOR	Transmission type	Propulsive efficiency	Manoeuvring capabilities	Robustness	Extra space on a vessel	Cost of a propulsor	Built-in cost	Maturity (developing phase)	Environmental pollution
1	Naked FPP	M-Hor	-	- -	+ +	- -	+ +	+	+ +	-
2	Naked FPP	M-Ver	- -	+	-	- -	-	+ +	+	-
3	Naked CPP	M-Hor	+	-	-	- -	-	+	-	+
4	Naked CPP	M-Ver	-	+ +	- -	- -	- -	+ +	-	+
5	Ducted FPP	M-Hor	+	- -	+ +	- -	+	-	+ +	-
6	Ducted FPP	M-Ver	-	+	-	- -	-	+ +	+	-
7	Ducted CPP	M-Hor	+ +	-	-	- -	-	-	-	+
8	Ducted CPP	M-Ver	+	+ +	- -	- -	- -	+ +	-	+
9	Steerable nozzle FPP	M-Hor	-	+	+	- -	-	- -	-	-
10	Tandem propeller	M-Ver	+	+ +	-	- -	-	+ +	+	-
11	CRP	M-Ver	+ +	+ +	- -	- -	- -	+ +	-	-
12	Pod propulsor FPP	E-Ver	- -	+ +	-	+ +	- -	+ +	-	+ +
13	Hydrostatic FPP	H-Ver	- -	+	- -	+ +	- -	+ +	- -	+
14	Tip-driven, shaft-less FPP	E-Hor	-	- -	-	+ +	?	+	- -	+ +
15	Tip-driven, hub-less FPP	E-Hor	?	- -	-	+ +	?	+	- -	+ +
16	Pump jet	M-Ver	- -	+ +	+ +	- -	- -	+	+	-
17	Vertical propeller	M-Ver	- -	+ +	-	- -	- -	-	+	+

Abbreviations:

Rating

++	very good
+	good
-	average
- -	bad
?	not known

Transmission types

M	mechanical
E	electrical
H	hydraulic
HOR	horizontal
VER	vertical

Note: HOR arrangements need a rudder (except in case 9), while VER arrangements, just by rotating the propeller about the vertical axis, generate thrust in all directions and thus do not need a rudder.

partly loaded condition). This can be realized with an adjustable or retractable tunnel (Fig. 9.47), which is—depending on the draught—aligned with the hull (Fig. 9.47a), or with fins folded downwards (Fig. 9.47b) forming a tunnel to prevent entrance of incoming air (ventilation) into the propeller at a low draught. Thus, the propulsion efficiency is significantly increased and resistance reduced at higher draughts as an installment of somewhat bigger propeller is allowed. On the other hand, the same propeller would work satisfactorily at lower draughts thanks to the retractable tunnel. In other words, a ship with the adjustable tunnel can operate at lower draughts without jeopardizing the propeller efficiency caused by the air intake. Without it, a smaller propeller (probably with fixed skirts, see Fig. 8.5) would have to be applied to satisfy both draughts. It essentially enables a ship without a conventional tunnel to operate efficiently under a partly loaded condition. An adjustable tunnel in full-scale

(FLEX tunnel—Van der Velden) is shown in Fig. 9.47. Efficiency improvements of approximately 10% are expected, depending on the draught.

a b

Fig. 9.47 Adjustable (FLEX) tunnel for an inland vessel. *Source* DAMEN (2019b)

Energy Saving Devices (ESD)

Various *Energy Saving Devices* (ESD), also called *thrust augmentation devices*, are available for seagoing ships. They may be divided into ESD which are placed[17]:

(a) *Before* the propeller (*pre-swirl* or *wake equalizing* devices)
(b) *After* the propeller (*post-swirl* devices—utilize the lost rotational energy of a propeller slipstream).

Most of ESD do not have any rotational parts (those which have are not of interest here) and on the seagoing ships can save approximately $2 \div 6\%$ of energy. Note that CFD can play a very important role in the design of both, the pre- and post-swirl devices. Concerning the IW vessels, the uneven waterway depth is the main obstacle for the wider application of ESD.

The IW vessels, on the other hand, use DPs. They (nozzle and propeller shaft) need to be supported by the struts positioned before the propeller. It thus seems obvious that these struts should be streamlined so that they can become a pre-swirl stator. Moreover, the stator may be integrated with a controllable pitch pre-swirl stator, which would then serve as a kind of CPP, see for instance Hughes (1991). The same idea is presented in Oosterveld (1972) with 'multiple nozzle arrangement with ring-propeller and ring-stator'.

Pre-Swirl Stators, in general, improve water inflow to the propeller and thus recover the rotational losses increasing the efficiency, probably by some 5%. The

[17]For the seagoing ships, beside ESD's placed before and after the propeller, there is also a third type placed *at the propeller* (e.g. propellers with end plates, contra-rotating propellers etc.) but these are treated here as different propulsor types and belong to Sects. 9.4.4 or 9.4.5.

Nautican, for instance, promotes an *integrated propulsion unit* that consists of a pre-swirl stator, nozzle and triple rudders (see Figs. 9.48 and 9.28) and reports fuel efficiency gains for the pushboats of up $10 \div 30\%$.

Fig. 9.48 Nautican pre-swirl stator. *Source* Nautican stators (2021)

Peters (1994) gives examples how to increase the efficiency of very shallow draught pushboats (having draughts of 0.84 and 1.3 m) by installing the pre-swirl fin; power savings of some 15% were reported. Similar fins in combination with the asymmetric stern form were tested on a single propeller selfpropelled ship ($L = 82$ m, $B = 9.5$ m, $T = 2.5$ m, $D = 1$ m) and a power reduction of some 10% was reported. It was concluded that the savings depend mainly on the propeller pitch ratio (P/D) and thrust coefficient of a DP.

The simplest *post-swirl devices* are the rudders. They can be further improved with the aim to recover lost rotational energy of a propeller slipstream, which then results in twisted rudders, see Sect. 9.4.2.

A *Flow Extending (or deflecting) Plate* (FEP) promoted by DST is a kind of ESD[18] for IW vessels positioned after the propeller, see Fig. 9.49. Its purpose is to separate the high velocities in the propeller jet from the low-speed wake of the hull, which has a drag-reducing effect. At the same time, it reduces the induced drag at the upper ends of the rudders and prevents ventilation at lower draughts and in stopping manoeuvres (see Zöllner 2009). The final results are improved propulsive efficiency and manoeuvring characteristics.

Summarizing this section, it may be concluded that DP with ESD placed in a tunnel (see for instance the recommendations for design of a pushboat tunnel, Sect. 8.2.2) leads to the design of a new propulsor concept called *Integrated Ducted Propeller* (IDP). IDP essentially integrates the propulsor with the ship stern (see Chen et al. 1991), while the conventional design approach assumes that a propeller is an add-on device to the hull.

[18]Flow extender plate is after the propeller ESD, but hardly can be called a post-swirl device, thus name *thrust augmentation device* is probably more appropriate.

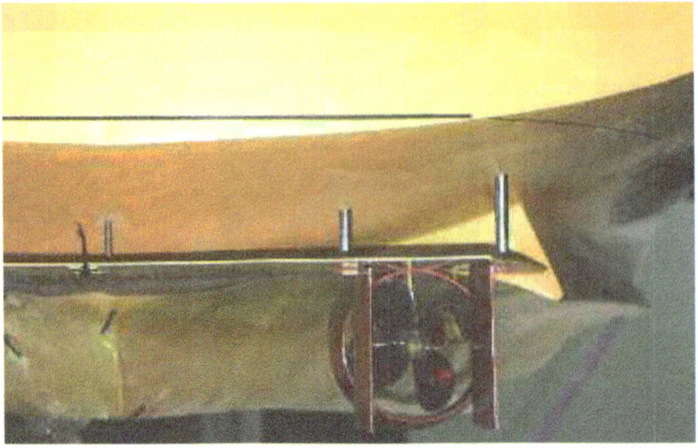

Fig. 9.49 Flow Extending Plate applied to a twin-propeller container ship. *Source* Zöllner (2009)

9.5 Innovations in Ship Structure

As illustrated in Fig. 9.1, ship lightweight and payload efficiency have multiple impacts on its overall efficiency. Lightweight can be reduced by choice of materials and optimization of hull structure scantlings. However, it should be noted that the decrease of lightweight does not necessarily lead to a decrease in ship draught and resistance. Shipowners are almost always tempted to counter-ballast lightweight savings by an increase of cargo weight. Note that the payload efficiency can also be improved by innovative design that leads to better utilization of cargo space, particularly in the case of container ships.

9.5.1 Reduction of Ship Weight

Low-speed IW vessels are almost exclusively made of mild shipbuilding steel due to its relatively low cost and high durability compared to other materials, especially having in mind that the exploitation period of these vessels is 50 years, often more (see Sect. 1.1.3 ad u and v). Therefore, the hull weight has not changed much over time. However, there are some weight reduction potentials, although most of them require a significant cost increase.

Use of High Tensile Steel

High Tensile Steel (HTS) is generally used in deck and hatch coaming structures areas where large stresses need to be endured. It is used instead of standard mild shipbuilding steel in order to achieve the same strength with lower thickness. However, consequent weight savings are far from sensational. For instance, considering the seagoing ships, a hull made of 10% of HTS can reduce the steel weight up to around 2%, while 60% HTS is expected to provide roughly 10% of hull weight savings, according to ABS (2014). For inland vessels, total savings would be even less. Moreover, excessive use of HTS and respective scantlings decrease, raises problems in the context of fatigue, buckling and corrosion issues.

Probably relatively large (for the Danube) Serbian selfpropelled bulkcarriers Sava Mala and Dorcol ($L_{OA} = 96.5$ m, $B = 13.8$ m, $H = 3.4$ m, $T = 2.9$ m, $m_{DWT} = 2600$ t) should be mentioned here. To increase the capacity, HTS was used for coamings and gangways, thus the vessels are elastic with unusually large sagging up to 25 cm.

Other Than Steel Materials

Alternative "other than steel" shipbuilding materials include *aluminium alloys* and *composites.* As a choice for hull structure, these are generally used where the speed is the most significant parameter (smaller high-speed units). Nevertheless, inland cargo vessels could use aluminium alloys and composite materials at least for the part of the hull (bridge, superstructure). So far, this showed not to be significant in the reduction of total hull weight.

Principally, "other than steel" materials can be used but the equivalent strength has to be proved as in the case of steel structure. Regarding the inland vessels, any lightweight material which is not standard would be considered on a case by case basis and should comply with general rules for materials (BV 2019a). The same "equivalent strength" and "case by case" principles are directed by DNV-GL (2015a). One should note that most of the classification societies provide guidance notes or recommended practices in aim to help designer when dealing with uncommon materials and "not-so-proven" technologies which are not considered in fully developed rules.

When it comes to material solutions, some other innovations have been developed over recent years, besides the use of aluminium and composites. Those lightweight materials remain mostly applicable to elements of the superstructure and do not provide a significant reduction of the total weight. Here, more attention will be made to the *Sandwich Plate System (SPS).*

Sandwich Plate System (SPS)

SPS seems to be able to replace the traditional steel plate with its secondary stiffeners. It consists of two plates with an elastomer injected in between to form a solid unit (see Fig. 9.50). The scantlings of SPS plating are generally in the range of 3 mm to 8 mm for steel face plates, and 15–50 mm for core thickness. The new structure is simpler (no secondary members), and hence, could provide some weight reduction.

Immediately after the development of SPS, scantlings could be determined only by direct calculations. In 2006, Lloyd's Register revealed "Provisional Rules for the Application of Sandwich Panel Construction to Ship Structure" (LR, 2006). These rules cover the construction procedures, scantling determination and framing arrangements. The rules are, in general, applicable to mono-hull ships of conventional forms, speed and proportions. Application of SPS in any area that is not specified by the rules requires special consideration by LR. More recently, in 2019, LR published

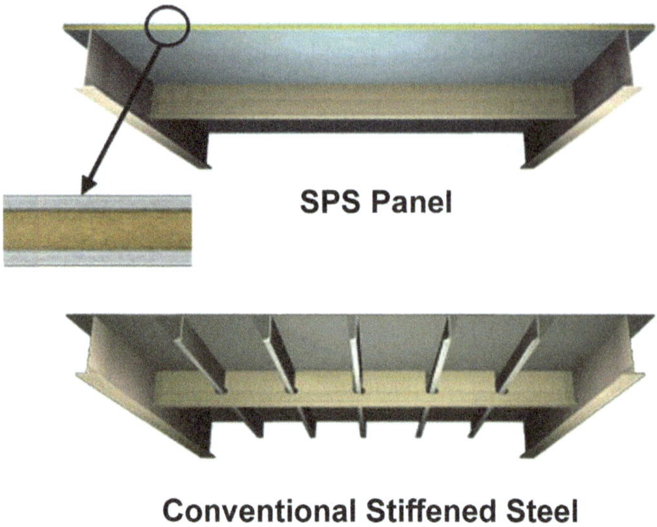

Fig. 9.50 SPS versus conventional structure. *Source* SPS Technology (2020)

"Rules for the Application of Sandwich Panel Construction to Ship Structure" (LR 2019), where construction procedures are presented, as well as the design basis for scantling determination. Moreover, BV (2018) presented general guidance notes (for all ships) for the construction and design principles for hull and superstructure made fully or partially of composite materials, in which the sandwich plates exist as one of the composite solutions.

Numerous projects were conducted with the purpose to investigate the weight saving possibilities of sandwich plates. According to INBAT project, a structural weight savings of around 40% were reported (see Guesnet et al. 2004 and Jastrzebski et al. 2003) if steel sandwich panels (labelled as I-core^R type) would be used for a small barge of 32.5 m (see Fig. 9.51). It is stated that the use of I-core^R panels simplifies barge production as well as maintenance; (note that the I-core^R panel is somewhat different than the SPS). Other projects in which various kinds of SPS were investigated are for instance DE-LIGHT (EC 2010) and LASS (Hertzberg 2009). Recent studies, e.g. Ramakrishnan and Kumar (2016), are addressed to seagoing ships and their repair, emphasising other SPS features, such as corrosion resistance, simplicity, blast resistance, etc.

A Case Study of Application of SPS for Danube Barge Hull Structure

As the abovementioned savings of 40% appear attractive and keeping in mind that LR Provisional Rules (LR 2006) were the only existing at the time of the examinations, custom calculations were performed on the application of an SPS construction for a typical general cargo Danube barge ($L = 77$ m, $B = 11$ m, $T = 2.8$ m, see Fig. 9.52).

The overall philosophy of the rules, as mentioned, is to ensure that designs utilising steel sandwich construction is equivalent in strength and safety to conventional

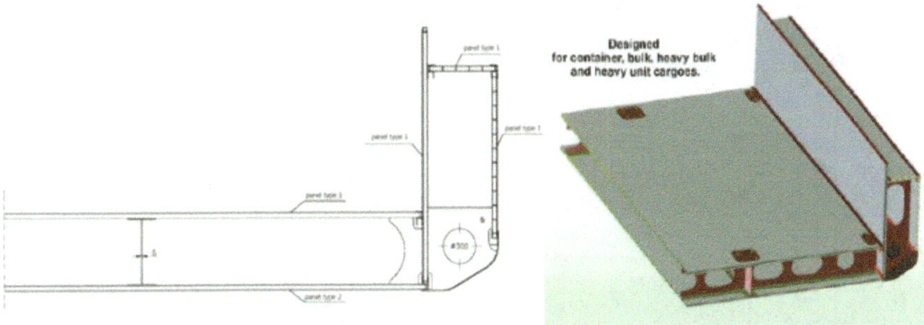

Fig. 9.51 Typical frame cross-section of an I-core^R panel barge. *Source* Guesnet et al. (2004)

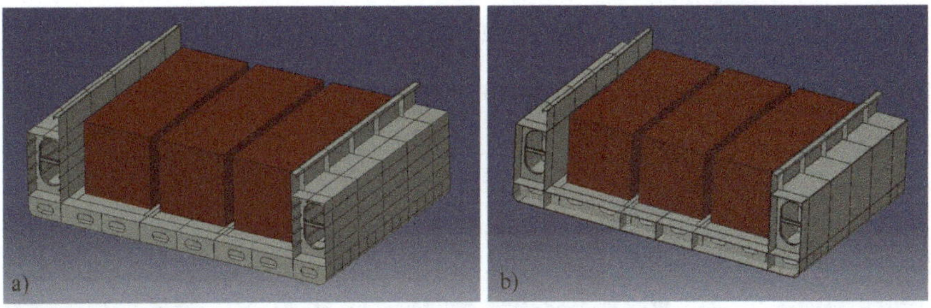

Fig. 9.52 Cross-section of conventionally structured, (**a**) and innovative SPS container barge (**b**). *Source* Radojčić (2009)

steel construction. The thickness of the top and bottom plates and core of the SPS are determined on basis of the scantlings given for the equivalent ordinary steel construction. The assumed scantlings of SPS construction are checked for strength by the formula given in the rules. If the strength is not satisfied, the chosen thickness has to be increased. The process is iterative. Welding is conducted via the perimeter bar, see connection details in Fig. 9.53.

Moreover, weight comparison for general cargo Danube barge ($L = 77$ m, $B = 11$ m, $T = 2.8$ m) was made for the following concepts:

- *existing steel barge,* conventionally-built (mixed framed), built 30 years ago according to Yugoslav Register of Shipping Rules,
- *mixed framing system, steel barge* (calculated according to LR), Fig. 9.54a,
- *longitudinal framing system, steel barge* (calculated according to LR), Fig. 9.54b,
- *SPS structure* (calculated according to the Provisional LR Rules*)*, Fig. 9.55.

A longitudinal framing system is the type of ship structure in which all secondary structure stiffeners are set up longitudinally, while in a transversally framing system all secondary structure stiffeners are positioned transversally. A mixed framing system denotes here double bottom which is to be transversally framed, while the double side and deck are longitudinally framed. The existing barge is built with

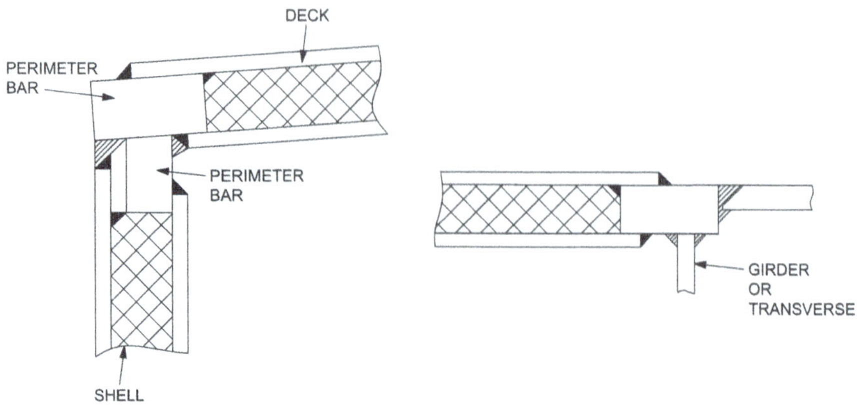

Fig. 9.53 SPS structural details. *Source* LR (2006)

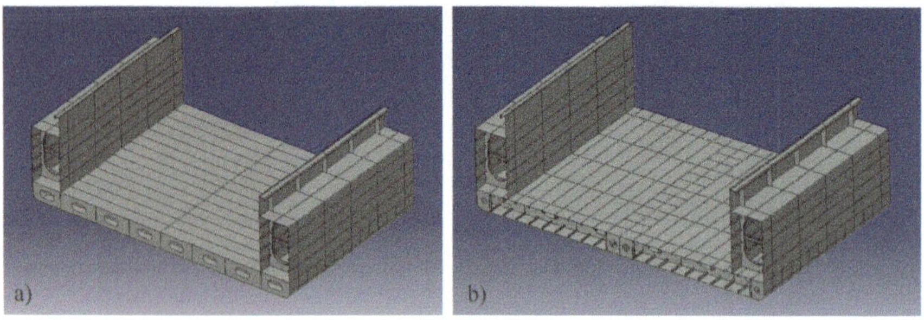

Fig. 9.54 Mixed, **a** and longitudinally, **b** framed barge. *Source* Radojčić (2009)

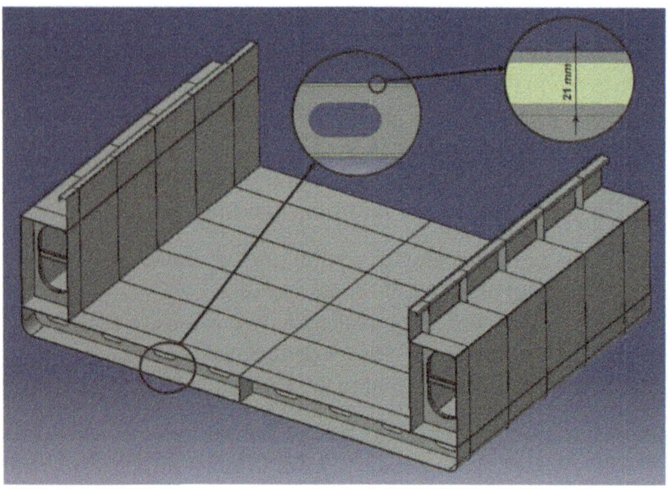

Fig. 9.55 SPS barge. *Source* Radojčić (2009)

a mixed framing system, which is nowadays typically applied for inland barge structures. Considering the local strength requirements, stern and bow structures are assumed to be conventionally built and therefore remain the same in all cases. Weight comparison for all four considered cases is shown in Fig. 9.56.

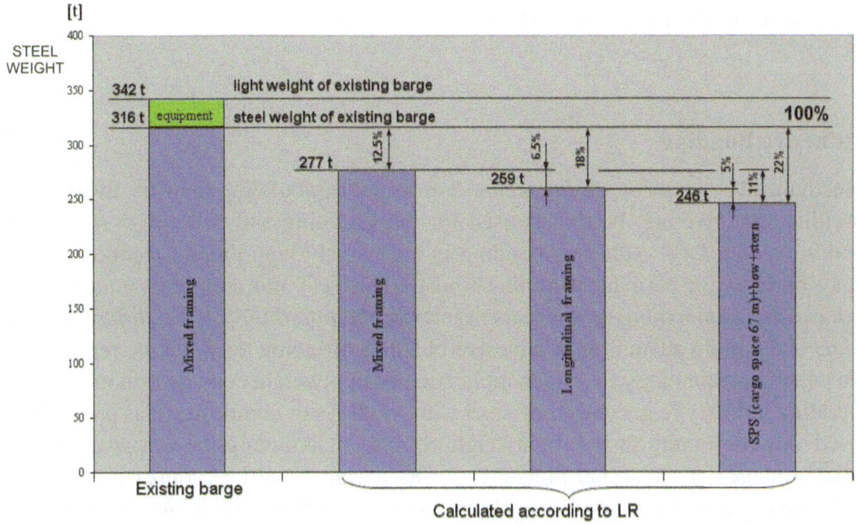

Fig. 9.56 Steel weight comparisons. *Source* Radojčić (2009)

Due to the various exploitation reasons, the hull structure scantlings of the calculated barges are often increased, allowing them to meet strength requirements over a prolonged period (more than 50 years). Consequently, the existing barge is heavier by around 12.5% than an equivalently framed calculated barge. The innovative SPS structure would be lighter by 22%, 11% and 5% respectively, compared to the existing, calculated mixed and calculated longitudinal framed barge respectively. It is worth mentioning that container and bulk-cargo barges (that were also analysed, besides this general cargo barge) are heavier by around 10% than the general cargo barges and that a savings due to the application of SPS construction would be smaller—around 2 ÷ 5% only.

Nevertheless, SPS construction may have some other advantages for specialized vessels (besides just weight savings), e.g. cheaper production and additional safety. For instance, the inner skin of an IW chemical tanker inner plate could be built of stainless steel and the outer plate of conventional steel, using only 50% of the expensive stainless steel in current designs. More on SPS use in IW vessels structures can be found in (Momčilović and Motok 2009).

Summarising, the innovative SPS barge can be lighter by only up to 10% than a conventional (calculated) barge, which is in the same range as just transferring the structure from the mixed to longitudinally framed (see below). Consequently, it seems that weight savings should first be examined within the conventional steel construction approach, and afterwards through the innovative approaches, like SPS.

Adhesive Bonding

Recognized *permanent joining technology* in the shipbuilding industry is primarily welding and riveting; bolting is used for the fastening and belongs to *temporary joining technology*. Adhesive bonding is a relatively new joining technology and may be used for permanent joining (most adhesives) and temporary joining (few adhesives). Concerning the permanent joining, the *possibility to join different materials* is the main advantage of adhesive bonding. Bonding thus may be regarded as an enabling technology for wider application of lightweight construction in the shipbuilding industry (e.g. combining steel with Al alloys or composites), as presently is used only for joining various lightweight equipment. Amongst the adhesive bonding disadvantages are heat and moisture sensitivity, approval procedures by the classification societies etc. For more about the use of adhesive bonding in the marine industry see for instance Weitzenböck (2012).

Optimization of Framing System of the Danube Barge

The diagram in Fig. 9.56 reveals some weight reduction possibilities other than the change of material. Although they are not a structural innovation in any sense, it is obvious that longitudinally framed vessels offer a lighter structure. Moreover, a change from existing mixed to longitudinally framed structure (as calculated) would reduce the steel weight almost in the same amount as the application of SPS—around 18%. Moreover, the change from mixed to longitudinal framing would decrease the weight up to 6.5%. However, one should bear in mind that most of the Danube vessels have an inner bottom thickness greater than prescribed by the rules for the sake of prolonged life (see Sect. 10.3.2). That explains the discrepancy between the existing and calculated weight of the barge.

Barges are usually below 100 m in length and therefore, mostly mixed or transversely framed. For more lengthy vessels, designers do not have a choice—due to longitudinal strength issues these vessels have to be longitudinally framed.

Optimization by Means of Direct Calculations

Most of the classification societies require direct longitudinal strength calculations and following buckling assessments for the vessels having a length over 60 m, and especially if the length to height ratio exceeds 35. More importantly, rules allow reduction of hull structure scantlings prescribed by their deterministic formulae if direct calculations can prove that strength of the structure is satisfied. This provides designers to have a greater role in weight optimization. However, direct calculations can "work in both ways", i.e. they can also reveal the need for additional strengthening. More on this particular topic can be found in Sect. 10.3.2.

The *Finite Element Method (FEM)* analysis is nowadays the standard tool for structural design analyses. Guidance for the FEM model development are incorporated within the classification societies' rules. They provide a set of specifications for making a respectable 3D model of the hull structure. For instance:

- Detailed recommendations, requirements, and general guidance for all ships, are provided in (DNV-GL 2015b).
- General FEM requirements and guidance for IW vessels are given in Rules for Inland Navigation Vessels, BV (2019b).
- BV rules covering the application of FEM are also associated with the use of *Veristar*—a company developed software for various types of finite element analyses (including fatigue, buckling etc.).

Similar guidances for FEM assessment procedures are provided by other classification societies.

Just as an illustration, in the following is presented FEM analyses required and performed for one unconventional (beamy) IW vessel in Stefanović (2019). The global model of complete hull structure is generated using standard mesh. Von Mises stress fields, as a consequence of fully loaded vessel condition, are given in Fig. 9.57.

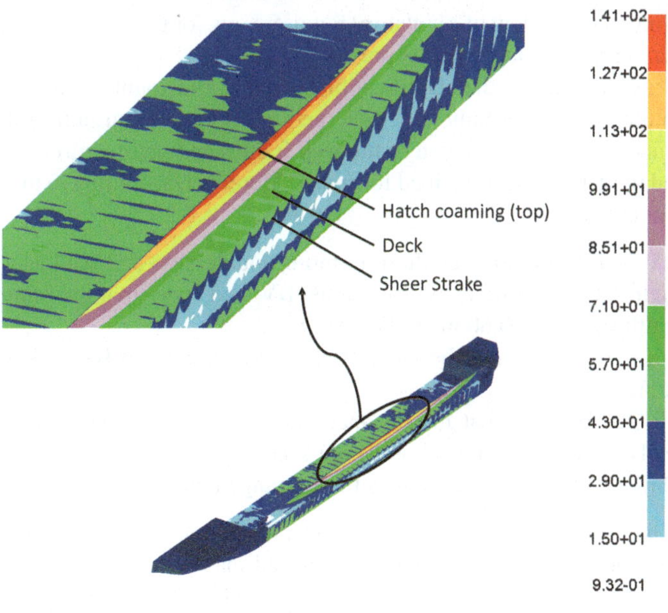

Fig. 9.57 FEM analyses of one unconventional IW vessel: Von Mises stress, in N/mm^2. *Source* Stefanović (2019)

High-stress zones are recognized in the sheer strake, deck and hatch coaming structure. Still, they are below the allowable 192 N/mm^2 specified by the rules (GL 2011)—suggesting that the particular "unconventional" vessel design passed this additional strength check.

9.5.2 Increase of Payload Efficiency by Enlarging the Inner Hull Volume

Besides the lightweight solutions, the payload can also be increased by enlarging the capacity of the cargo hold. That becomes important particularly in the case of container vessels.

Double Side Width

The Danube waterway constraints (see Sect. 2.4) govern the width of the vessels, and therefore, their cargo capacity. Due to the Upper Danube locks limits, the maximum width of the vessels can be 11.45 m. Therefore, the potential to increase the width of the cargo hold is driven by the decrease of the double side width (see Sect. 3.2 and Fig. 3.3). The minimum width of the double side is an issue that could be investigated from the damage stability and structural (crashworthiness, torsion, etc.) analyses. However, the rules and statutory limitations are quite deterministic. Most important are the ADN requirements (for more on the regulatory framework, see Chap. 5). In the following, an overview of regulatory requirements for the double-side minimum width of dry cargo inland vessels is presented.

Regarding the rules of the classification societies, the double side width and the double bottom height are indirectly governed by the minimum hull girder section modulus, and cross-sectional area of the web frames and bottom girders, which are determined upon the rule prescribed formulae. Moreover, according to the following rules for inland vessels:

- BV (BV 2019b) determines the minimum double side and walkway width as 0.60 m, to facilitate safe crew operations. BV also states that the minimum side tank width should be 0.60 m, at its lowermost.
- DNV-GL (2015c) requires the same width of stringer plate as BV, and thus, directs the double side space.
- LR (LR 2020) states that the distance between the side shell and cargo hold bulkhead (inner side plating) should not be less than 0.80 m. However, a reduction to 0.60 m is allowed if specific structural arrangements are provided.

Above mentioned arrangements are achievable and are not disturbing the vessel design since they are mostly related to the required spacing of the structural elements. Requirements in LR are derived from ADN (UNECE 2019), which relates to the carriage of dangerous goods. Nevertheless, one should not disregard ES-TRIN

(CESNI 2019) regulations, where the minimum distance between the outer edge of the hull and the inner edge of the hold should not be less than 0.65 m.

Therefore, when it comes to the cargo space increase by reduction of the double side width, according to rules for IW dry cargo vessels: *minimum double side width is constrained to 0.60 ÷ 0.65 m*. ADN rules govern the same constraint for container vessels if they are intended to carry dangerous goods. Note that the share of dangerous goods in overall container transport is significant.

Double Bottom Height

Rules of classification societies usually guide the double bottom height by the section modulus of double bottom structural members: floors and girders. ADN also states it should not be less than 0.50, or 0.60 m if intended to be used as a fuel oil tank.

Single Side Hull

All abovementioned is related to the double-hull vessels, while the largest payload increase can be obtained with a single side hull. Classification societies provide rules for single hull dry cargo vessels but those vessels are of significantly lower safety (lower strength and damage stability). What's more important they do not satisfy the ADN rules for the carriage of dangerous goods, which considerably lowers their all life payload efficiency.

Payload Increase by Unconventional Cargo Hold Design

Various projects, amongst other issues, offer to increase the payload by cargo hold design improvements. Some of them are briefly described in the following.

The ECCONET project (ECCONET 2012), although it was not its prime objective, also discussed the Rhine vessel (11.45 m wide) weight saving potential based on the slight reduction of the double-side width where only one side would be used as a gangway. It is estimated that such a design would deliver 60 t of weight savings as the steel structure would be lighter. Similar investigations, along with some other measures, may lead to weight savings in the case of the Danube vessels, but lower than in the case of Rhine vessels, i.e. the Danube vessels offer lower weight saving potential due to their smaller lengths and draughts compared to Rhine vessels. According to ECCONET, these measures include: HTS instead of mild steel (where possible), reduced stiffener spacings to ensure the lower thickness of the plating, sandwich plates and innovative cargo hold considerations (as mentioned above). Considering the abovementioned measures, the weight saving potential for a typical European dry bulk vessel is presented in Table 9.3.

NEWS project (NEWS 2014) offered a solution with one gangway positioned in the deck centerline and therefore splitting the cargo hold in two to accommodate more payload. However, this raises problems in satisfying the ADN rules for the carriage of dangerous goods (see above Single Side Hull).

NOVIMAR project (NOVIMAR 2017) discusses the possibility of stowing all containers on the deck and placing only one gangway in the centerline. Apart from enlarged deck space for placing containers, loading and unloading is much easier. Namely, within the project a special vehicle for horizontal loading/unloading is developed; it needs the centerline (rather than side) platform for its manoeuvring.

Table 9.3 Typical European vessels weight savings and potential increase of carrying capacity—shaded cells. *Source* ECCONET (2012)

	Max. savings [t]	Unit immersion [t/cm]	Draught decrease [cm]							
			5	6	7	8	9	10	15	
Gustav Koenigs extended	45	6.1	31	37	43	45	-	-	-	Corresponding potential increase of carrying capacity [t]
Johann Welker extended	55	7.5	38	45	53	55	-	-	-	
GMS 110	90	11.1	56	67	78	89	90	-	-	
GMS 135	150	14.3	72	86	100	115	129	143	150	
JOWI	200	21.1	105	127	148	169	190	200	-	
Europe II barge	70	8.0	40	48	56	64	70	-	-	
Danube barge	40	8.0	40	-	-	-	-	-	-	

9.6 Innovations Important for Better Ship Utilization

Innovations that can be used for improved ship utilization and navigation are a result of global technological development and therefore related to various aspects of digitalization and automatization of the vessel itself and its navigation (e.g., see SAFETY4SEA 2019). These innovations are not necessarily developed for the shipping industry but can be applied in it. Their core is the application and integration of communication technologies between all parties involved in navigation (vessels and shore authorities). Accordingly, the EU Commission is developing a regulatory framework in order to facilitate the introduction of digital technologies in inland navigation (see EU Commission 2017).

A side effect of the abovementioned modernization is that the vessels and navigation are becoming vulnerable to cyber-attacks, thus cybersecurity is introduced as a new task in the digital domain. The classification societies have already issued cybersecurity regulations in the maritime industry and are in process of developing the same for IW navigation.

In the following, innovations that have been already applied in IW shipping are highlighted in more details (River Information Services in Sect. 9.6.1) or more deeply investigated in related research projects (Smart and Energy-Efficient Navigation projects in Sect. 9.6.2 and Autonomous Ships in Sect. 9.6.3).

9.6.1 River Information Services

RIS[19] provide integrated and easy to use electronic data for better utilization of the fairway, voyage planning, optimization of the schedule and speed, and therefore improved safety, efficiency and reliability of IW navigation. Its application leads to a reduction of fuel consumption, improved environmental friendliness and rises IWT competitiveness compared to other transport modes. RIS includes various technologies and harmonizes their application, especially through the guidelines and standards issued by CCNR. According to "RIS Directive", see EU Commission (2005), the implementation of RIS is standardized through these four technologies:

- Inland Electronic Chart Display and Information System (Inland ECDIS)
- Notices to Skippers (NtS)
- Inland AIS
- Electronic Reporting International (ERI).

Inland ECDIS (Electronic Chart Display and Information System)

Inland ECDIS is the chart of geographical data gathered from Inland Electronic Navigational Charts (IENC) and can be called the "electronic atlas" of the waterway (see Fig. 9.58). Data include water depth, shore geography, structures and obstacles (bridges), other hydrographic data, availability of navigational aids, etc. and replaces classical nautical charts. This allows choosing the best route and navigation with an optimum speed. Consequently, it leads to better voyage planning and increased safety and efficiency of the navigation. Inland ECDIS has been mainly developed within the German project ARGO (ARGO 2000) some 20 years ago, and later adopted by CCNR. ECDIS is a base technology of RIS. It is used along with inland AIS and radar.

NtS (Notice to Skippers)

NtS is a set of information prepared for skippers that mainly include: vessel identification, fairway and traffic-related messages, ice data, and weather report. All these data are part of the so-called Fairway Information Services (FIS).

Inland AIS (Inland Automatic Identification System)

Inland AIS is an automatic tracking system that provides identification, position, course, speed and additional specific data of the vessel (a transponder is installed onboard to broadcast this information). Inland AIS is used by waterway authorities to locate and track the vessels and by vessels themselves to track the other vessels. It also enables calculation of the Estimated Time of Arrival (ETA) and lock management. AIS can be used along with ECDIS to provide better tracking and updating of the system, therefore enhancing safety.

[19]The main source of information in this section is RIS (2021).

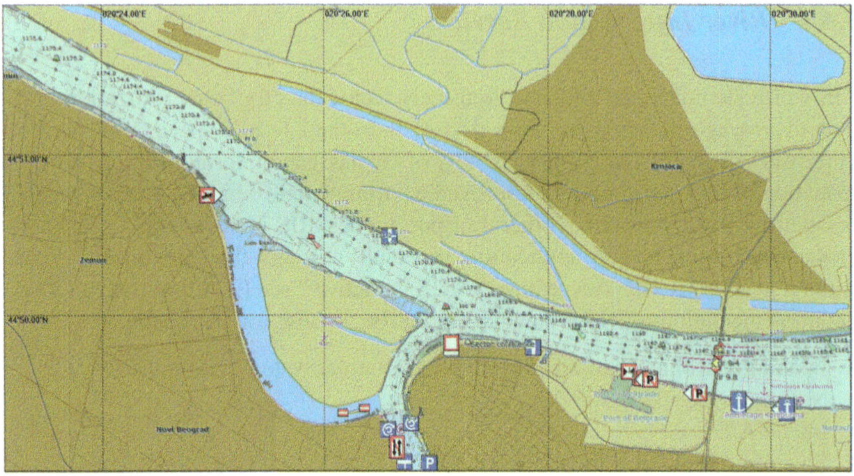

Fig. 9.58 ECDIS display, Danube km 1166 ÷ 1175. *Source* Plovput (2021)

ERI (Electronic Reporting International)

ERI is a feature used in the RIS system for strategic traffic information (e.g. transport of dangerous goods), traffic management and accident preventions. Electronic reporting can replace paper and voice interaction between the vessel and shore centers. Besides, the ERI is also used with the vessel's partners in the multimodal transport chain.

Generally, all mentioned technologies are based on the digitalization of the vessel‘s navigation data. However, they do not exclude the skipper‘s visual monitoring. They are helping the "traditional" communication aids, which still remain in use: radar and radiotelephone. Therefore, radar and radiotelephone functions are integrated into RIS.

RIS standards and the abovementioned technologies are adopted by the CCNR and EU Commission, which propelled their use in inland navigation. Since 2005, the RIS technical standards development is under the regulations of the EU Commission. Consequently, RIS has also been recognized and established by countries outside the EU (see Sect. 5.3). Guidelines and recommendations for the use of RIS are reviewed in Braunroth and Haupt (2018).

To summarize, the main output of the RIS are information such as geographical and hydrological data on waterways, weather conditions, vessels position, data related to the accidents, statistical data on navigation etc. These data help skipper decisions about the vessel's route, planning and timing of the operations (in the short and long term), and are heavily used in Energy-Efficient Navigation (EEN) policy (see Sect. 9.6.2). They can allow adjusting the speed and estimation of the time of arrival, which in overall, ensures safety and increases the vessel's efficiency.

9.6.2 Smart and Energy-Efficient Navigation (EEN)

The onboard computerization and RIS application, in addition to crew training, can *lead to Energy Efficient Navigation (EEN)*, i.e., optimal navigation concerning waterway configurations, engine performance, lock availability etc. For instance, according to Dutch Programme "Saving While Sailing" (part of a Dutch Air Quality Action Plan, see Gille and de Vries (2011)) initiated to reduce fuel consumption and pollutant emissions from IW vessels, fuel savings of $5 \div 10\%$ may be obtained. Some other studies reported reductions in air emissions of up to 15% from ships, just by using available technology on today's ships without incurring additional costs! According to Friedhoff et al. (2018), sailing *with* and *with no* EEN policy can lead to a fuel consumption difference of around $4 \div 14\%$. An example of EEN optimization is presented in Suvačarov (2019), where depending on the river sector and chosen optimization criterion (minimization of environmental impact, maximization of the profit rate, etc.) an optimal speed is evaluated by the application of Artificial Neural Network (ANN).

Note that an increase of efficiency and emission reduction can be achieved by *carefully analyzing and optimizing all aspects of vessel navigation*, such as engine performance, trim for all draughts and speeds, the propulsion system efficiency etc., and by *improving the voyage management and crew skills.*

In Table 9.4 given is a short list of notable projects, studies, training programs, tools, and commercial services related to EEN. The table contains basic information on their objectives, results and/or features. For more details, the reader is addressed to web sites given in the references.

9.6.3 Autonomous Ships

One of smart navigation concept's goals (see also Sect. 10.3.1) is to pursue autonomous shipping, i.e. introduce vessels that are partly or fully remotely controlled. Such vessels can sail with reduced or no crew at all, which can significantly reduce shipping costs. All such concepts are leaning on modern technologies for communication between the vessels themselves and between the vessels and shore authorities, which also leads to safety improvements.

Autonomous vehicles are not a new concept, as this technology is already state-of-the-art in some land-based transport modes. However, the development of an autonomous cargo ship may be considered as an innovation. Most of the current developments in this field are done within large European research projects and these will be highlighted in the following. They all, generally, include feasibility studies, investigations of challenges in safety and socio-economic impacts, identification of major threats and development of new technical standards. According to Bačkalov (2020), a review of the technical standards for inland vessels, contained in ES-TRIN and ADN regulations (see Chap. 5), revealed that there is a considerable number of

Table 9.4 Projects related to smart and energy–efficient navigation

Project name, framework, dates, ref.	Objectives	Results/Conclusions/Specific outcomes
VOORTVAREND BESPAREN Dutch Gov. program, from 2011 managed by (EICB) (VVB, 2011)	Stimulation of inland skippers and shipping companies to adopt efficient sailing behaviour.	E-learning courses and use of smartphone application for monitoring fuel consumption.
MUNIN EU: FP7, 2012 - 2015 (MUNIN, 2012)	Use of smart technologies for autonomous shipping.	Developed own EEN
PROMINENT EU: HORIZON 2020 2015 - 2018 (PROMINENT, 2015)	- Reduction of energy consumption by adaptation of speed (power) to the waterway. - Preparing technologies and concepts applicable to 70% of the fleet with a 30% reduction of implementation costs. - Certification and monitoring of emission performance.	The greatest impact on the reduction of fuel consumption can be achieved by combining: - speed (power) adaptation to the fairway - choice of optimum sailing track - provision of needed information to skipper.
SYNCHRO-NET EU: HORIZON 2020 2015 - 2018 (SYNCHRO-NET, 2015)	- Optimization of ship operations and integrated hinterland logistics. - Reduction of fuel usage by smart steaming approach.	- Vessel speed is dynamically optimized based on the real-time data on sea state, weather and the destination port. - Reported savings in fuel usage are up to 30%.
VIRTUAL REALITY TOOLS LR 2017 (LR, 2017)	Development of smart systems and tools and their application into the EEN policy in order to optimize navigation efficiency.	- Digital replica/twin of a wheelhouse or engine room used for crew training. - Virtual reality devices for staff to monitor their health parameters.
GRENDEL EU: Danube Transnational Programme (DTP) 2018 - 2020 (GRENDEL, 2018)	- Modernization of the Danube fleet: Strategy and Recommendations. - Development of innovative vessel concepts; improvement of logistics. - Regulatory framework and well-designed public support measures.	- Increase of sailing time by 10% may lead to fuel savings of 30%. - Improvements in logistics chains are important - Boatmasters skills are important - Ship operator can improve sailing behaviour.
COVADEM Dutch Gov. initiative 2019 – (COVADEM, 2019)	- Safe and sustainable navigation with maximum efficiency. - Development of smart technologies for real-time data collection that can be used by skippers as an insight for navigation optimization.	Commercial real-time service platform that "processes data in the cloud"; accessible by the user-friendly onboard system.
TOPOFAHRT DST EU-NRW Ziel-2 (EFRE) (TOPOFAHRT, 2021)	Promotion of traffic safety and efficiency, reduction of fuel consumption by changing skippers' driving style.	Development of training courses and manuals.

design and construction requirements affecting safety that ought to be adapted to the concept of autonomous inland navigation in order to properly take into account safety challenges. Nonetheless, IMO is currently investigating how the existing regulations relating to ships with a certain level of automation. In that regard, the Maritime Safety Committee in 2019 approved Interim guidelines for Maritime Autonomous Surface Ships trials. Moreover, considering inland navigation, CCNR (2018) defined the following levels of automation:

- No automation (the vessel is full time governed by the human);
- Steering assistance (steering automation system is employed);
- Partial automation (both steering and propulsion systems have a level of automation assistance);
- Conditional automation (automation system is used in almost all dynamic navigational tasks, including collision avoidance while expecting that human would intervene in case of failures);

- High automation (somewhat similar as above but without expecting the human to intervene);
- Full automation (unconditional performance by navigation automation system without expecting human intervention).

It should be noted that the issue of crew reduction by the increase of automatization appears not to be so straightforward. For example, a study performed on a 750 TEU container vessel (Kooj and Hekkenberg 2020) showed that the increase of automation can decrease the required number of the crew only in certain phases of the navigation. On the other hand, another interesting consequence of increased automatization examined in the same paper is the potential reduction of the workload of the crew members. The benefit of that could be a reduction of crew fatigue and therefore more safety in navigation. This topic is directly connected to the EEN policy as the crew's performance directly influences optimal navigation of the vessel (see EEN in Sect. 9.6.2).

Notable projects, not necessarily aimed to IW but dealing with technologies that can be applied to enhance autonomous shipping generally, are presented in Table 9.5. Although some of these vessel concepts are also mentioned in Sect. 10.1.3, their key innovations are given here as well. A similar, but an extended list of pilot and research projects in the field of automation in inland navigation is presented in CCNR (2020b).

Table 9.5 Notable projects related to autonomous ships

Project name, framework, dates, ref.	Objectives	Results/Conclusions/Specific outcomes
MUNIN EU: FP7 2012 – 2015 (MUNIN, 2012)	- Development and verification of an autonomous ship concept. - See Table 9.4 - MUNIN project.	- Case study performed on a cargo ship for deep-sea voyage. - See Table 9.4 - MUNIN project.
WATERTRUCK+ EU + Group De Cloedt 2014 – 2019 – 2020 (WATERTRUCK, 2014)	Development of a system that can monitor vessels sailing in convoys, or ones decoupled for sailing in small waterways	Experiment with unmanned barges controlled from the shore.
AUTONOMOUS SHIPPING Dutch project 2017 – 2019 (AUTONOMOUS SHIPPING, 2017)	Development of autonomous shipping in order to reduce operating costs.	Trails performed on 26m fast crew supplier vessel.
NOVIMAR EU: HORIZON 2020 2017 – 2021 (NOVIMAR, 2017)	Development of a "vessel train" concept – a fleet of several digitally connected IW vessels.	Vessel train concept is in the process of development.
AUTOSHIP EU: HORIZON 2020 2019 – 2022 (AUTOSHIP, 2019)	- Reduction of costs. - Development of autonomous vessels for short-sea shipping and inland waterways.	Currently, an analysis is performed regarding the use of novel technologies such as "artificial captain and chief engineer".

9.7 Concluding Remarks

Possible IW vessel greening technologies are summarized in Tables 9.6 and 9.7. Namely, Table 9.6 refers to prime movers and fuels (treated in Sect. 9.2) and is *focused on the emissions* (as the engines are *direct emitters* of GHG and air pollutants, the first abridged by the quantity of CO_2 and the second by NO_x and PM emissions). If the efficiency is represented with *EEDI* (see Sect. 6.1) then it is directly correlated to the quantity of emitted CO_2. All other greening technologies (Sects. 9.3–9.6) directly influence the power needed for moving a certain quantity of cargo at a certain speed, thus *indirectly influence emissions*.

Table 9.6 Greening measures—propulsion plants and fuels (see Sect. 9.2)

MEASURE	REDUCTION POTENTIAL compared to CCNR II with fossil Diesel			FEASIBILITY to be applied	MATURITY of technology	COMMENTS/REMARKS
	CO_{2e}	NO_x	PM			
SCR	0-4%	75%	0%	O – 4 N – 5	4	-
DPF	0%	0%	85%	O – 4 N – 5	4	-
SCR + DPF	0-4%	75%	85%	O – 4 N – 5	4	For St. V or equivalent emission performance
HVO	≤100%	≤13%	≤60%	O – 4 N – 4	5	Currently no supplier active in the Danube region
LNG	10%	75%	100%	O – 2 N – 3	4	CO_{2e} including moderate methane slip. Gas engines have to fulfil the St. V limits.
LBM or LMG from PTL	100%	75%	100%	O – 2 N – 2	3	Negligible availability today
Methanol ICE	100%	75%	85%	O – 1 N – 2	3	For green Methanol
Methanol FC	100%	100%	100%	O – 1 N – 2	2	For green Methanol
Hydrogen ICE	100%	75%	100%	O – 1 N – 2	2	For green Hydrogen
Hydrogen FC	100%	100%	100%	O – 1 N – 2	2	For green Hydrogen
Batteries	100%	100%	100%	O – 2 N – 3	3	-

Legend: O - Old vessels (retrofitting); N - New vessels
Grading: 1-low, 5-high

The increase of vessels efficiency may be represented through the *savings* (i.e. percentage reductions of power or fuel consumption as a result of applied greening measure), see Table 9.7.[20] The energy efficiency measures discussed in Sects. 9.2–9.5 are often referred to as *technical*, while the ones from Sect. 9.6 belong to a group of *operational* measures (the third group are *logistical* measures, but these are out of the book's scope).[21] Therefore, discussed measures for efficiency improvement are far from being complete.

[20]Table 9.7 contains *air hull lubrication* which as a greening measure reduces energy consumption, although that topic is treated in Sect. 10.1.3.

[21]These three are the transport issues. Greening technologies may also involve the infrastructure and education, amongst other.

Note, however, that discussed energy efficiency measures (summarized in Fig. 9.1 and Tables 9.6 and 9.7) are *not suitable for every ship; the individual case must always be considered.*

Greening Measures

Most greening measures are associated with significant investments. The complexity of the systems rises, which also increases maintenance costs and requires additional qualification of the crew. The energy density of alternative energy carriers (fuels) is usually lower compared to diesel and requires more space and/or more frequent bunkering. Bunkering infrastructure for alternative energy carriers hardly exists. Future developments of costs for energy carriers and technologies, as well as the development of infrastructure, are extremely uncertain. Today most emission reduction technologies increase operating costs. No Return On Investment (ROI) can be achieved compared to the current cost structure of conventional diesel drives. For most zero-emission technologies the maturity of the technologies themselves can possibly also lead to a barrier as they are not yet broadly in use. Last but not least, the production of green fuels requires sufficient resources of sustainable feedstocks and/or large amounts of cheap renewable electricity.

Complete avoidance of local exhaust emissions in ship operation is possible through the use of electric drives with energy supply from batteries and/or fuel cells. While electric drives are very widespread in rail transport with supply via overhead lines or conductor rails, electromobility on the roads has started to gain importance in the last few years. Electric cars and buses with lithium-ion batteries are becoming technically and economically competitive, even if the range is limiting

Table 9.7 Technical measures (excluding propulsion plants and fuels which are given in Table 9.6) for reducing energy consumption (compilation based on GRENDEL, PROMINENT and other sources)

MEASURE (Sect.)	SAVINGS / EFFECTS from the greening viewpoint	FEASIBILITY to be applied	MATURITY of technology	COMMENTS/REMARKS
The Right Choice of Principal Dimensions (9.3.1)	Right sizing (of pushed convoys too) significantly influence economic effectiveness	O[1] – 2 N – 5	5	- Large vessels are economically effective in deep water only. - Formation type of pushed/couple convoys is extremely important
Hull form improvements (9.3.2)	May reduce resistance up to 10%	O – 2 N – 5	5	- CFD and model experiments are recommended - Focus is on stern and prop. tunnel
Air Lubrication (10.1.3)	Up to 20% (extra power is required for air pumping)	O – 2 N – 3	3	- Different, not equally efficient, air lubrication systems exist - May be expensive - Increases draught
Diesel Direct (DD) (9.4.1)	Robust, simple, cheap	N – 4	5	- Rudders & steering gear is necessary - For all kinds of IW vessels
Diesel Electric (DE) (9.4.1)	Up to 10 – 20%[2] (compared to DD) Reduces emissions	O – 2 N – 4	5	- EM inside the hull or in a pod - Mainly for selfpropelled vessels
Hybrid Propulsion (HP) (9.4.1)	Up to 10 – 20%[2] (compared to DD) Reduces emissions	O – 2 N – 4	4	- There are several kinds of HP - May be expensive - Mainly for selfpropelled vessels

(continued)

Table 9.7 (continued)

MEASURE (Sect.)	SAVINGS / EFFECTS from the greening viewpoint	FEASIBILITY to be applied	MATURITY of technology	COMMENTS/REMARKS
Better rudders (9.4.2)	Can increase efficiency for up to 5 %	O – 3 N – 5	5	- Maneuverability is of primary importance. - Flanking rudders increase resistance
Bow thrusters (9.4.2)	Increase efficiency for approx. 5% if replace flanking rudders	O – 3 N – 5	5	- Obligatory on larger selfpropelled vessels, usual on new pushboats (better than flanking rudders)
Ducted Propeller (9.4.3)	10 – 20% and more (compared to the naked propeller)	O – 4 N – 5	5	- For heavily loaded propellers - Nozzle type is important.
CRP (rudder propeller) (9.4.4)	Approximately 10% (compared to naked propeller)	O – 1 N – 4	5	- For lightly loaded propellers (passenger ships)
Tip Driven Prop. (9.4.5)	Not known	O – 1 N – 1	3	- Seems promising - May be used with DE or HP
Side-by-Side Prop. (9.4.5)	Not known	O – 1 N – 1	2	- Seems promising
Pump-Jet (as a propulsor) (9.4.6)	Lower efficiency than of ordinary prop.	O – 2 N – 3	4	- For extremely shallow draught vessels - Very good maneuvering characteristics
Adjustable Tunnel (9.4.8)	Power savings at large draught; up to 10% at low draught.	O – 3 N – 3	4	- For selfpropelled ships (without tunnels) which navigate at reduced draught
Energy Saving Devices (ESD) (9.4.8)	5 – 10%	O – 3 N – 3	3	- Different design for each vessel - Relatively cheap for installation
Reduction of Ship Weight (9.5.1)	5 – 10%	O – 1 N – 3	4	- Some lightweight structures are considered to be frail - Draught reduction or payload increase - May be expensive
Increase of Payload efficiency (9.5.2)	Less than 2 – 3%	O – 1 N – 4	5	- Unconventional design approach often limited by various regulations - Reduces EEDI
RIS (9.6.1)	Influence fuel consumption but quantification is not possible	O – 5 N – 5	5	- Increase safety, efficiency and reliability of IW navigation - Basis for EEN
Smart and Energy Eff. Navigation (EEN) (9.6.2)	Speed reduction reduces fuel consumption for 10 – 20% and more	O – 3 N – 5	4	- Relatively low cost. - Increases cost-effectiveness

Legend: O - Old vessels (retrofitting); N - New vessels; Grading: 1-low, 5-high
[1] Lengthening; [2] Depend on operational profile

Note: Savings expressed in percentages are not cumulative

for many applications. Compared to road vehicles, the limitations in inland navigation in terms of space and mass are lower, but the requirements from the power and especially the usually high energy demand are considerable. Effects such as recuperation or the possibility of short-term inductive charging at stops, which can contribute significantly to economic efficiency, are largely absent in shipping.

Before a mass roll-out of advanced greening technologies can start, these barriers need to be alleviated. International coordination to regulate fleet modernisation, while maintaining a level playing field, has been initiated. With the long life cycles of ships and the pressing time, diesel-electric propulsion systems play an important role in the long-term conversion of the fleet.

Electric motors allow high torque at low rpm and are well suited to the characteristics of the propeller. With appropriate power management, the electric drives are future proof and can be supplied with energy from different sources. It is not expected that in the next ten to twenty years a single energy source and a single energy conversion technology will prevail as a replacement for today's diesel propulsion systems in IW navigation.

However, the emerging trend towards very different systems that are adapted to the respective application also has advantages. Diversification reduces the risk of dependencies and scarcity of raw materials.

References

ABS (2014) Ship energy efficiency measures, status and guidance. Ship energy efficiency measures advisory. https://ww2.eagle.org/content/dam/eagle/advisories-and-debriefs/ABS_Energy_Efficiency_Advisory.pdf. Accessed 19 Dec 2020

ABS (2019) EEDI phase 3 compliance for large bulk carriers and tankers

ARGO (2000) http://www.german-argo.de. Accessed 01 Jan 2021

AUTONOMOUS SHIPPING (2017) First autonomous maneuvering vessel trials held on North Sea. https://autonomousshipping.nl. Accessed 22 Dec 2020

AUTOSHIP (2019) AUTOSHIP—autonomous shipping initiative for european waters. www.autoship-project.eu. Accessed 01 Nov 2020

Bach R (2015) Reduzierung des Dieselverbrauchs und der CO_2-Emission durch Diesel-direkten Hybridantrieb, Hybrideantrieb.pages (bonapart.de). Accessed 24 Feb 2021

Bačkalov I (2020) Safety of autonomous inland vessels: an analysis of regulatory barriers in the present technical standards in Europe. Saf Sci 128

Baroni C, Legittimo CM, Reina C, Zotti I (2016) Recent research development on a bivortix vertical axis propeller. In: 22nd symposium SORTA, Trogir, 2016

Becker Marine Systems (2020) Becker flap rudder. http://www.becker-marine-invest.de/05_downloads/zzpdf/product_pdf/Becker_Flap_Rudder_US_Waterways.pdf. Accessed 21 Dec 2020

Braunroth N, Haupt W (2018) River information services (RIS) in Germany. PIANC-World Congress, Panama City

Brix J (1993) Manoeuvring technical manual. Seehafen Verlag GmbH, Hamburg. ISBN 3-87743-902-0

BRUNVOLL (2014) Complete thruster systems. https://issuu.com/brunvoll/docs/brunvoll_complete_thruster_systems_. Accessed 25 Dec 2020

BV (2018) Hull in composite materials and plywood, material approval, design principles, construction and surveys. rules and guidance notes, Bureau Veritas

BV (2019a) Rules on materials and welding for the classification of marine units. Rule Note, Bureau Veritas

BV (2019b) Rules for classification and construction of inland navigation vessels, Bureau Veritas

Carlton J (2018) Marine propellers and propulsion. Butterworth-Heinemann. ISBN 9780081003664

CCNR (2018) Automated navigation definitions of levels of automation in inland navigation. https://www.ccr-zkr.org/files/documents/AutomatisationNav/NoteAutomatisation_en.pdf. Accessed 21 Dec 2020

CCNR (2020a) Financing the energy transition—studies on energy transition towards a zero-emission inland navigation sector. https://www.ccr-zkr.org/12080000-en.html. Accessed 06 Mar 2021

CCNR (2020b) Listing of pilot and research projects in the field in automation in inland navigation. https://automation.ccr-zkr.org/1000-en.html. Accessed 01 April 2021
CESNI (2019) European Standard laying down technical requirements for Inland Navigation vessels (ES-TRIN)
Chen B, Peterson F, Valentine D (1991) Integrated ducted propulsor concept. In: SNAME symposium of propellers/shafting, Virginia Beach 1991
CREATING (2005) WP5 State of the art existing hull- and propulsor concepts (propulsor concepts of IW vessels)
COVADEM (2019) https://www.covadem.com. Accessed 03 Jan 2021
Dahlke-Wallat F, Friedhoff B, Martens S, Rafael R (2020a) Report on applicable most promising greening technologies & efficient fleet management solutions. Project GRENDEL (D 3-1-1 Greening technologies and solutions)
Dahlke-Wallat F, Friedhoff B, Martens S (2020b) Assessment of technologies in view of zero-emission IWT – Part of the overarching study "Financing the energy transition towards a zero-emission European IWT sector". Report DST 2293
DAMEN (2019a) Inland rudder and steering system. https://www.damenmc.com/-/media/damen/vandervelden/products/rudders/downloads/inland_rudder_and_steering_systems.pdf. Accessed 19 Jan 2021
DAMEN (2019b) Flex tunnel system. https://www.damenmc.com/-/media/damen/vandervelden/products/energy-saving/flex-tunnel/downloads/flex_tunnel_system.pdf. Accessed 21 Dec 2020
DNV-GL (2015a) Rules for classification of inland navigation vessels, part 2: material and welding. Det Norske Veritas-Germanisher Lloyd
DNV-GL (2015b) Finite element analysis. Class Guideline, Det Norske Veritas-Germanisher Lloyd
DNV-GL (2015c) Rules for classification of inland navigation vessels. Det Norske Veritas-Germanisher Lloyd
DNV-GL (2019) 2019 Energy transition outlook—a global and regional forecast to 2050
DST (2021) Project 20545 Rudder systems—development of a web-based decision support for the optimized rudder design of inland vessels based on the latest measurement and computational methods » DST – Entwicklungszentrum für Schiffstechnik und Transportsysteme e.V. (dst-org.de). Accessed 29 March 2021
EC (1997) Council Directive 97/68/EC on the approximation of the laws of the Member States relating to measures against the emission of gaseous and particulate pollutants from internal combustion engines to be installed in non-road mobile machinery
EC (2009a) Council Regulation (EC) No 595/2009 on type-approval of motor vehicles and engines with respect to emissions from heavy duty vehicles (Euro VI) and on access to vehicle repair and maintenance information and amending Regulation (EC) No 715/2007 and Directive 2007/46/EC and repealing Directives 80/1269/EEC, 2005/55/EC and 2005/78/EC
EC (2009b) Council Directive 2009/30/EC amending Directive 98/70/EC as regards the specification of petrol, diesel and gas-oil and introducing a mechanism to monitor and reduce greenhouse gas emissions and amending Council Directive 1999/32/EC as regards the specification of fuel used by inland waterway vessels and repealing Directive 93/12/EEC
EC (2010) DE-LIGHT—developing lightweight modules for transport systems featuring efficient production and lifecycle benefits at structural and functional integrity using risk-based design
EC (2016) Council Regulation (EU) 2016/1628 on requirements relating to gaseous and particulate pollutant emission limits and type-approval for internal combustion engines for non-road mobile machinery, amending Regulations (EU) No 1024/2012 and (EU) No 167/2013, and amending and repealing Directive 97/68/EC
ECCONET (2012) Deliverable 2.1.1: IWT fleet and operation. Report for European commission directorate-general for energy and transport, Brussels
Eichsleder H, Klell M (2020) Wasserstoff in der Fahrzeugtechnik, https://doi.org/10.1007/978-3-8348-9674-2
EU Commission (2005) European Commission Directive 2005/44/EC

EU Commission (2017) Digital inland waterway area. Study on support measures for the implementation of the TEN-T core network related to sea ports, inland ports and inland waterway transport. https://doi.org/10.2832/26584. p 109

EUROMOT (2020) FAQ/Guidance notes https://www.euromot.eu/publication-and-events/publications. Accessed 06 Mar 2021

Friedhoff B, Abma D, van Mensch P, Verbeek R, Schulz AC, Lutz A (2018) Digital solutions for environmental performance and efficient navigation using on-board monitoring and river modeling. Transport Research Arena—TRA conference, Viena, 2018

Gille J, de Vries A (2011) VoortVarend Besparen. Antwerpen, 2011

GL (2011) Rules for classification and construction of inland navigation vessels. Germanisher Lloyd

Godjevac M and Drijver M (2014) Power configuration of an inland pusher. ICMT 7th-9th July, Glasgow 2014

GRENDEL (2018) Energy efficient navigation. Fact sheet no. 8. http://www.interreg-danube.eu/approved-projects/grendel. Accessed 21 Dec 2020

GRENDEL (2020) Interreg—danube transnational programme green and efficient danube fleet—innovation fact sheets. http://www.interreg-danube.eu/approved-projects/grendel/section/technological-factsheets. Accessed 06 Mar 2021

Guesnet T, Delius A, Jastrzepski T (2004) Efficient freight transport on shallow inland waterways. In: 9th symposium PRADS 2004, Luebeck-Travemuende, results of the INBAT R&D project, 2004

Hertzberg T (2009) (Ed) LASS—Lightweight construction applications at sea. Project report, SP Technical Research Institute of Sweden

Heuser H (1990) Binnenschiff der Zukunft. VBD Report 1260

Hughes M (1991) An experimental analysis of adjustable pitch pre-swirl stators in combination with a ducted Propeller. SNAME's New England Section

IMO (2013) MEPC 1/Circ.815 - 2013 guidance on treatment of innovative energy efficiency technologies for calculation and verification of the attained EEDI

Jastrzebski T, Sekulski Z, Taczala M, Graczyk T, Banasiak W, Zurawski T (2003) A concept of the inland waterway barge base on the I-coreR steel panel. In: EIWN 2003 conference, Gyor, 2003

KOOIMAN (2020) Kooiman bow rudders. https://kooimanmarinegroup.com/assets/flyer-a4-bow-rudders-en.pdf. Accessed 28 Dec 2020

Kooj C, Hekkenberg R (2020) The effects of autonomous systems on the crew size of ships—a case study. Maritime Policy Manage. https://doi.org/10.1080/03088839.2020.1805645

Kuiper G (1992) The wageningen propeller series. MARIN Publication 92-001. ISBN 90-900 7247-0

Lasco RD, Ogle S, Raison J, Verchot L, Wassmann R, Yagi K, Bhattacharya S, Brenner JS, Partson Daka J, González SP, Krug T, Li Y, Martino DL, McConkey BG, Smith P, Tyler SC, Zhakata W (2006) IPCC guidelines for national greenhouse gas inventories, volume 4 agriculture, forestry and other land use, Chap. 5 cropland

Lassesson H, Andersson K (2009) Energy efficiency in shipping—review and evaluation of the state of knowledge. Chalmers University of Technology, Göteborg, Report No. 09:115. ISSN 1652-9189

Liu J (2020) Mathematical modeling of inland vessel maneuverability onsidering rudder hydrodynamics. ISBN 978-3-030-47475-1

Liu J, Hekkenberg R (2015) Impacts of rudder profiles on ship manoeuvrability. MARSIM, 2015

Liu L, Hekkenberg R (2017) Sixty years of research on ship rudders: effects of design choices on rudder performance. Ships Offshore Struct. https://doi.org/10.1080/17445302.2016.1178205

Liu J, Hekkenberg R, Rotteveel E (2014) A proposal for standard manoeuvres and parameters for the evaluation of inland ship manoeuvrability. In: EIWN 2014 conference, Budapest 2014

LR (2006) Provisional rules for the application of sandwich plate construction to ship structure. Lloyd's Register of Shipping

LR (2017) Announcing new Virtual Reality (VR) safety simulation and gaming technology. https://www.lr.org/en/latest-news/virtual-reality-safety-training-technology. Accessed 21 Dec 2020

LR (2019) Application of sandwich panel construction to ship structure. Lloyd's Register of Shipping
LR (2020) Rules and regulations for the classification of inland waterways ships. Lloyd's Register of Shipping
MariGreen (2018) Perspectives for the use of hydrogen as fuel in inlad shipping. https://www.dst-org.de/wp-content/uploads/2018/11/Hydrogen-Feasibility-Study-MariGreen.pdf. Accessed 06 Mar 2021
Molland AF, Turnock SR (2007) Marine rudders and control surfaces—principles, data, design and applications. Butterworth-Heinemann. ISBN 978-0-75-066944-3
Molland AF, Turnock SR, Hudson DA (2017) Ship resistance and propulsion – practical estimation of ship propulsion power. Cambridge University Press. ISBN 9781316494196
Momčilović N, Motok M (2009) Estimation of ship lightweight reduction by means of application of sandwich plate system. FME Transactions 37(3)
Müller E (2003) Innovative types of inland ships and their use on the river Rhine, its tributaries and the adjacent canals. SPIN, Version 1, VBD Duisburg
MUNIN (2012) Maritime unmanned navigation through intelligence in networks. Funded by EU commission (7th Research Programme). www.unmanned-ship.org. Accessed 22 Dec 2020
Nautican rudders (2021) Nautican product sheet triple rudders. https://nautican.com/wp-content/uploads/2015/06/Nautican-product-sheet-triple-rudders.pdf. Accessed 02 Feb 2021
Nautican stators (2021) Nautican product sheet pre-swirl stators. https://nautican.com/wp-content/uploads/2015/06/Nautican-product-sheet-pre-swirl-stators.pdf. Accessed 02 Feb 2021
Neste Corporation (2016) Neste renewable diesel handbook
NEWS (2014) Concept and specification of a novel inland container vessel. Development of a next generation european inland waterway. Ship and logistics system, report, Vienna, 2014
NOVIMAR (2017) https://novimar.eu. Accessed 20 Dec 2020
Oosterveld MWC (1972) Ducted propeller systems suitable for tugs and pushboats. ISP 19(219)
Peeters G, Afzal MR, Vanierschot M, Boonen R, Slaets P (2020) Model structures and identification for fully embedded thrusters: 360-degrees steerable steering grid and four channel thrusters. J Mar Sci Eng 8. https://doi.org/10.3390/jmse8030220
Peters HE (1994) Propulsion improvements of inland going ships with ducted propellers. WEGEMT 21st graduate school, Gerhard Mercator University, Duisburg
Plovput (2021) http://www.plovput.gov.rs. Accessed 18 Jan 2021
PROMINENT (2015) Sustainable, safe and economically feasible energy concepts and technologies for European inland shipping. https://www.prominent-iwt.eu/wp-content/uploads/2018/04/2015_09_26_PROMINENT_Mid-Term_Review.pdf. Accessed 01 Jan 2021
PROMINENT (2018) D2.8/D2.9—Standardized model and cost/benefit assessment for right-size engines and hybrid configurations, TNO
Radojčić D (1997) Tip-driven marine propellers and impellers—a novel propulsion concept. In: SNAME propeller/shafting symposium, Virginia Beach, 1997
Radojčić D (2009) Environmentally friendly inland waterway ship design for the Danube river. World wide fund for nature international, Danube-Carpathian programme (WWF-DCP)
Ramakrishnan KV, Kumar PGS (2016) Applications of sandwich plate system for ship structures. In: International conference on emerging trends in engineering and management—ICETEM 2016
REINTRIEB (2019) Ambient water transmissions and side-by-side propeller. Presentation at working group meeting on fleet modernisation, Vienna, 2019
RIS (2021) https://ris.cesni.eu. Accessed 01 Jan 2021
Rolls Royce (2020) https://www.rolls-royce.com/~/media/Images/R/Rolls-Royce/content-images/marine-product-finder/hero-images/mpf-hero-az-pm-990x481.jpg?h=481&la=en&w=990. Accessed 19 Dec 2020
SAFETY4SEA (2019) 5 digital innovations that are changing the maritime industry. https://safety4sea.com/5-digital-innovations-that-are-changing-the-maritime-industry. Accessed 21 Dec 2020
Schneekluth H, Bertram V (1998) Ship design for efficiency and economy. Butterworth-Heinemann, ISBN 0 7506 4133 9

Schot J, Rotteveel E (2020) Comparing azimuthing thrusters and conventional shaftlines for propulsive efficiency of pushboat-barge convoys. MARIN Report, No, p 128
SHOTTEL (2020) Schottel TwinPropeller. https://www.schottel.de/en/portfolio/products/product-details/stp-schottel-twinpropeller. Accessed 21 Dec 2020
Schottel Pump Jet (2020) Schotel PumpJet. https://www.schottel.de/en/portfolio/products/product-details/spj-schottel-pumpjet. Accesed 25 Dec 2020
SPS Technology (2020) What is SPS? www.spstechnology.com/what-is-sps. Accessed 29 Jan 2021
Stefanović A (2019) Structural analysis of an unconventional inland container vessel using finite element method. MSc thesis, University of Belgrade, Faculty of Mechanical Engineering
Stellmach H (2020) Deutz baut Wasserstoff-Verbrennungsmotor, https://bi-medien.de/fachzeitschriften/baumagazin/tiefbau-strassenbau/deutz-baut-wasserstoff-verbrennungsmotor. Accessed 06 Mar 2021
Suvačarov A (2019) Analysis of the navigation of the IWW selfpropelled ship in the optimal way (in Serbian). MSc Thesis, University of Belgrade, Faculty of Mechanical Engineering, Department of Naval Architecture, Belgrade
SYNCHRO-NET (2015) De-stressing the Supply Chain. https://www.mjc2.com/synchro-net-smart-steaming-supply-chain.htm. Accessed 20 Jan 2021
TOPOFAHRT (2021) https://www.dst-org.de/topofahrt. Accessed 01 Jan 2021
UNECE (2019) European agreement concerning the international carriage of dangerous goods by inland waterways (ADN). UNECE Committee on Inland Transport, New York
van Meij K, Raven C (2014) Promising hydrodynamic improvements for inland vessels. EIWN 2014 conference. Budapest, 2014
Van der Velden (2021) Van der Velden Marine Systems. https://www.nauticexpo.com/prod/van-der-velden-marine-systems/product-23779-456947.html. Accessed 19 Jan 2021
Verhaar Omega (2020) Verhaar Omega V-POD series. http://www.gulermarine.com/resimler/20111102_Leaflet_V-Pod.pdf. Accessed 21 Dec 2020
VETH (2020) Counter rotating. https://www.vethpropulsion.com/products/rudder-propellers/propeller-type/counter-rotating. Accessed 19 Dec 2020
Virendra DP, Abhishek A, Bhushan M, Sushmita B (2015) A review of azimuth thruster. SSRG—IJME 2(10)
VOITH (2020) Exactly on course. Marine products and services. https://d2euiryrvxi8z1.cloudfront.net/asset/445934742530/d35bd377a147f0af71906e17b7208e09. Accessed 19 Dec 2020
VVB (2011) Project VoortVarend Besparen. https://www.eicb.nl/projecten/voortvarend-besparen. Accessed 01 Jan 2021
Waldron C D, Harnisch J, Lucon O, Mckibbon R S, Saile S B, Wagner F, Walsh MP, Maurice LQ, Hockstad L, Höhne N, Hupe J, Lee DS, Rypdal K (2006) IPCC guidelines for national greenhouse gas inventories, vol 2 Energy, (Chap. 3 Mobile combustion)
WATERTRUCK + (2014) http://www.watertruckplus.eu/. Accessed 01 Jan 2021
Weitzenböck J (2012) Adhesives in marine engineering. Woodhead Publishing, ISBN 9780857096159
Zibell HG, Müller E (1996) Binnenschiffe fur extrem flaches Wasser. Ergebnisse des VEBIS-Projektes", Handb. D. Werften, vol XXIII
Zöllner J (2003) Vortriebstechnische Entwicklungen in der Binnenschiffart. IST Symposium, New and Further Development, Duisburg, 2003
Zöllner J (2005) Neue Containerschiffe zur Hinterlandanbindung. 26th Duisburg colloquium: The Ship in Intermodal Traffic. Duisburg, 2005
Zöllner J (2009) Strömungstechnische Möglichkeiten zur Reduzierung des Kraftstoffverbrauchs und der CO2-Emissionen von Binnenschiffen. Vortrag beim ZKR Kongress Rheinschifffahrt und Klimawandel, Bonn, 2009
Zondervan GJ, Hoekstra M, Holtrop J (2006) Propellers and shafting—flow analysis, design and testing of ducted propellers, MARIN

Chapter 10
Concepts of Contemporary and Innovative Vessels

10.1 Vessels Concepts in Various Research Projects

Some carried out researched projects concern the *innovative vessels* whose time, seems, did not come yet due to various reasons; or did, but the vessels never became fully operational. Beside necessary costs for the development of a new concept, amongst the reasons are that innovations are not so easily accepted by the rather conservative IW shipping society. Traditionally, it accepts only well-proven, durable and safe products.

Some of the concepts of researched innovative vessels that "never came to be" are described in this section. These concepts cover the time span of over two last decades. They are presented according to their main aspect of design at the time they were conceived, whether it is the use of LNG and/or electric powering, autonomous navigation, air lubrication, extremely shallow draught or some other innovative feature. Their main characteristics, stated in the following, are taken from the referenced projects, thus the authors of the book reserve the right not to be responsible for all information given below.

Note that some ex-concepts are realized by now. A very good example is the ship developed within the INBISHIP project at the very end of the last century (see Sect. 10.1.2) and realized 20 years later (Sendo Liner, presented in Sect. 10.2.2).

10.1.1 Concepts of LNG Fueled Vessels

The LNG fueled ships primarily produce lower emissions, which is extremely important in view of emerging environmental issues. Furthermore, although compatible with the conventional vessels of the same type in most of the specifics, presented concepts also tend to improve the overall performance as much as possible. As

D. Radojčić et al., *Design of Contemporary Inland Waterway Vessels*,
https://doi.org/10.1007/978-3-030-77325-0_10

a consequence for high power output, lower total operational costs are expected although the investment and crew costs of LNG vessels are higher due to the application of novel technologies.

Presented concepts include results from the IDV and GRENDEL projects, both developed for the Danube navigation; these are one selfpropelled vessel and two pushboat concepts. As explained in Sect. 9.2.5, two types of LNG fueled ships exist: (a) those that are powered solely by LNG, and (b) those with dual fuel arrangements—LNG and standard diesel fuel. In recent years similar vessel types came into life on the Rhine and became fully operational (see examples of inland tankers in Sect. 10.2.1).

LNG Fueled Selfpropelled Ship (IDV)

The aim of the Innovative Danube Vessel project (2012–2013), see IDV (2014), was to pursue modernization of the Danube fleet using state-of-the-art technology and innovative solutions. In that sense, IDV identified solutions that would lead to cost savings and lower environmental impacts. Proposed LNG vessel concepts include a selfpropelled vessel and a pushboat.

The selfpropelled vessel, designed by DST, was conceived to be used as a single vessel or in a coupled convoy with one to three barges. Two hull concepts were offered, differing only in bow shape (see Sect. 8.2.1):

- A hull with a *blunt bow* (intended for permanent operation in coupling convoy with a conventional barge in-front)
- A hull with a *conventional bow* (in case of convoy operation it should be coupled with a dedicated barge in-front, with a specially adapted stern shape).

Design requirements follow:

- LNG should be used as a fuel, but the engine should be able to run on diesel fuel as well
- Application of modern and innovative technologies for propulsion
- Good comfort onboard (all premises regarding noise and vibrations)
- Adaptation to the current and expected Danube infrastructure
- Full compliance with present and foreseen European IW rules and regulations
- Cargo: containers and bulk cargo.

Main characteristics are the following:

Length overall	105 m (min. lock chamber length of 190 m allows the coupled barge length of up to 85 m)
Breadth	11.4 m
Height	3.1 m
Max. draught	2.8 m (min. operational draught of 1.6 m)
Air draught	7.5 m
Lightship mass	670 t
Main engine	2 × 750 kW, Wartsila 6L20DF, dual fuel (98% gas; 2% diesel)

Propulsion	Direct drive, FP propellers of 1.75 m in nozzles, adjustable retractable tunnels
Bow thruster	350 kW diesel driven
Harbor generator	35 kW on the deck, air cooled
Water ballast	867 m^3
Diesel oil	90 m^3
LNG	140 m^3 in total (LNG storage tanks at the aft part of deck)
Operational range	Diesel only—120 h i.e. 1400 km in still water LNG only—140 h i.e. 1800 km in still water
Fresh water capacity	15 days

The wheelhouse is located at the fore end of the vessel, while LNG storage tanks are on the aft end deck. Cargo hold width is 10.05 m, allowing four ISO containers to be positioned abreast, whereas 13 TEU bays can be carried lengthwise. The vessel is designed to carry 208 TEU containers or 2200 t of bulk cargo. When operating in a 190 m long convoy with one barge ahead, 400 TEU or 4180 t of bulk can be transported. The cargo capacity for a convoy with three barges is expected to be 784 TEU or 8140 t of bulk cargo. The draught in all cases is designed to be 2.8 m.

Amongst the conclusions is that the application of LNG leads to reduced emissions and fuel costs, while the crew and fixed costs are expected to be higher. However, the overall transport costs would be smaller than for conventional IW selfpropelled vessel.

LNG Fueled Pushboat (IDV)

The pushboat concept was designed under the IDV project along with the abovementioned LNG selfpropelled vessel. The following main requirements and specifics are given in IDV (2014) for the dual-fuel-electric pushboat concept designed by SDG, see Figs. 10.1 and 10.2.

Design requirements follow:

- LNG should be used as a fuel, but the engine should be able to run on diesel fuel as well
- Application of modern and innovative technologies for propulsion
- Good comfort onboard (all premises regarding noise and vibrations)
- Adaptation to the current and expected Danube infrastructure
- Coupling with the existing fleet of barges
- Full compliance with present and foreseen European IW rules and regulations.

Main characteristics are the following:

Length overall	39 m
Breadth	11.4 m
Height	3 m
Design draught	1.7 m
Full load displacement	520 t
Lightship mass	370 t

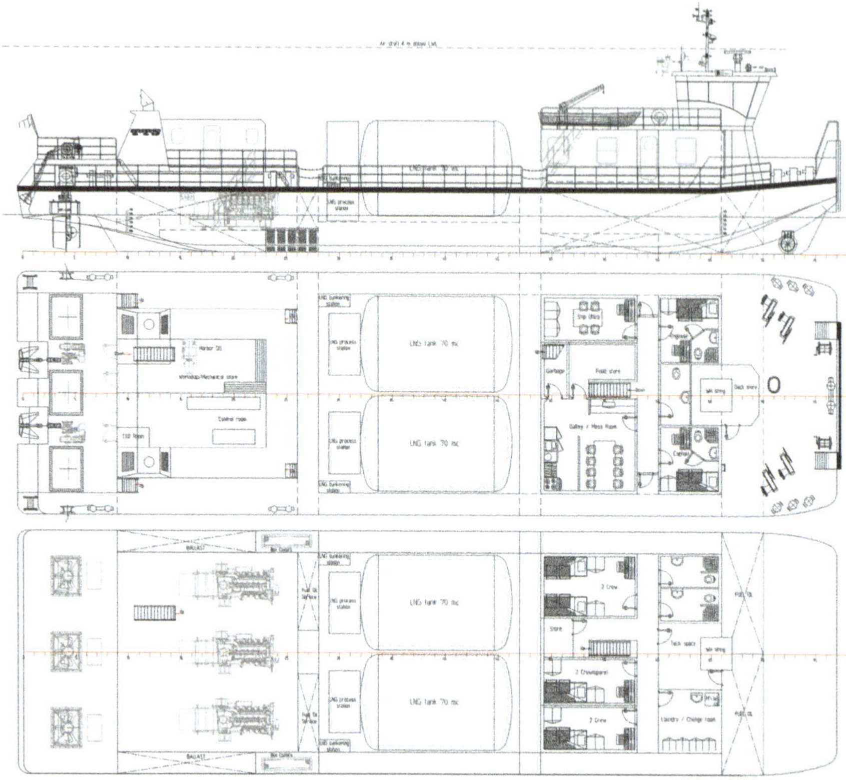

Fig. 10.1 General arrangement plan of LNG fueled pushboat concept (IDV). *Source* SDG

Fig. 10.2 LNG fueled pushboat concept (IDV). *Source* SDG

Main engine	3 × 800 kW/1000 RPM, Wartsila 6L20DF, dual fuel (98% gas; 2% diesel)
Propulsion	Electrically driven 3 rudder propulsors (azimuth thrusters), FP propellers of 1.45 m in nozzles
Bow thruster	1 × 120 kW electric driven (optional)
Thrust	270 kN at cruise speed of 12 km/h
Water ballast	100 m^3
Diesel oil	70 m^3
LNG	140 m^3 in total (2 × LNG storage tanks)
Operational range	Diesel only—120 h i.e. 1400 km in still water LNG only—120 h i.e. 1400 km in still water
Fresh water	15 days

The main advantage of the pushboat, as expected, is reduced fuel consumption. The crew costs (especially in terms of specific training) and fixed costs are higher than for the conventional designs. However, the total transport costs and especially emissions should be significantly lower.

LNG Fueled Pushboat (GRENDEL)

The Romanian Ship Design Group (SDG), the same that designed the IDV pushboat, and shipping company NAVROM, designed a new LNG fueled pushboat concept within the GRENDEL project (see GRENDEL 2020). The GRENDEL's aim was to modernize the Danube navigation by making it "greener" and more efficient, stimulating, amongst other, two research subjects: (a) retrofitting of inland pushboats (see Chap. 11) and (b) development of LNG powered inland pushboat.

The next generation of LNG fueled pushboat concept is given in the following, see GRENDEL (2020) and Ship & Boat Int (2020). Expected savings in emissions are 25% reduction of CO_2, 90% reduction of NO_X, 100% reduction of PM.

Design requirements follow:

- LNG fuel only (no dual fuel), Power > 4000 HP, Range > 1000 km upstream
- $L < 42$ m, $B < 23$ m (limitation due to lock chamber sizes), $T < 2$ m (preferably 1.85 m), Air draft <7.70 m (bridge height at Passau)
- The navigation area is the Danube up to Passau and Danube-Black Sea Canal
- Convoy configurations: 3 + 3 + 3 and 2 + 2 + 2 (230 × 33 m and 230 × 22 m, respectively, without the pushboat).

According to the above requirements, this concept is supposed to be one of the most powerful pushboats on the Danube (in this respect see Fig. 4.11). The design should be in line with all current emission limits, general rules for inland vessels and specific rules for LNG-powered vessels. It is envisaged, however, that in 20–30 years the hydrogen, as a cleaner fuel than LNG, may be employed. Special attention is paid to safety taking into consideration the LNG is a fuel, i.e. to certified materials for cryogenic temperatures (−160 °C), placement of the LNG tank and processing units, the possibility of leaks, specifics of the ventilation system etc. (see Sects. 9.2.4 and 9.2.5). Placement of the LNG tanks (which are huge compared to the pushboat size), planning of zones of hazardous areas and other unconventional requirements that follow LNG application, *dictated the layout of the vessel*, see Figs. 10.3, 10.4 and 10.5. A detailed technical description of the concept is given in the Project GRENDEL (2020).

Main characteristics are the following:

Length overall	42.00 m
Breadth	13.5 m
Height/Depth	3 m
Design draught	1.85 m
Air draft above B.L.	9.40 m (vertically movable wheelhouse)
Propulsion engines	3 × 1460 kW @ 900 RPM, gearbox ratio 1:2.55
Propellers	2 × FPP, 1 × CPP, in nozzles, diameter 1.8 m
Bow thrusters	2 × 250 kW, diameter 1067 mm
LNG (total/net)	220/190 m^3 (2 LNG tanks)
Capacities	Ballast/Fresh water/Sewage/Lub.oil 83/30/26/8 m^3
Generators	Shaft/Gas 100/100 ekW
Hydraulic unit	600 kW
Crew	8
Max thrust	3 engines—408 kN (at 4050 kW and 13 km/h)
	Side engines only—270 kN (at 2700 kW and 11 km/h)
Range	Upstream/Downstream—1200/3150 km
Speed	Upstream/Downstream—8/14 km/h (average current 4 km/h)

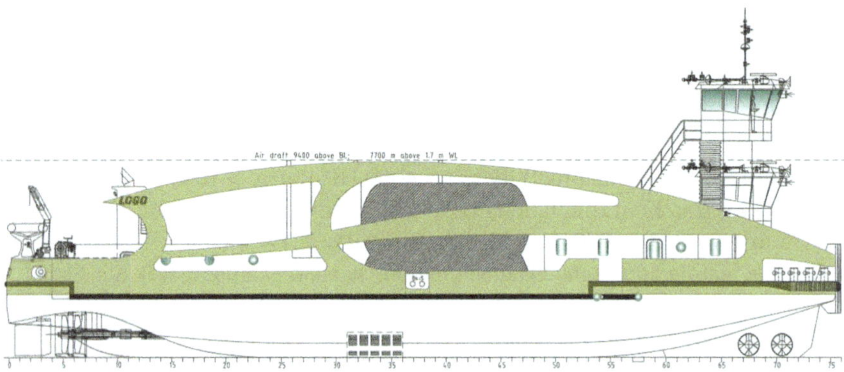

Fig. 10.3 LNG fueled pushboat concept—side view. (*Source* Grendel (2020)

Some noteworthy follow-up features of the innovative approach are:

- Three pure-gas engines. Their engagement depends on the load.[1] When the central engine is turned off, CPP feathering feature minimizes resistance.

[1]Similarity with a completely different diesel-hydraulic concept with 2-engines and 3-propulsors—Fig. 10.11, should be noticed. The idea in both cases is the same: *flexible power distribution* and *separate engagement of engines* according to need.

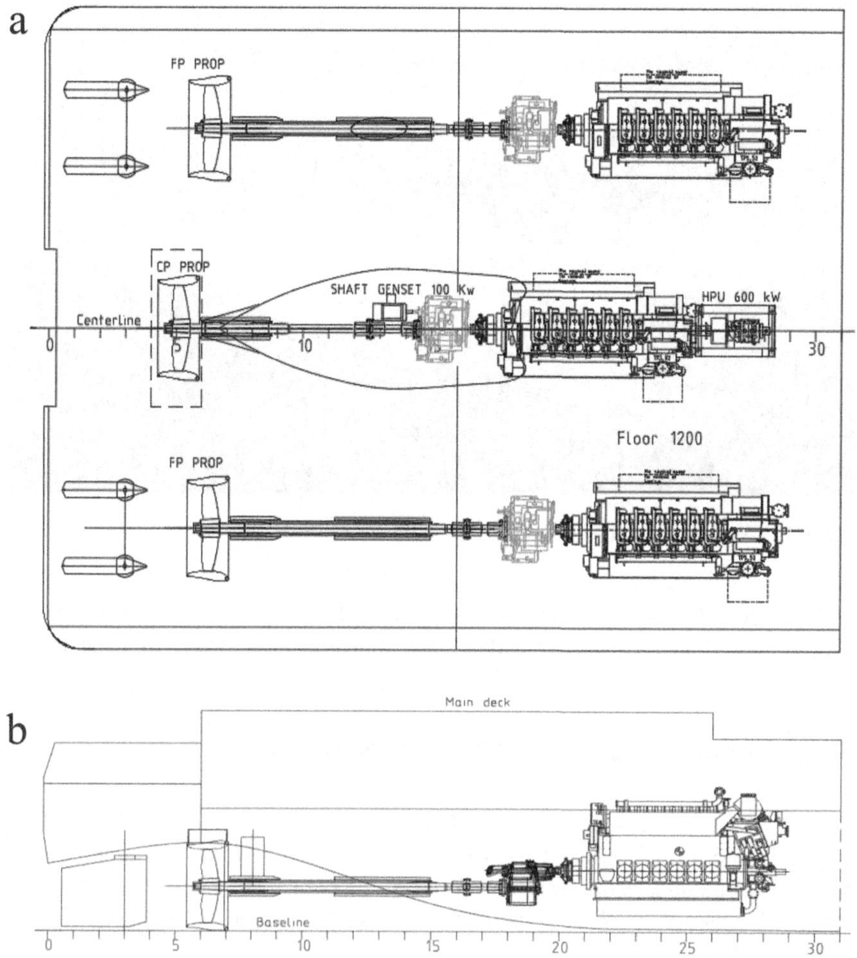

Fig. 10.4 LNG fueled pushboat concept GRENDEL. *Source* Grendel (2020); **a** Arrangement of main engines, **b** Side shaftline

- The central engine works at constant RPM, power is distributed through a power take-off (PTO) to a generator and CPP, without rudder behind (see Fig. 10.4a)
- Side engines work at variable RPM driving FPPs with two rudders per propeller (see Fig. 10.4a, b)

- FPPs enable reliability, while CPP enables flexibility. Note that the application of CPP is unusual on the IW vessels
    - A shaft generator is used in navigation only when the central engine is running
    - A separate gas generator is in use when the vessel is stationary and the power requirement exceeds the output of the solar panels (15 ekW)

Fig. 10.5 LNG fueled pushboat concept. *Source* Grendel (2020)

- A protective composite side and the top shell is equipped with solar panels (150 m^2) giving an attractive and unusual appearance (see Figs. 10.3 and 10.4).

10.1.2 Concepts of Electrically Driven Vessels

Electric vessel concepts presume propellers *driven by electric motors* (EM). Energy supply can be: (a) *diesel-electric,* (b) from the *batteries as the only source*, or c) hybrid, i.e. in a *dual arrangement with diesel, LNG or hydrogen.* Certain concepts include the use of *Fuel Cells (FC)* which are still regarded as innovative and technology of the future.

Generally, the use of electric propulsion tends to reduce fuel consumption and particularly emissions (see Sects. 9.2.5 and 9.4.1). However, due to the introduction of new technologies, the cost of the investments, as in the case of LNG fueled ships, is significantly higher. Moreover, over the years, the electric concepts included mostly diesel-electric or gas-electric propulsion and new tendencies are oriented towards *fully electric* or *hydrogen-electric* concepts. Fully electric vessels are already operational, as one can see in Sect. 10.2.2.

Selfpropelled Vessel Concept (INBISHIP)

Amongst the first projects of the kind is Project INBISHIP (INBISHIP 1998). It offered an innovative approach to IW ship design proposing: diesel-electric powering, pod propulsion, optimization of hull lines, excelled manoeuvrability and increased economic characteristics. Recently, a similar vessel but with a diesel direct drivetrain was actually built, see Sendo Liner in Sect. 10.2.2.

Amongst the novelties in the INBISHIP concept is that the engine room can be hypothetically placed anywhere along the ship (even at the bow if necessary) as there is no direct coupling of diesel engines and propellers. As a consequence, the machinery compartment requires less space and the cargo space may be increased (in particular, on a 110 × 11.4 m ship, one container bay more can be loaded!). Overall ship design leads to lower fuel consumption, emission levels and maintenance costs.

Inland Container Vessel (NEWS)

The aim of this project (2013–2014), see NEWS (2014), was to develop an inland container vessel for the Danube (labelled as NEWS Mark II). The design options included diesel-electric or gas-electric propulsion. The main design objectives and particulars, as stated in NEWS (2014) are:

- Capability to carry three layers of containers, stacked four abreast, on at least 80% of the European IWW
- A deckhouse and a liftable wheelhouse at the fore end
- Completely new aft ship design for optimized propulsion efficiency
- Diesel or gas-electric propulsion for clean and economical operation.

Main particulars:

Length	110 m
Breadth	11.45 m
Height	4.5 m
Max. payload	1907 t
TEU capacity	156 (can load up to 208 TEU with reduced average cont. mass)
Propulsion	LNG-electric power station—4 × 290 kW resulting in 1160 kW, or Diesel-electric power station—3 × 480 kW resulting in 1440 kW + 1 × 480 kW in spare

The vessel was designed to have two pod propulsors or a conventional transmission system. Both concepts were aimed to reduce a carbon footprint, with an expected fuel cost reduction of around 10%. The vessel was also designed with active ballast system to manage large draught variations (up to 80 cm) to cope with the low bridges.

Elektra

Elektra is a hybrid hydrogen-electric pushboat developed upon the German agenda on climate policy, thus the vessel is expected to be emission-free. Anticipated investments cost is considerably higher than in the case of conventional pushboats. The vessel is designed by TU Berlin and reported as the first *hydrogen-fueled pushboat* concept in the world, see Fig. 10.6. Contrary to other electric arrangements that consist of gas or diesel engine with electric drive (INBISHIP and NEWS vessels) Elektra uses hydrogen as a fuel located in six containers at the aft part of the deck. Hydrogen feeds the fuel cells (see Sect. 9.2.5) that supply the electric power when the batteries are run out (see Innovation Origins 2020). The vessel has solar cells on the roof of the wheelhouse and can use the shore-to-ship electricity supply to charge the batteries. Batteries are positioned below the deck (behind the wheelhouse); their share in the total weight is around 15%. This vessel specifics are presented in the following.

Elektra's main particulars:

Length	20 m
Breadth	8.2 m
Draught	1.25 m
Operational	range more than 400 km
Propulsion	Electric motors 2 × 200 kW, Battery packs 2.5 MWh (total capacity)
Hydrogen	6 containers (750 kg in total) under a pressure of 500 bar to power the fuel cells of 3 × 100 kW
Thrusters	2 fully rotating rudder propellers

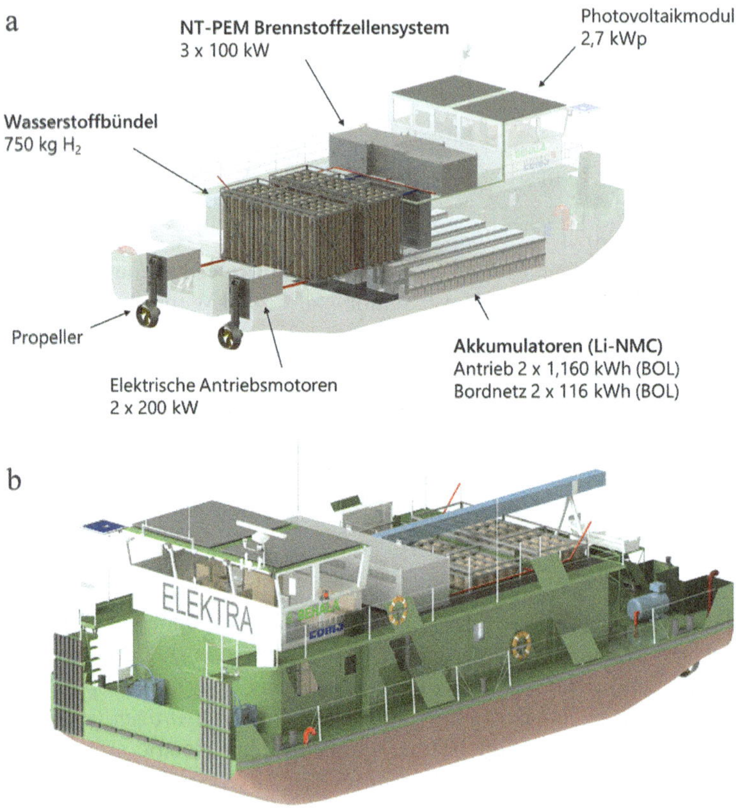

Fig. 10.6 Pushboat Elektra: **a** *Source* e4ships (2021), **b** *Source* EST-Floattech (2021)

Besides the pushboat Elektra, the world's first *IW commercial cargo vessel operating on hydrogen* (1 MW of on-board fuel cell power) is developed within the FLAGSHIPS project (FLAGSHIPS 2021). Commercial operations are set to commence on the river Seine in Paris in 2021. Although these particular concepts did not become operational yet, some hydrogen-powered vessels were delivered, as for instance Emeli and Hydroville (see Sect. 10.2.3).

10.1.3 Other Notable Concepts

Some of the concepts given in the following are developed decades ago, but are still considered novel in regard to their specific design. They include features as: operation in shallow waters, modular hull concept, hybrid propulsion, air lubrication, air-supported hull and autonomous navigation. Some projects, thus their concepts too, include more than just one of the abovementioned (e.g. INBAT project) which will be discussed in the following as shallow draught, modular and hybrid concept.

Shallow Draught Concepts

The shallow draught concepts are designed with an aim of efficient operation at small or reduced draught enabling flexible navigation in the shallow waterways. As explained below, for shallow water navigation it is necessary to obtain (a) a small draught (by weight reduction, e.g. use of lightweight materials), and (b) propulsors that are adapted to shallow water navigation. Notable projects in that regard are VEBIS and INBAT.

R&D project VEBIS (German acronym for improvement of the efficiency of inland water transportation) was initiated with a twofold goal: (a) to increase the transport capacity on Elbe and Odra rivers that allow ship draughts of only 1.0–1.4 m, and (b) to enable efficient operation at larger draughts also. So, amongst others, 4 types of pushboats with draughts between 0.8 and 1.7 m with the pump-jets and propellers were developed (see Zibell and Müller 1996). VEBIS is an instructive and useful study; its conclusions and recommendations are certainly valid for the Danube and particularly for the vessels intended for its tributaries and canals.

An EU project under the acronym INBAT (INnovative BArge Train) that lasted from 2001 to 2004, was initiated after the VEBIS project, with a similar goal, i.e. development of a barge train that will operate efficiently at draughts ranging from 0.6 to 1.7 m, see Guesnet et al. (2004). Within INBAT, amongst others, the application of new lightweight construction materials and structural designs were investigated. Furthermore, within the project a shallow draught pushboat was developed. Its specific is a retractable middle rudder-propeller of 1.2 m (with the purpose to enhance manoeuvrability and to reduce resistance when not in use) together with two classical horizontally driven shaft propellers of 0.675 m (see Fig. 10.7). During the operation in very shallow water, only the side propellers are supposed to be used (the central propeller would be retracted). In deeper waters the central propeller would be used

Fig. 10.7 The pushboat propulsion arrangement developed within INBAT project. *Source* Guesnet et al (2004)

too, to increase the barge train speed. Various power plants were investigated: (a) 3 diesel engines (mechanical transmission), (b) 2 diesels driving 2 generators (diesel-electrical transmission), (c) 2 diesels driving hydraulic pumps (hydraulic transmission), (d) 3 fuel cell modules supplying electric power (see Zenczak et al. 2003). In all four cases propulsors were the same, i.e. as explained above and shown in Fig. 10.7.

Another design suited particularly for operation in shallow waters is an inland ship with retractable side buoyancy bodies (blisters); their application alters the ship breadth from 9 to 12.6 m enabling the draught reduction without reducing the cargo quantity. Side bodies are supposed to be operated by hydraulic cylinders, Fig. 10.8. There is a series of other technical solutions that are based on the same principle (e.g. inflatable and steel blisters). Expected effects and installation costs for various ship types are discussed in ECCONET (2012).

Modular Vessels

A modular pushboat design was also investigated within the INBAT project, see Fig. 10.9. Amongst the objectives was to reduce production costs and to increase the lifetime. Main particulars of INBAT pushboat are: $L = 25$ m, $B = 11.4$ m, $T = 1.1$ m.

Another modular vessel prototype was built in Serbia, see Fig. 10.10 and Bilen and Bilen-Katić (1997); the vessel was named MMPO (Modular Multi-Purpose Vessel; $L = 13.75$ m, $B = 7.6$ m, $T = 2.4$ m). It consisted of two propulsive modules with a driving complex (each module had 2 diesel engines of 90 kW which were powering a single hydrostatic rudder-propeller), connecting modules (pallets) which provide stiffness, accommodation modules and a wheelhouse module. The idea

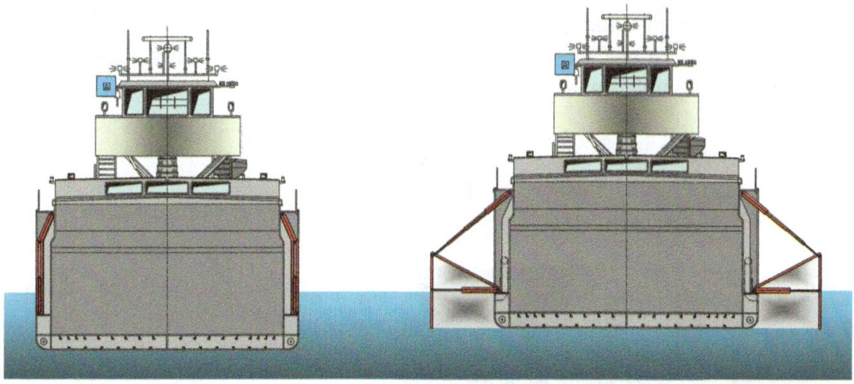

Fig. 10.8 Cross-section of the IW ship with laterally extractable buoyancy elements. *Source* Müller (2003)

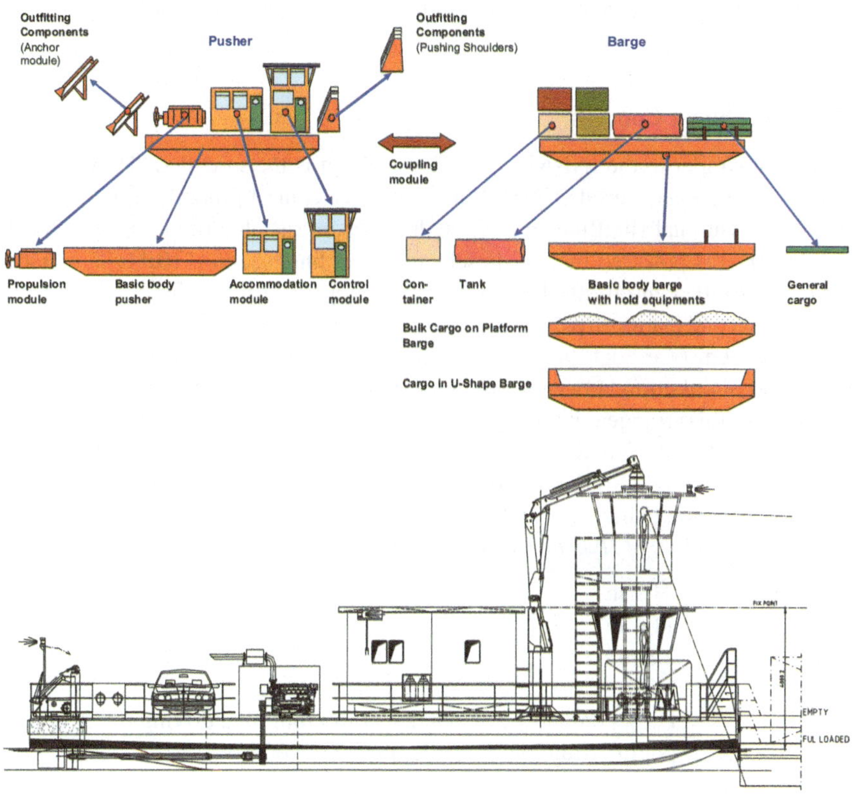

Fig. 10.9 Pushboat modules. *Source* Guesnet et al. (2004)

behind MMPO was to enable transport of modules by trucks, which would be easily joined on the spot of use, i.e. to be a kind of so-called "truckable tug". Modules are not welded, but coupled with special wedges.

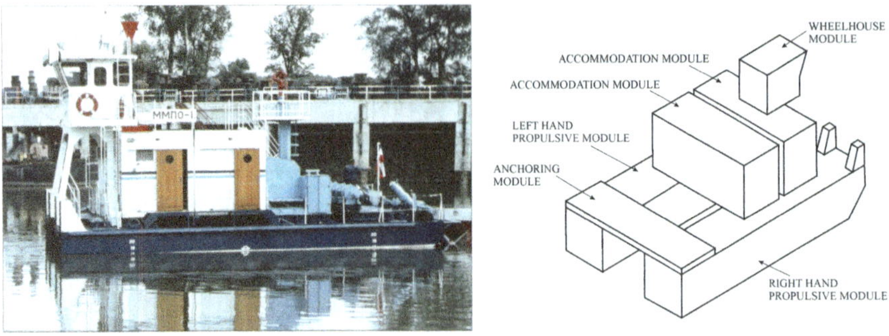

Fig. 10.10 Modular multi-purpose vessel—MMPO. *Source* Shipyard "Belgrade"

Hybrid Pushboat Concept

Somewhat similar to the INBAT propulsion concept is the so-called "Hybrid Pushboat", see Bilen and Žerjal (1998). It has one large central propeller ($D = 1.85$ m) with a conventional shaftline and two hydrostatic, retractable rudder propellers ($D = 1.35$ m). It is powered by two diesel engines in line, see Fig. 10.11. The first is connected to the central, mechanically-driven propeller (diesel direct propulsion), while the second engine drives two hydrostatic side propulsors via a hydraulic transmission system (see Sect. 9.4.1) which enables independent and flexible control. Thus, load distribution between central and side propellers is optimized. The main advantage of this arrangement is the possibility to draw nominal power for a particular convoy also enabling good manoeuvring characteristics. Note that this propulsion concept (as well as INBAT's) inherently eliminates the rudders (thus the steering gear too) increasing efficiency and safety.

Main particulars of Hybrid Pushboat are:

Length overall	24.2 m
Breadth	11.4 m
Height	2.8 m
Draught	1.9 m
Engine	Diesel engine—2 × 600 kW/1800 rpm (one driving the hydraulic pump), (nominal propeller power is 960 kW)

Research that not conceive a particular ship concept as does not treat the vessel itself, but is informative from the viewpoint that various pushboat power configurations have been examined, is given in Godjevac and Drijver (2014). Seven selected power configurations, all based on the same direct drive, 4-propeller pushboat and operational profile given in Fig. 9.20, were examined and benchmarked against the

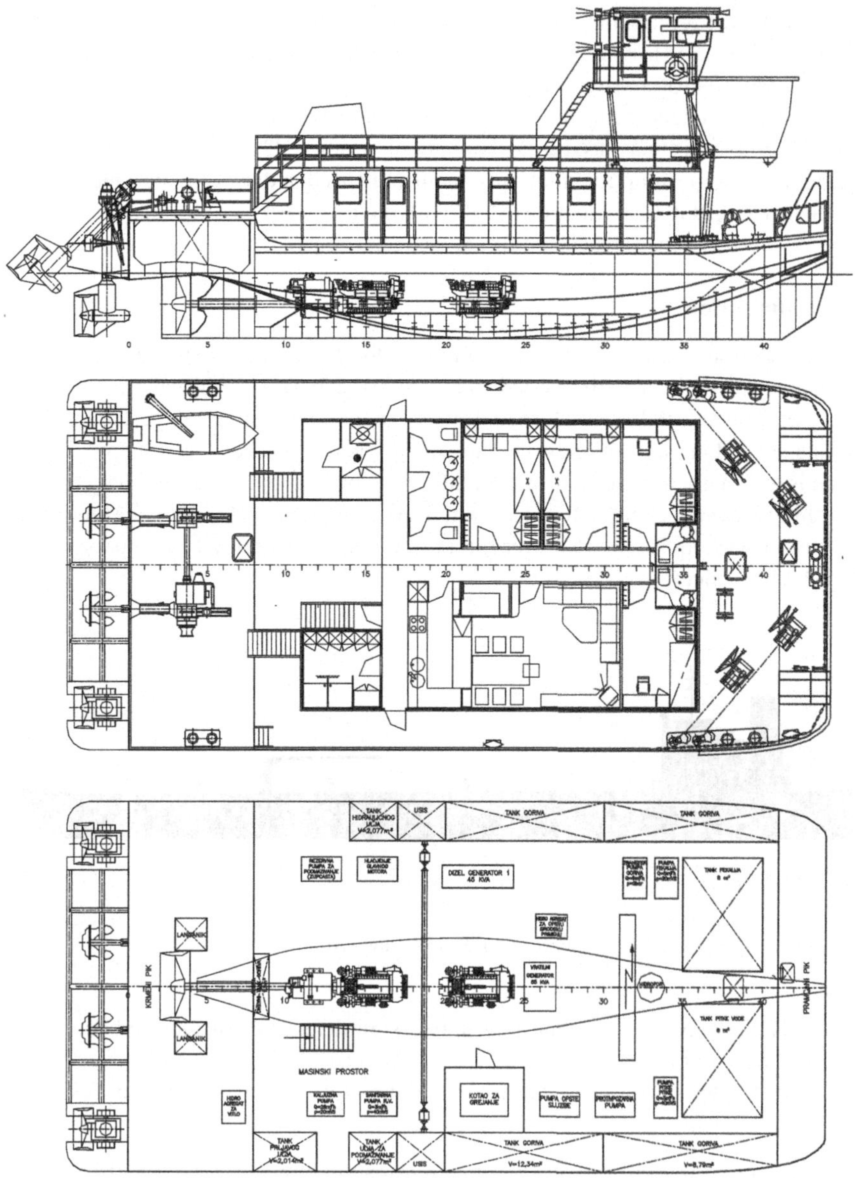

Fig. 10.11 Hybrid pushboat concept. *Source* Bilen and Žerjal (1998)

state-of-the-art 3-propeller Rhine pushboat Verhaven X (sistership Verhaven IV is shown in Fig. 4.14). Examined power configurations include diesel-direct, diesel-electric, gas-electric and hybrid (diesel-direct and diesel-electric combination). Superior configuration regarding the energy consumption was diesel-direct, while for NO_X and PM emissions that was the gas-based configuration. Gas engines, however, have higher energy consumption, thus the conclusion was that diesel engine supplied with the Selective Catalytic Reduction (SCR) will remain to be the prime mover workhorse for IW vessels in the foreseeable future (see also Sects. 9.2.3 and 9.2.6).

Air Lubrication

Air lubrication of the hull bottom can reduce the *frictional resistance* of the vessel (which is proportional to the wetted area), while the *wave-making* and *pressure resistance* may be reduced by hull form optimization. For lower speed range, the frictional resistance is dominant amongst the three mentioned resistance components. Thus, it sounds logical to investigate the potentials of air lubrication, as it is probably the only currently known technique for local skin friction reduction. However, air lubrication (air injection) requires extra power for air pumping (powering of a compressor), see Fig. 10.12, hence:

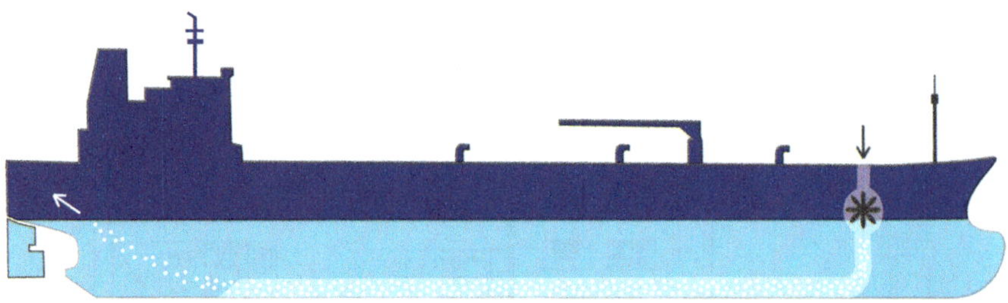

Fig. 10.12 The principle of air lubrication. *Source* Naval Architect

(a) for very low speeds, that power is a significant part of required propulsive power, and
(b) for higher ship speeds, the wave-making resistance is a dominant resistance component.

Consequently, there is a speed range (corresponding to $F_n = 0.05$–0.15) where air lubrication may be a feasible technique for power and emissions reductions. That speed range matches normal IW vessels' speed range.[2] Moreover, hull form requirements for air lubrication also match those of the IW vessels (e.g. large parallel middle section, small bilge radius, etc., see Fig. 10.13).

---

[2]Some types of high speed craft (speed range above $Fn_\nabla > 1$) use air lubrication. These are air supported vessels e.i. hovercrafts and SES.

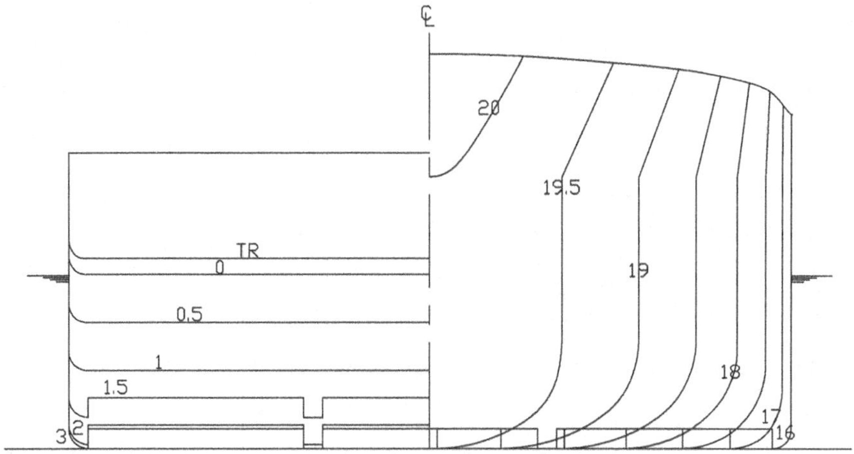

Fig. 10.13 Body plan of ship model with air cavities. *Source* Thill et al. (2005)

Three air lubrication techniques exist (see Thill et al. 2005; Pavlov et al. 2020):

- Air cavities in the bottom of the hull
- Injection of micro air bubbles along the hull
- Formation of air films under the hull.

Amongst the projects which are of interest for IW vessels are Dutch projects initiated in 1999. That is PELS project (see Thill et al. 2005 and Pavlov et al. 2020) that researched the air lubrication capabilities on theoretical and numerical grounds; the follow-up project was PELS 2 which was focused on the air cavity ships. Follows the EU funded SMOOTH project (see Foeth et al. 2010) which was focused on air bubbles and air films in combination with special coatings. The project‘s researched themes were resistance, propulsion, manoeuvring and seakeeping within which model- and full-scale experiments were performed. It was concluded that the net power savings depend on the air lubrication technique, while manoeuvring, in general, does not depend much on air lubrication.

The abovementioned R&D projects were conducted by the research centres MARIN and DST. The industry followed, in particular Dutch DAMEN Group together with DUT developed DACS (Damen Air Cavity System). It can be used on the seagoing, and particularly on the IW vessels by forming stable air cavities by the so-called cavitators—transversal profiles on the ship's bottom, (see DAMEN 2020). Fuel savings and emission reductions of 10–20% are expected. Moreover, according to model test results, the highest savings may be expected in shallow water navigation. Disadvantages are that DACS increases the draught by 50–100 mm (due to the cavitators, pipes etc.) and is vulnerable to grounding.

Actually, Damen Group built IW tanker Ecoliner in 2013 (see Sect. 10.2.1) with *Air Chamber Energy Saving* (ACES) system that enabled 10–15% resistance reduction at a speed range of 13–16 km/h (see Provoost 2018; Pavlov et al. 2020). ACES system's technical characteristics were: 64 air chambers, 2 air blowers of 5.5 kW, 150 m^3, 0.4 bar—see Fig. 10.14. After encouraging results and further developments, the ACES system was renamed to Damen Air Cavity System DACS (see the previous paragraph).

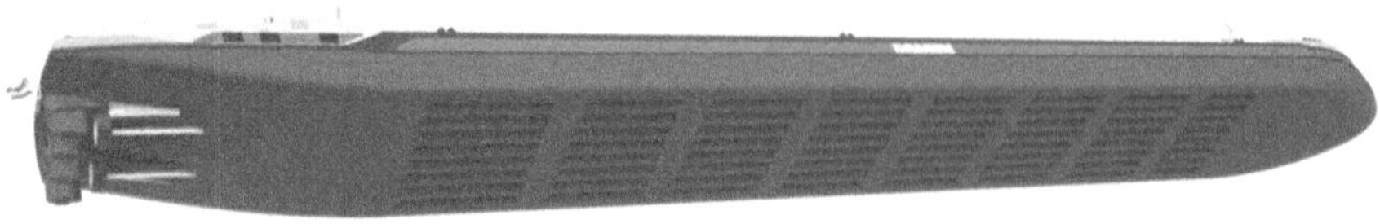

Fig. 10.14 Ecoliner's air lubrication system. *Source* Provoost (2018)

Note that model- and full-scale tests were also performed with the Futura-type vessel Till Deymann (see Sect. 10.2.3) supplied with the *air bubble lubrication system*, but that technique seemed not to be successful.

Similar savings and disadvantages were reported in Gorbachev (1993) and Pavlov et al. (2020), where full-scale results representative for the IW vessels were given. Specifically:

- For a river-sea ship of Volga-Don type (135 × 16.3 m, 5300 tdw) power saving of 16–17% was reported for a speed of 20 km/h in deep water (10–12% at lower speeds). Necessary compressor power was 3% of the main engine power.
- A pushboat and two barges (84 × 14 m of 3000 t capacity each) were tested in two pushed convoy formations: (a) convoy with a single barge, and (b) 1 + 1 convoy. Reported pushboat power-saving were 20% and 10% respectively. Note, however, that the mechanism for bubble formation reduces the carrying capacity and robustness, and increase resistance for some 10% when the compressor is not working.
- Full-scale shallow water tests were performed for $h/T = 1.67$ (results for $h/T = 3.5$ are practically the same as in deep water) with a 1000 tdw barge. Reported power savings were 20% at 8.5 km/h (15% at 6 km/h). In this case, the mean increase of draught was only 4 cm (compared to conventional barge) and a compressor of 6 kW was sufficient.

More about air lubrication can be found in Pavlov et al. (2020), as well as in IMO MEPC.1/Circ.815 (IMO 2013) as it belongs to innovative energy efficiency technology of Category B, see Sect. 9.1.

Air Supported Vessel Concept

PACSCAT (Partial Air Cushion Supported CATamaran) is a catamaran vessel with *hovercraft/SES capabilities* that has advantages at relatively higher speeds, but still requires very large power to be installed compared to conventional ships. The design started in 2006 and was coordinated between several EU research institutions and companies with the aim to facilitate river freight operations for Rhine and Danube.

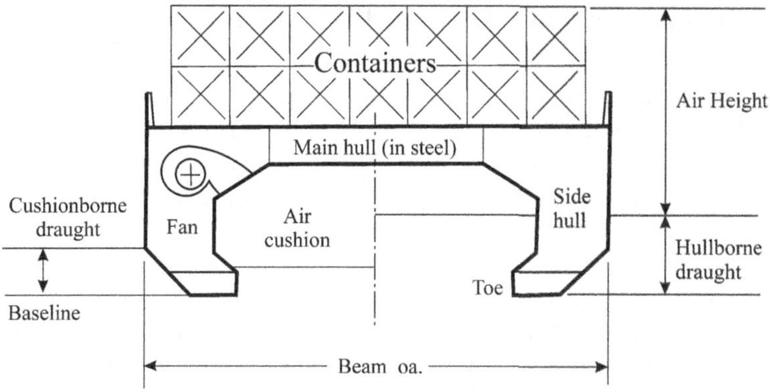

Fig. 10.15 PASCAT concept. *Source* IMAA (2019)

The payload was expected to be around 2000 t with a service draught of 1.5 m to suit shallow waters. A cross-section of this vessel is shown in Fig. 10.15.

Autonomous Vessel Concepts

Autonomous vessels are those that are partly or completely remotely controlled. The autonomy level raises the questions of vessels' safety, manoeuvrability, human factor, etc. Most of the autonomous vessels are still the subject of ongoing projects. Three, which are applied in IW or short sea shipping, are listed below.

NOVIMAR project is currently running under the HORIZON 2020 framework (see NOVIMAR 2017). It introduces the so-called "vessel trains". The Vessel train consists of one lead vessel which fully governs several *digitally connected* following vessels. The followers are able to exit and enter the train in order to facilitate different port locations. The vessel train can be partially autonomous with the future aim to achieve full remote control (unmanned followers).

WATERTRUCK + project's goal is to propose a fleet of vessels consisting of a pushboat and barges, that would be coupled conventionally (not digitally as in NOVIMAR); however, the barges are supposed to be adapted for easy decoupling and remotely controlled navigation afterwards (having just a single skipper onboard or even no crew at all) to enter smaller IWWs. Watertruck + project is in the phase of testing remotely controlled barges.

AUTOSHIP project (AUTOSHIP 2019) is also expected to develop and test two different autonomous vessels and ensure their remote control.

Note that some aspects of autonomous vessel innovations are discussed in Sect. 9.6.3.

10.2 Concepts of Built Advanced IW Vessels

The IW vessels built in the recent past and *innovative at the time they were built*, are reviewed in this section. These vessels may also be regarded as the concept vessels, or demonstrators as they are unique and often experimental as they do not completely comply with the regulations. Sections 10.1 and 10.3 *treat the concepts* which are researched or are proposed but are not yet built. On the other side, Sects. 4.2 and 4.3 *treat the representative ship types*, out of which many are *state-of-the-art vessels*. Consequently, the *advanced vessels* are somewhere in between these two. There is often some overlapping in the stated classification, because the difference between the state-of-the-art vessel and the advanced vessel may be arguable. Moreover, it may be expected that present advanced vessels will become state-of-the-art vessels in the foreseeable future. The term *state-of-the-art* refers here to modern vessels but built with already accepted, matured technology and hence may be delivered in series. The advanced vessels, however, have some notable feature, in most cases more efficient and/or environmentally friendly machinery, equipment or hull form, not yet completely accepted in today's IW vessel designs. Most of the following examples of advanced vessels involve several innovative features, but the emphasis is on LNG or some other clean fuel type, electric propulsion or unusual propulsion arrangements. Note, however, that chosen examples are not necessarily cost-effective vessels, regardless of their (high) building cost. Also, some of their innovative features will probably not be realized in the foreseeable future.

10.2.1 LNG Fueled Vessels

Although used in maritime shipping for decades, the LNG technology started to be applied in IW navigation just in recent years. These vessels are still more expensive than the conventional ones and thus are often subsidized. On the other side, the novel technology is environmentally friendly, as poisonous emissions are significantly reduced, often the operational costs too. In the following, presented are operational vessels designed to use LNG as a fuel. Some of them have additional novel features as well (e.g. electric drive).

MT Argonon

MT Argonon, delivered in 2011, is a dual-fuel powered chemical tanker, reported to be the first IW newbuilding to navigate on LNG as the main fuel (see HARTENERGY 2011). LNG and diesel are used as fuel with an 80:20 ratio. The main engines are two dual-fuel Caterpillars of 1115 kW each; natural gas is the main fuel and diesel is used for ignition. Dual-fuel feature results in reduced CO_2 and NO_2 emissions to a large extent and PM emission to nearly zero (see Sect. 9.2.5). Moreover, taking into account the vessel's lifespan, significant financial savings were expected since the LNG is generally cheaper than diesel. LNG storage tank and the gas handling

Fig. 10.16 Argonon, first newbuilt LNG powered IW vessel *Source* Argonon (2020)

equipment are located on the deck at the midship, see Fig. 10.16. Ship dimensions are $L = 110$ m, $B = 16.2$ m, $T = 4.95$ m. The vessel is built by a Dutch Shipyard Trico as the main contractor.

Ecotanker II and III (Ex Greenstream and Greenrhine)

Greenstream (sister ship Greenrhine, renamed to Ecotanker II and Ecotanker III) is 110×11.4 m, $T = 3.46$ m, 2877 tdw (3124 m^3), IW and coastal double-hull tanker. She is a fully LNG-electric powered ship delivered by Dutch Peters Shipyards in 2012, see Fig. 10.17. Propulsion consists of four LNG fueled gen-sets of 285 kW each powering two electric motors of 500 kW with Z-drive rudder propulsors and a Veth-Jet bow thruster of 380 kW, see Fig. 10.18. Greenstream/Ecotanker II has two LNG tanks with a total capacity of 80 m^3 (size of a 40-feet container each) at a pressure of 3 bar and -165 °C (Ship Technology 2020). Emission savings are reported to be 40% and 80–90% for CO_2 and NO_X respectively, while SO_2 and PM emissions are almost zero. With the vertically movable wheelhouse positioned at the fore-end the ship has a striking appearance.

Ecoliner

The Ecoliner is an IW tanker (110×11.45 m, 2635 t capacity, Fig. 10.19) produced by Damen and developed as a part of the *LNG Masterplan for Rhine-Main-Danube* project (DAMEN 2015). The Ecoliner has a full LNG propulsion plant (see Fig. 10.20). The gas-electric propulsion system consists of two LNG tanks (2×26 m^3) positioned transversely (Note: Ecotanker II tanks are larger and positioned longitudinally), which supply four gas engines; they power four generator sets of 285 kW each (see Provoost 2018). Moreover, the ship has an air lubricated hull and Flex tunnels, see Sect. 10.1.3 and 9.3.8, respectively.

It should be mentioned, however, that the Ecoliner was not abolished of problems (see Provoost 2018). Namely, the original design (when launched in 2016) had two

Fig. 10.17 Greenstream. *Source* Greenstream (2021)

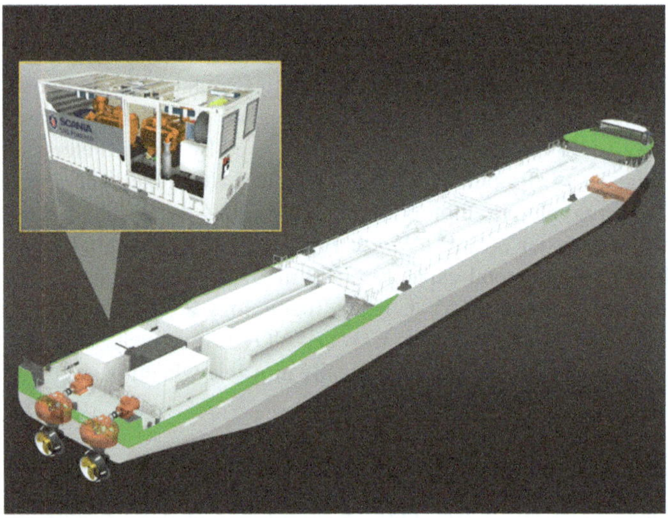

Fig. 10.18 Greenstream's drive components. *Source* Pospiech (2013)

rudder propellers with relatively small propellers in nozzles of 1200 mm. Their efficiency was low (which has nothing to do with other Ecoliner's innovations). Consequently, they have been replaced with electric-direct propulsion with four rudders, i.e. two conventional horizontal shaft lines and propellers in nozzles of 1500 mm. Then the Flex tunnels were added too, thus the efficiency grew by some 10%. Several other deficiencies concerning the LNG technology were also noticed, amongst them

Fig. 10.19 Ecoliner. *Source* Provoost (2018)

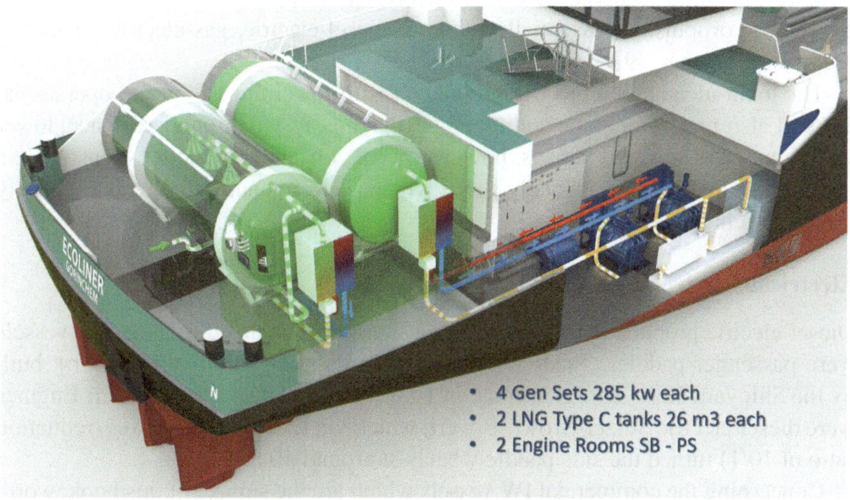

Fig. 10.20 LNG system on Ecoliner. *Source* Provoost (2018)

is the LNG quality, problems with LNG tanks (etc. see Provoost 2018), so the ship returned to service in the middle of 2018, almost two years after the launching. Taking into account the abovementioned, as well as high initial cost and a need for qualified crew, it is questionable whether Ecoliner's original goals—low fuel cost, low emissions and low operational costs—are fulfilled.

Seems, however, that the trend towards the widespread implementation of LNG in inland navigation has lost momentum. The additional investment costs are too high to allow for a business case based on cheaper fuel with given energy throughput in most applications. Nevertheless, Damen Concordia Shipyard received an order in late 2020 for a total of 40 gas-electric type C tankers to be delivered between late 2021 and the end of 2024. They will be chartered by Shell and operated by VT Group/Marlow (see Concordia Damen 2020).

10.2.2 Electrically Driven Vessels

This section presents the vessels powered by an electric motor (EM), i.e. cases where EM is transmitting the torque and delivering power to the propeller. However, the source of electric power may be the generator powered by a diesel engine running on diesel or gas as fuel, batteries (which only store the electricity), fuel cells, etc. Whichever is the case, some other type of energy (e.g. from fuel, electrochemical etc.) has to be converted into electricity. Therefore, the ships as autonomous vehicles, essentially contain (and carry) the energy source and some kind of gen-sets for energy conversion into electrical power. The EM which is powering the propeller belongs to a well-matured technology. Consequently, the term *electrically driven vessels* (or electric propulsion) essentially relate to diesel-electric, gas-electric, or hybrid propulsion, see Sect. 9.4.1.

The main idea behind electric propulsion is to lower emissions and operational costs. It also may offer some additional advantages, as larger onboard space, lower vibrations etc. Delivered vessels, however, were considerably more expensive than the conventional ones, especially those which were intended to navigate as fully electric (on batteries).

Electrically Driven Vessels on the Danube

Diesel-electric propulsion on the Danube is not new. The first diesel-electric vessels were passenger paddlers Stadt Wien and Stadt Passau (77.7 $\times$ 8.4/15.2 m), built by the Shipyard Korneuburg, Vienna, in 1939 and 1940 respectively. Their Engines were diesel/electric Sulzer/Brown-Boveri, which via DEMAG gearboxes (reduction ratio of 10:1) turned the side paddle wheels at around 40 RPM.

Concerning the commercial IW vessels which are the subject of this book, worth mentioning is a towboat Goliath built in 1914, originally powered by a 1000 HP triple expansion steam engine. In 1956 it was converted by MAN/Elin Weiz to diesel-electric propulsion in the Shipyard in Linz, see Fig. 10.21. This remarkable but expensive conversion was very effective and Goliath was not decommissioned until 1974. During her second life, Goliath was successfully towing barges along the whole Danube, with a cargo of up to 10,000 t.

Note that Goliath, thanks to old but relatively efficient paddle wheals, hydrodynamically may be very efficient in shallow water, probably more than contemporary vessels with the propellers. Diesel engines, however, are nowadays much cleaner and more efficient than 50 years ago.

Roethelstein is an Austrian diesel-electric, shallow draught, river icebreaker built in 1995, (Ship & Boat International 1995). Trials proved that the vessel was capable

Fig. 10.21 DDSG diesel electric towboat Goliath 66.0 × 8.52/16.5 m, 2 × 800 HP (2 × 597 kW) and 500 RPM. *Source* Franz Dosch

of penetrating 4 m thick ice ridges and breaking 0.7 m level ice at the speed of 1.5–2 km/h. The hull has a cylindrical bow, parallel mid-body and a stern designed for podded azimuth propulsion units, see Fig. 10.22. Roethelstein is an interesting vessel because of its propulsion system, i.e. application of relatively small Azipods. The diesel-electric propulsion consists of two diesel generator sets (2 × 700 kW at 1500 RPM) which are powering two Azipods (2 × 560 kW at 550 RPM).

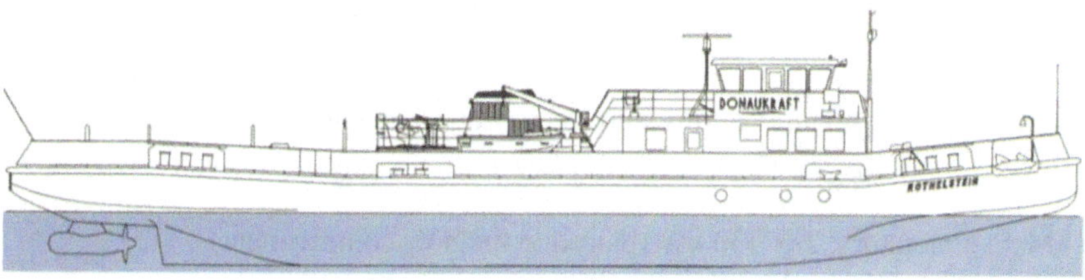

Fig. 10.22 River icebreaker Roethelstein 42.3 × 10.3 × 3.35 m, $T = 2$ m, bollard pull 125 kN. *Source* Ship & Boat Int, 1995

Semper Fi

Semper Fi, delivered in 2012, is IW vessel of 110 × 11.4 m with hybrid propulsion, developed by MARIN institute and Dutch company Carpe Diem Shipping (Semper Fi 2012). Propulsion consists of two diesel engines of 512 kW (diesel-direct propulsion) and two electric motors of 330 kW (diesel-electric propulsion). The core of this specific hybrid propulsion are two VETH Z-drives (rudder-propeller units of 845 kW with contra-rotating propellers[3]) that enable diesel-direct or diesel-electric propulsion, or a combination of both. Owing to the installation of VETH Z-drives with CRP, a compact engine room and smaller propeller diameter than on comparable IW vessels was obtained. Therefore, the fully loaded vessel is intended to be propelled by diesel engines, while in empty condition or when sailing downstream, electric motors are employed. Diesel engines are supplied with an integrated exhaust post-treatment system (catalytic converters and soot treatment) that additionally minimize noxious emissions. Summarizing, around 10% savings in terms of energy consumption are expected to be achieved.

Sendo Liner

Sendo Liner was awarded the "ship of the year" prize (see SWZ Maritime 2019) for the design and innovations (see Fig. 10.23). Its main advantage is propulsion configuration (diesel-electric supplemented with a battery pack). Hull structure consists of double skin over the entire length of the ship, including the engine rooms, for increased safety. The owner and builder were Sendo Shipping and Concordia Damen, both from the Netherlands.

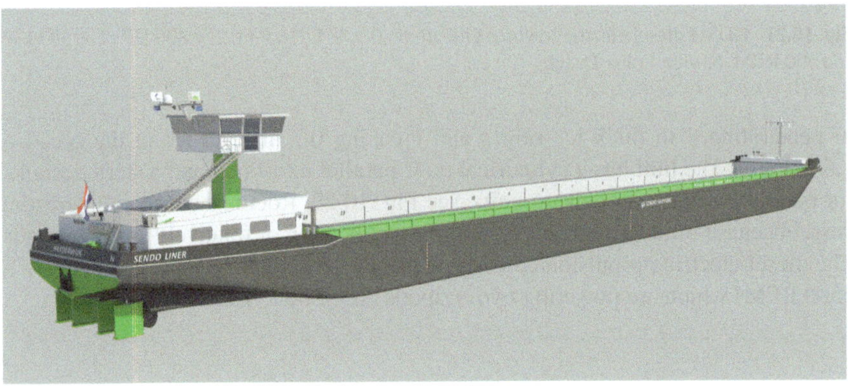

Fig. 10.23 Sendo Liner 110 × 11.4 × 4.25 m, $T = 3.65$ m. *Source* Concordia Damen

The following features of this ship are chosen to be highlighted here:

The Propulsion plant

- Main generators: 2 × 430 kW (placed in the bow compartment)
- Battery bank 564 kWh (placed in the stern compartment)
- Two electric motors, each of 350 kW, driving the propellers (max speed 21 km/h)
- Most of the time ship operates with only one diesel generator working
- Battery capacity is sufficient for 2–3 h of sailing with zero emissions
- Batteries are charged by the onboard generators or shore power
- Two bow thrusters of 300 kW each, provide thrust 360° around.

[3]Note that CRP are usually used on relatively fast and light vessels (e.g. passenger ships). Selfpropelled IW ships usually have propellers in (short) nozzles.

The Hull

- Optimization of the hull form performed by CFD
- Stern part (shape of the tunnels) optimized for the most common operating condition: shallow water of just 1.5 m below the keel and speed of 14 km/h (instead for maximal speed of 18 km/h in deep water, as is usually done)
- Increased payload by approximately 8% (by stowing 14 instead of 13 TEU bays longitudinally). This means the reduction of fuel consumption per cargo unit by 32%, and CO_2 emissions by 40% per container transported.

De Alphenaar

An enterprise called Zero Emission Services B.V. (ZES 2020) planned to start leasing of *exchangeable battery containers* (ZES-Packs) to inland shipping in 2020, to pursue emission-free navigation. Members of ZES consortium, which is supported by the Dutch Ministry of Infrastructure and Water Management, are Port of Rotterdam, ING and Wärtsilä, among others. The aim of the project is to completely electrify the IW transport and thus make it a more acceptable transport mode.[4] The batteries are intended to be charged from the renewable energy resources located at open-access charging points – loading stations. The intention is to replace the battery containers (ZES-Packs) with charged ones in a short time. The capacity of a single ZES-Pack is about 2 MWh, hence a typical barge of 500–1000 kW will have to swap a ZES-Pack every 2–4 h (EV Barge 2020).

The electrically powered IW vessel De Alphenaar (see Fig. 10.24) is supposed to use ZES-Packs. She is carrying beer from the Heineken brewery (Alphen aan den Rijn) to Port of Moerdijk in the Nederlands. The vessel's autonomy should be 50–100 km on two ZES-Packs, resulting in the expected reduction of the carbon emissions by 1000 t per year per ship. Another five vessels with exchangeable battery containers are expected to be employed in 2021 (by 2030 even 150), as loading stations expand along the waterways.

[4] See discussion on energy carriers and energy storage devices production in Sect. 9.2.7.

Fig. 10.24 Battery powered vessel De Alphenaar, 90 × 10.5 m. *Source* ZES (2020)

E-ferry Ellen

E-ferry Ellen (navigating since 2019) was developed as an EU project (under the HORIZON 2020 framework) called E-Ferry (E-Ferry 2020). The project's aim was to emphasize "green" sailing without polluting and CO_2 emissions, thus establishing an energy-efficient design. Although aimed for the seagoing navigation (with dimensions compatible with the IWW restrictions $T_{MAX} = 2.5$ m) this ship is completely running on the batteries. Ellen's operating costs are said to be 75% lower, but the vessel is about 40% more expensive compared to the conventional vessels of the same type.

Ellen's technical characteristics are the following: 59.4 × 13.4 × 3.7 m; lightweight 650 t, capacity 31 cars or 5 trucks on the open deck and 200 passengers; power 750/1000/1500 kW for speeds 13/14.5/15.5 kn. The hull is made of steel and the bridge of aluminium, while deck furniture was made of recycled paper! Batteries are located in two battery rooms below the main deck and are intended to be charged in the harbour. Their capacity of 4.3 MWh out-passes any other electric vessel. The range of navigation is 40 km and designed area of operation is a coastal area in Southern Denmark.

10.2.3 Demonstrators and Other Advanced Vessels

In the following, presented are vessels that have some other features besides LNG or electric propulsion as the main aspect of the design, e.g. air lubrication, optimized hull form, specific propulsion or hydrogen used as a fuel.

Futura Carrier

Futura Carrier was delivered in 2007, along with a few other "Futura type" vessels based on the same innovations (Renner 2007; NEW-LOGISTICS 2007). German Federal Ministry for the Environment, Nature Conservation and Nuclear Safety co-funded the project. The ship has an innovative semi-catamaran hull form (fully catamaran shape at the bow only, see Fig. 10.25), with two rudder propellers at the stern and identical two at the bow placed in the specially designed tunnels, see Figs. 10.26 and 10.27. Power is generated by four diesel engines having an output of 600 kW each. Furthermore, air lubrication is applied (for reduced frictional resistance, see Fig. 10.27 and Sect. 10.1.3) for the first time on a European IW vessel. The hull form is also optimized for operation in shallow water.

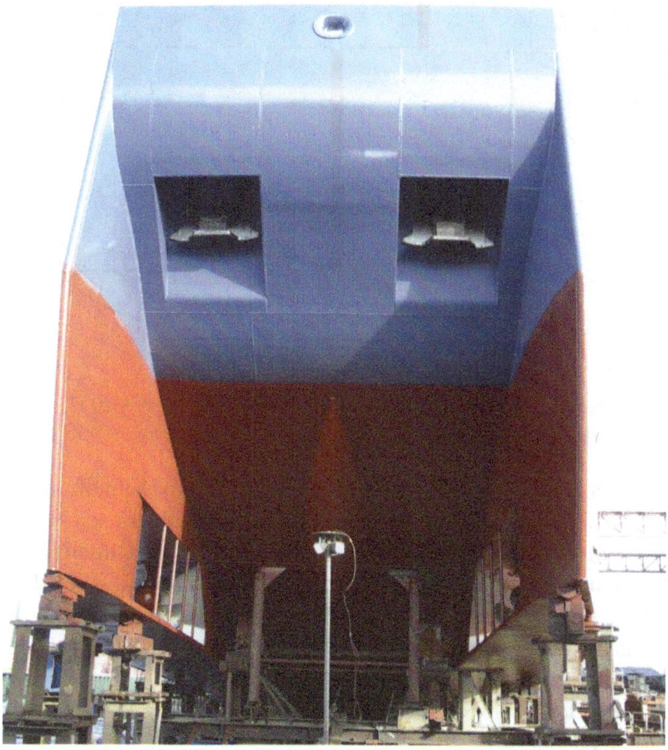

Fig. 10.25 Futura Carrier bow. *Source* NEW-LOGISTICS (2007)

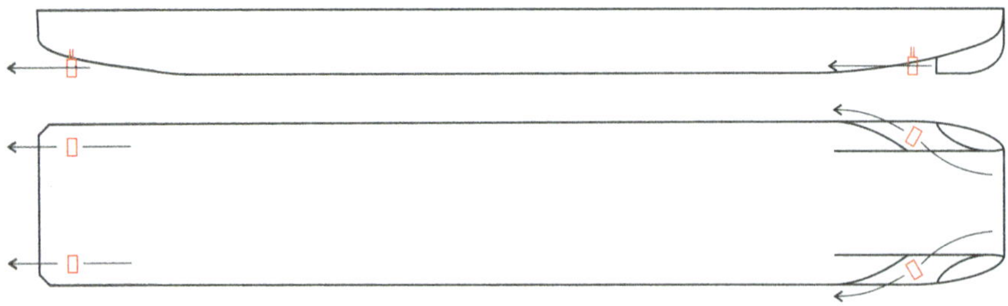

Fig. 10.26 Futura Carrier drives. *Source* NEW-LOGISTICS (2007)

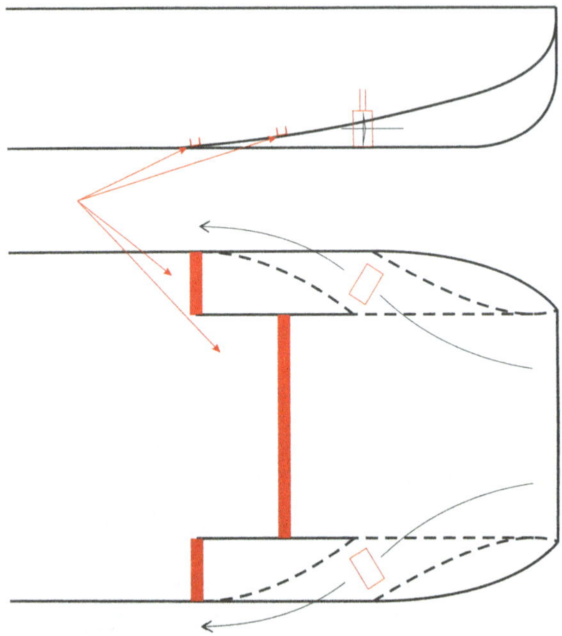

Fig. 10.27 Futura carrier air bubble injection. *Source* NEW-LOGISTICS 2007

Futura Carrier is a river-sea vessel and as such cannot be soundly compared with the river vessels (some design compromises had to be made for sea sailing). Consequently, several innovations that were employed are not as attractive for the river vessels as they may seem at the first sight. The main dimensions of the vessel are 97.5 × 13.6 m. Other already built Futura-type vessels have different dimensions, while a similar vessel of 134.6 × 20.8 × 6 m was anticipated. Concerning air lubrication (bubble lubrication system, see Sect. 10.1.3), extensive model- and full-scale tests were conducted with Futura-type vessel Till Deymann, but obtained results seams not to be encouraging.

Hydrogen Propulsion

Emeli is a training cargo vessel with main dimensions 55 × 7.2 m, used by the staff of Maritime Academy Harlingen, Netherlands. She uses both diesel-electric and hydrogen propulsion. The latter consists of hydrogen storage supplying the 30 kW fuel cells to produce electricity for the EM, along with a 60 kWh battery pack (INE 2019). This can allow 10 h of sailing on fuel cells with the emission reduction (CO_2, NO_2, PM) of 100%. The hydrogen innovations were made through the retrofit of the vessel originally built in 1961.

Hydroville (Internationales Verkehrswesen 2018) is a passenger vessel, with dimensions 14 × 4.2 m, having two dual-fuel hydrogen-diesel co-combustion engines with a total power of 441 kW. Fuel storage consists of 12 hydrogen tanks (205 l at 200 bar each) and two diesel fuel tanks (2 × 265 l) for backup fuel. This kind of propulsion allowed the following savings: 58–73% of CO_2, 65% of NO_X and 57% of PM emission. Same as Emeli, Hydroville was initially built as a demonstrator, but currently is in service as a shuttle vessel for CMB—a maritime group from Belgium. The vessel was delivered in 2017 and given a "Port of Antwerp Sustainability Award" in 2018.

10.3 Proposed Concepts for the Danube River

It is generally regarded that decisions made in the early design phases are by far the most influential from the viewpoint of cost-effectiveness. Namely, of all achievable savings, the good design probably contributes 3/4, while all other savings (reconstruction, retrofitting and modernisation of existing vessel) contribute up to 1/4, usually much less. Consequently, three new designs are the subject of this section, while conversions and retrofitting options are considered in Chap. 11.

To demonstrate how a contemporary, safe, cost-effective, shallow draught vessel considered particularly for the Danube waterway should look like, some of the energy-efficiency technologies discussed in previous sections will be incorporated into the design of the following three specific ship types:

- Selfpropelled, narrow body general cargo and container vessel
- Selfpropelled, wide body (beamy) general cargo and container vessel
- Pushboat (for bulk cargo transport by conventional barge convoys).

These three distinct ship type concepts are chosen because they are good representatives of typical present and near-future vessels for the Danube waterway. Tankers of all kinds, passenger ships and other vessel types are out of the book's scope. However, this does not mean that proposed features relate just to the above three ship types, nor that innovations integrated into one concept cannot be applied in the other. Moreover, for all three concepts it is assumed that the *present-day technology* and *current Danube river infrastructure* are taken into consideration in their design.

Note however that the LNG technology is omitted, as is already discussed in Sects. 10.1.1, 10.2.1 and 9.2. It is a specific subject that requires a different approach from the outset, thus would draw too much attention from the other issues which are very important for IW vessels.

Needless to say, the new vessels have to be built (hence designed and operated through their lifetime) *in compliance with various international and national rules and regulations*. This is, per se, *a guarantee that the vessel will be safe and environmentally acceptable*. Also, it is usually underlined that ship and her equipment will be made *according to good shipbuilding practice and experience, standards of the yard* etc; those phrases, however, do not mean much as they are not so well-defined, although used worldwide.

10.3.1 Narrow Body Selfpropelled Vessel[5]

In this case, special attention is paid to existing navigation conditions on the Lower Danube, but as it is expected that the container vessel would operate on the whole Danube, restrictions on the Upper Danube are also taken into account. An optimal vessel is briefly explained in Sect. 2.2, restrictions of the Danube waterway in Sect. 2.4 and its implications on ship design in Sect. 2.5. The cargo that should be transported—intermodal loading units (ISO containers and EILUs), their size etc. are discussed in Sect. 3.2, while the design recommendations for conventional IW selfpropelled container vessels are given in Sect. 8.7.

Proposed Features of a Concept Ship

The objective here is to explain in words (general arrangement plan is discussed later) how successful shallow water general cargo and container vessel could look like. Consequently, the recommended vessel dimensions and other important characteristics to be incorporated into the design are the following:

- *Draught (T)—2.5 m maximum for three layers of loaded containers* of the average mass of 13 t. *With two layers of loaded containers, draught will be up to 1.85 m.* Nevertheless, according to transport statistics, on average 2/3 of all containers are loaded and 1/3 are empty (with a mass of approx. 2 t only). Therefore, it might be expected that in reality draught will be smaller than stated above.
- *Breadth (B)—11.65 m (cargo hold breadth just above 10.3 m)* allows abreast loading of four ISO containers or 2.50–2.55 m pallet-wide domestic containers—EILUs (see Sect. 3.2). Note that the usual ship breadth is up to 11.45 m (due to locks and gangway restrictions), but keeping in mind the extensive use of pallet-wise EILUs within Europe (other transport modes are using them) and that the competitiveness of IWT should be increased, a ship width of 11.65 m is suggested (although locks on the Upper Danube are 12 or 24 m wide, see

[5]The material presented in this section originates from Radojčić (2009).

Table 2.2). Note that for the Rhine corridor the breadth of 11.65 m was requested, but is not allowed yet. Concerning the Danube, downstream of Vilshofen the allowed breadth could be anything up to 23.4 m (see Sect. 2.5). Larger breadths, generally bring some drawbacks too, thus a breadth of 11.65 m was chosen as a good overall compromise.

- *Length (L)—104 m follows from the desired cargo hold length of around 80 m.* Within this length, *13 TEUs may be stowed longitudinally* with 20–50 mm clearance between them (this requires top-lift transhipment with a spreader). A hold length of 80 m allows also a wide variety of other stowing possibilities, for instance (6 × 40' + 1 × 20'), (4 × 45' + 4 × 20'), (4 × A1360 + 4 × 20'), (9 × C745 + 2 × 20'), etc. Discussion given in Sect. 8.7.2 (long or beamy vessel) and Sects. 9.4.1 and 10.3.2 (*L/H* ratio) explains why the longer ship is not recommended. Furthermore, with this ship length, a coupling train with a standard 77 m Danube barge would be shorter than 185 m, still acceptable for passing through Danube locks (see Table 4.1a).
- *Height (H)—3.1 m.* This is a discussible subject and is beyond the scope of this study. Namely, as stated in Sect. 8.3, according to UNECE, 2006 Resolution No 61 for Ship type C in Zone 3 minimal freeboard is 150 mm and minimal safety distance is 500 mm. This implicitly defines the minimal hatch coaming height of 350 mm. Commonly, due to longitudinal strength requirements, container stuck arrangement and desire that coaming stands also as a hatch railing—hatch coaming is much higher.
- Moreover, taking into account some recent disasters due to insufficient safety clearance (see Hofman et al. 2006) *F = 0.6 m and a coaming height of 1.1 m* is suggested, which is more than required by the rules. Besides, $H = 3.1$ m also satisfies DNV-GL requirement for $L/h < 35$.
- *Ship form—should be optimised for low resistance navigation in shallow water.* The form should be relatively full (both C_B and C_P around 0.9), with full fore- and after-body providing substantial buoyancy (hence allowing larger payload) at low draughts. This, however, inevitably increases resistance. After body is strongly influenced by propeller diameter and propulsor type (twin rudder-propulsors with propellers having a relatively small diameter are imagined here). Ship form should mirror longitudinal weight distribution (accommodation in the front, engines at the stern) which should reduce the trim of a partly loaded ship. The above-water bow form should be adapted for pushing (coupling train formation). Note, however, that hull form optimisation with the purpose to reduce resistance requires model testing (see Sects. 8.2.1 and 9.3.2).
- *Ship weight—should be reduced by around 10% compared to conventional designs* by applying state-of-the-art technologies, probably high tensile steel for the hull structure, SPS or aluminium for superstructure (see Sect. 9.5). Capital weight savings, however, should not be expected, but overall weight savings within the common steel-building approach might be obtained (see Sect. 9.5.1). Although somewhat opposite to the weight savings, ballasting is often necessary, so ballast tanks should be considered too.

- *Propulsion—two rudder (azimuthing) propellers in nozzles with* $D \approx 1.35$ *m* optimised for both low draught and full draught operation, see Sect. 9.4.4. Propulsors may be (a) of innovative design, for instance *diesel-electric* (see Figs. 9.38 and 9.39 for, respectively, electric podded and integrated L-drive propulsors), or (b) conventional mechanical *"Z drive" rudder-propellers* (see Fig. 9.37a. Both will eliminate the need for rudders and will also enable exceptional manoeuvring capabilities. If diesel-electric propulsion is envisaged then an innovative tip-driven propulsor might also be considered (see Sect. 9.4.5). A bow thruster of around 250 kW (with the ability to assist stopping and improve thrust—4-channel) either diesel or electrically driven, should be considered (see Sect. 9.4.6).
- *Engines*—Newly installed diesel engines must comply with the *limit values of the NRMM Regulation Stage V*. These engines are equipped with exhaust gas aftertreatment systems (see Sect. 9.2.3). If diesel-electric propulsion would be applied, then a power of around 4 × 400 kW is suggested; for mechanical rudder propellers around 2 × 700 kW or so would be sufficient. Power is estimated for an assumed speed of 16 km/h, as well as a coupling convoy formation with one, probably two Danube barges (depending on waterway conditions). Diesel-electric propulsion is ecologically very attractive, but is also more expensive than mechanical transmission.
- *Shore-to-ship-power supply* (of electricity while in port, often called *cold ironing*) should be considered with the aim to reduce onboard diesel emissions.
- *Accommodation, wheelhouse and engine room.* All crew premises should be dimensioned according to ES-TRIN recommendations (see Chapt. 5) based on six crew members and should be positioned in the bow (the wheelhouse too), while the engines (placed in well insulated spaces) should be at the stern. This enables good visibility (hence safety too), crew comfort (no vibrations and noise) and a well-balanced ship at low draughts.
- *Electronics* and computerisation should be of the latest generation, providing a *one-man bridge system* including a data-based ship monitoring system (engine and ship-system monitoring and recording, voyage optimization, etc.); see Sect. 9.6.
- *An onboard crane with a capacity of 35t/30 m* could be considered as this would allow transhipment at any port (see Fig. 4.20). However, this would reduce the number of containers that could be transported and require ballasting for stability.

It should be noted, however, that the shipowner, according to his own requirements, judgments of the market trends and costs, usually requires a specific ship to be designed/built (with particular dimensions, carrying capacity, engines, equipment etc.) and that "design freedom" as exercised above is very seldom. For instance, ship speed, which is amongst the most influential design parameters, was omitted in this discussion.

General Arrangement Plan of a Concept Ship

General Arrangement (GA) plans of two variants of a container ship concept are shown in Figs. 10.31 and 10.32, and its enlarged bow and stern parts, in Figs. 10.28, 10.29 and 10.30, respectively. Namely, diesel-electric propulsion and a conventional

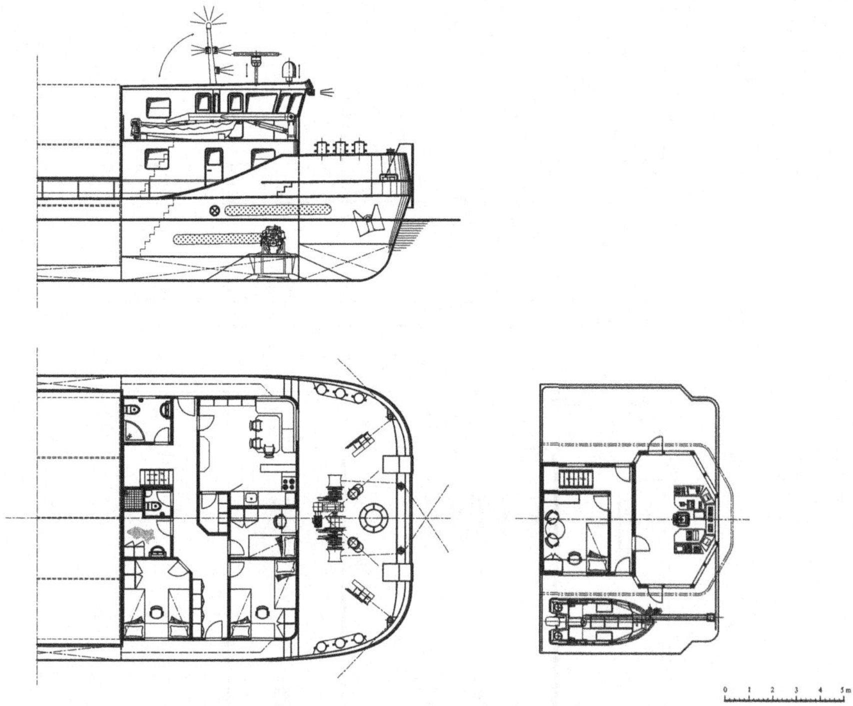

Fig. 10.28 Enlarged bow part (both variants)

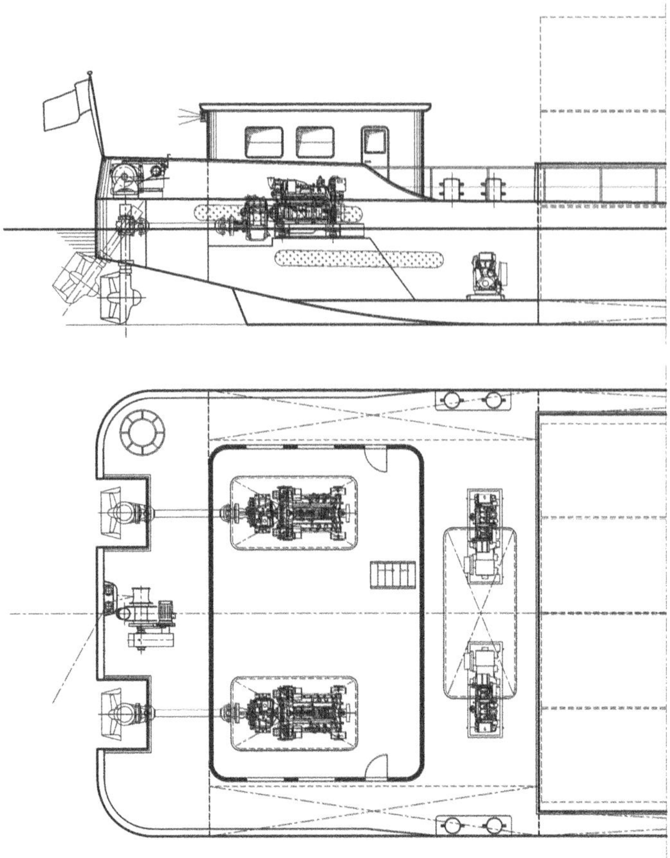

Fig. 10.29 Enlarged stern part—variant with conventional “Z drive” rudder-propellers

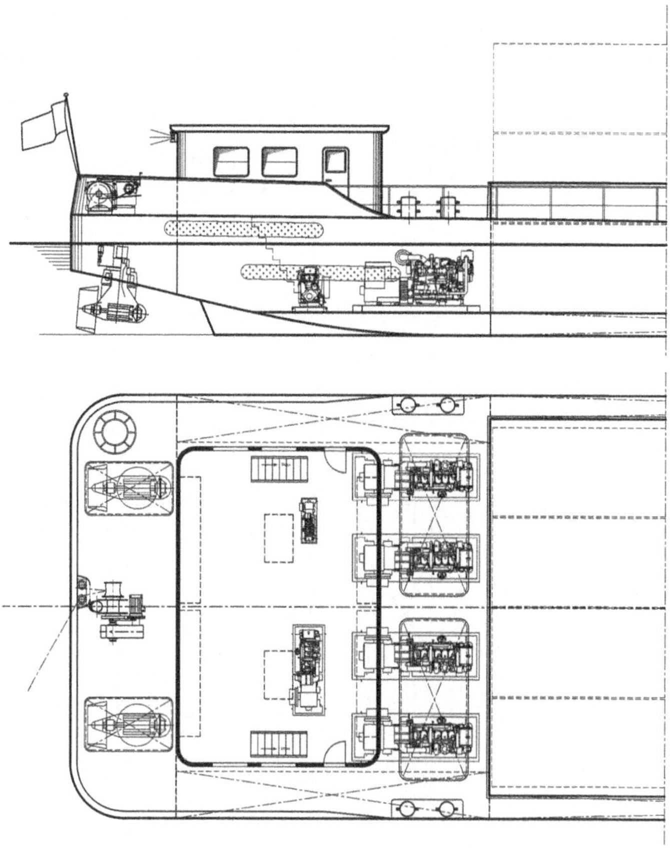

Fig. 10.30 Enlarged stern part—variant with diesel-electric propulsion

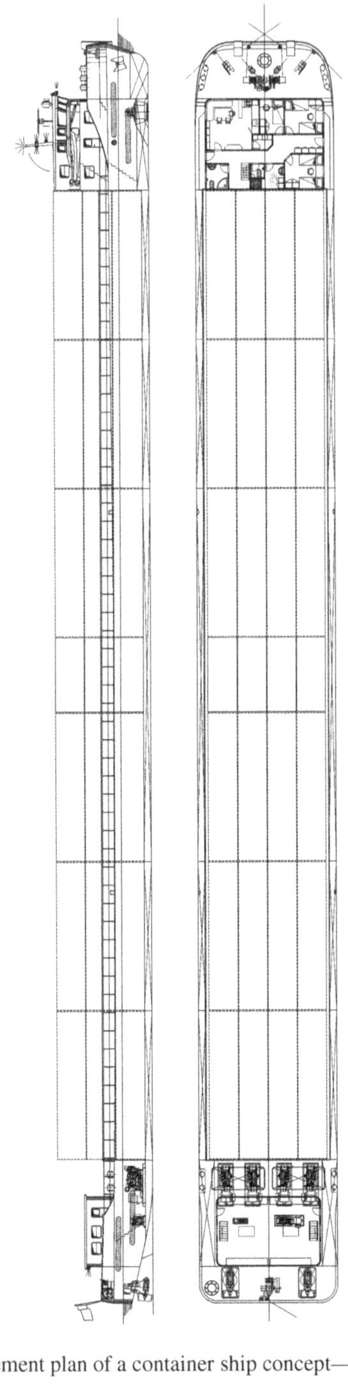

Fig. 10.31 General arrangement plan of a container ship concept—diesel-electric propulsion

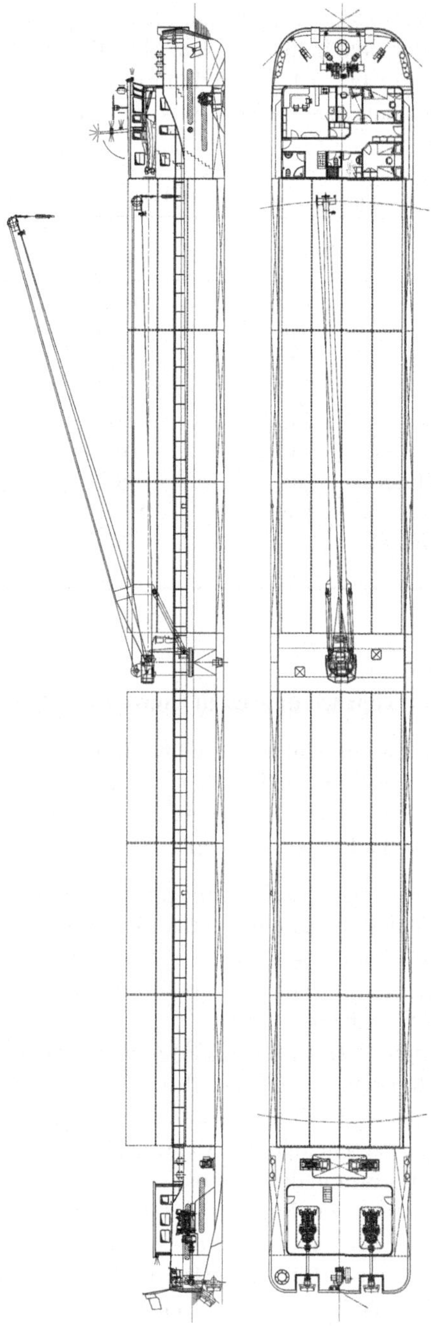

Fig. 10.32 GA plan of a container ship concept—variant with conventional rudder-propellers and onboard crane

mechanical (azimuthing) rudder-propeller may be employed (Figs. 10.29 and 10.30). An on-board crane is depicted in the GA plan, Fig. 10.32 (with a mechanical transmission already shown in Fig. 10.29) resulting in reduced carrying capacity. The main particulars of the abovementioned vessels are given in Table 10.1.

Table 10.1 Main particulars of proposed selfpropelled ship

Characteristics	Dimension	Configuration	
		Basic	With a Crane
Loa	m	104.0	102.5
Boa	m	11.65	11.65
H	m	3.1	3.1
T	m	2.5	2.5
Hold length	m	80.0	78.5
Hold width	m	10.34	10.34
Height above basis line	m	8.3	8.3
P_B	kW	4 x 400	2 x 700
TEU (3 layers / 4 layers)	-	156/208	134/172
Payload capacity	t	1950	1800

Advantages of this Concept Compared to Conventional Vessels

Some of the concept's features underlined below suggest an environmentally acceptable vessel with a large volume and payload capacity. At the same time, due to superior manoeuvring capabilities, the proposed concept should be safer than similar selfpropelled vessels on the Rhine and Danube. So:

- *Special attention was paid to low-draught performance.* Consequently, the proposed concept should be able to operate successfully and therefore be cost-effective at both low *draught of up to 1.7–1.8 m* (with two container layers) and *full draught of up to 2.5 m* (with 3 layers of full containers or even 4 layers of mixed full and empty containers).
- The chosen *hold length (80 m) and breadth (10.34 m) allow stowing of a variety of 2.50–2.55 m wide domestic containers* (EILUs of C745 and A1360 type), besides the usual ISO containers (TEUs and FEUs). The same hold length and somewhat narrower breadth has a contemporary MGS, which is 110 m long (vs. proposed 104 m). The concept's overall dimensions *104 × 11.65 m allow passage through all Danube locks,* even in coupling convoy formation.
- An *on-board crane would allow transhipment at any port* which would be an advantage particularly on the Middle and Lower Danube where adequate container ports (hubs) and dedicated container transhipment equipment are rare. However, this would reduce the number of containers that could be transported.
- *Rudder propulsors enable exceptional manoeuvring characteristics* (steering & stopping) even at low draughts, resulting in a safer ship. If diesel-electrical

propulsion is installed, additional benefits would be evident, i.e. better adaptation to various operation/sailing modes (upstream/downstream, speed and coupling train formation requiring the employment of 1, 2, 3 or all 4 diesel engines). This would also reduce fuel consumption (probably up to 10% in upstream and even more in downstream navigation); therefore emission levels would be lower as engines would run at optimal RPM/loading. For refrigerated containers and other large electricity consumers (a bow thruster for instance), the same electrical network could be used, eliminating the need for auxiliary units. Conventional, mechanically-driven rudder-propellers have the advantage of being cheaper and retractable (hence can better adapt to water depths).

- The latest engine generation *emits less air pollutants* (nitrogen oxides and particulate matter) and significantly less than the Stage IIIa or older engines. In addition to the engine classes IWA and IWP, marinised NRE engines of up to 560 kW and Euro VI truck engines (about 250–400 kW) are also approved for use on inland waterway vessels (see Table 9.1). They emit even fewer air pollutants and are generally lighter, more compact and can be economically advantageous due to their high efficiency and lower cost. Smaller engines can show their advantages particularly in diesel-electric propulsion systems adapted to the operating profile (see discussion on right-sizing in Sect. 11.2). Diesel-electric systems are more expensive than diesel-direct drives (due to the generators and electric motors) but offer other advantages, e.g. low noise and emissions. Moreover, if a ship is equipped with adequate batteries it can operate locally and temporarily without any emissions. This would be a *way towards gradual decarbonization* until the alternative technologies, such as FCs, become technically and economically mature technologies.
- The position of the engines/engine room at the stern and the crew premises at the bow offers *additional crew comfort* (reduced vibrations and noise). The application of contemporary equipment and electronics enables safer sailing and lower overall operational costs.

Final Notes on Narrow Body Concept Ship

An alternative to electric *rudder propellers* in nozzles (whether electric podded or integrated L-drive) would be *electric-direct* propulsion with conventional horizontal shafts and electric motors placed within the hull (see Sendo Liner and Ecoliner, Sects. 10.2.2 and 10.2.1, Figs. 9.21, 10.23 and 10.14). That would allow installation of larger ducted propellers and for low draught operation adjustable retractable tunnels would be beneficial. The efficiency of this installation would certainly be higher, though the manoeuvrability would be a bit worse. A 4-channel thruster is required in both cases to enhance the stopping ability at low draughts. This kind of drivetrain, however, requires the rudders, most probably two per propeller as well as adequate steering gear.

10.3.2 Concept of Wide Body (Beamy) Selfpropelled X-Type Vessel[6]

The proposed concept of a container vessel, named X-Type, is investigated in Bačkalov et al. (2016). The vessel development is governed by an idea to explore possibilities for design improvements of the Danube container vessels, especially taking into account that the climate changes could lead towards the lower river depths and interrupted navigation in the future. Therefore, the novel design, in terms of principal dimensions, should have *shallow draught and increased breadth* compared to conventional IW vessels. This should result in continuous year-round service, thus the number of carried containers would be maximized. Summarizing, the following four design objectives of the wide body X-Type concept should be emphasized:

- continuous year-round service on the Danube
- number of carried containers should be maximized
- deadweight should be maximized
- vessel speed should make the container service attractive and competitive.

The vessel should be designed to satisfy the requirements of the Directive 2006/87/EC (EC 2006), ADN (UNECE 2015) and DNV-GL (GL 2011), see Chapt. 5. Due to the larger length to height ratio, this kind of vessel would generally be regarded as unusual or unconventional from the standpoint of the rules of the classification societies, and hence, required to be additionally assessed by direct longitudinal strength calculations.

Genesis of X-Type Vessel

In Bačkalov et al. (2014) the draught and breadth of IW ship for the Danube waterway were varied systematically. The starting point were 4 vessel series (A- to D-Type) with a standard breadth of 11.4 m, while the draught was systematically changed from 2 to 2.8 m, freeboard from minimal recommended (0.15 m) to that evaluated from the ratio $L/H < 35$ valid for the "usual" design (see Sect. 9.5.1) and ship lengths 78.75–135 m. After thorough analysis, the E-Type vessel was derived with beam and length, and draught and height, of 15 and 103.75, and 2 and 2.15 m, respectively. It was concluded that in terms of cargo-carrying capacity and for water depths of up to 3.5 m, wide, shallow draught vessel is more cost-efficient than the standard selfpropelled IW ship. However, it was also surfaced that special attention should be paid to structural strength, i.e. *direct calculations*, as the ratio L/H is much larger than the usual 35. Consequently, the direct calculations of structural strength are placed in the focus of E-type, and as will be seen later, in X-Type vessel too. Essentially, the E-Type vessel ($L/B = 6.92$, $L/H = 48.26$) is a kind of prototype for the X-Type ($L/B = 7.46$, $L/H = 41.52$) vessel, although the second is breadth optimized from the container carrying capacity viewpoint. The design objective for both vessel types is to prolong a regular service on the Danube throughout most of the year. The analysis

[6]The main source of information in this section is Bačkalov et al. (2016).

also revealed that standard Danube IW vessels could be improved in the early design stage by rational selection of main dimensions.

Main Particulars of Wide Body (X-Type) Vessel

Low draught is the main feature here since the vessel is conceived to operate during most of the season, taking into account the Danube depth restrictions. Consequently, the *maximal draught is chosen to be 2.3 m*, as a compromise between prolonged navigation during the dry seasons and the possibility to carry four container layers during most of the service (for instance while sailing from the Black Sea to Vienna). In the case of lower depths, navigation with three container layers should be considered.

Freeboard of 0.2 m is adopted which is just slightly above the minimum required, resulting in the *side height of 2.5 m*. The length and breadth of the vessels are influenced by cargo capacity. Thus, to allow the fifth container row abreast, the *breadth of 13.9 m* is chosen. This allows navigation along the whole Danube, except on some upper sectors due to the presence of narrow locks (see Table 2.2). This restriction, however, may be regarded as the main disadvantage of this design.

The length is determined by the fact that the vessel should enter the locks when coupled with a standard 77 m barge. Consequently, a *length of 103.8 m* is proposed, which enables competitive cargo capacity by accommodating *12 container bays*. Further lengthening would have a negative effect on the longitudinal strength. The X-Type vessel is conceived to be diesel powered with twin ducted propellers (propeller diameter of 1.55 m). The proposed wide body ship (X-Type) is illustrated in Fig. 10.33, whereas her main particulars are given in Table 10.2.

Fig. 10.33 3D model of the X-Type vessel. *Source* Bačkalov et al. (2016)

Comparisons with Other Danube Vessels

The proposed X-Type vessel is compared with other Danube concepts presented in IDV (2014), NEWS (2014) and Radojčić (2009),[7] see Table 10.2. The Block coefficient of all vessels is assumed to be 0.88, as it is a typical value for IW ships. Summarizing, X-Type vessel has a higher deadweight coefficient and number of TEUs but assumed a smaller average container mass (m_{CON}) than for other concepts. This may also be considered as a disadvantage of the proposed design.

The study also compares the designs in terms of the number of days necessary to transport the reference amount of cargo, while taking into account the potential draught. The reference amount of cargo is actually the maximal annual cargo carrying capacity of the X-Type vessel. Since m_{DWT} is known, it is possible to estimate the number of days necessary to transport the same amount of cargo by each vessel. The final comparison is presented in Fig. 10.34, while the calculation procedure can be found in Bačkalov et al. (2016). Therefore, while navigating on their design draughts and in unrestricted water depts, the IDV and X-Type concepts need almost the same number of days to transport the reference cargo; NEWS vessel would need a week less, while Radojčić concept would need seven weeks more for the same cargo.

Table 10.2 Main particulars of the existing danube container vessel concepts and the proposed X-Type design

Concept	IDV (2014)	NEWS (2014)	Radojčić (2009)	X-Type (2016)
L [m]	105	110	104	103.8
B [m]	11.40	11.44	11.65	13.90
T [m]	2.80	2.97	2.50	2.30
H [m]	3.10	4.50	3.10	2.50
TEU	192	156	208	240
m_{DWT} [t]	2261	2299	1970	2248
m_{LWT} [t]	688	990	696	671
m_{DWT} [/]	0.767	0.699	0.737	0.770
m_{CON} [t]	10.6	13.3	8.5	8.4

Note: 1) m_{DWT}, m_{LWT}, m_{CON} and $\eta_{DWT}= m_{DWT}/\Delta$ are mass of deadweight, mass of lightweight, average mass of containers and deadweight coefficient, respectively
2) For Radojčić (2009) concept the actual number of TEU is 208, not 192 (see Sect. 10.3.1) as given in the original Bačkalov et al. (2016) paper. This is corrected throughout

Source Bačkalov et al. (2016)

[7]Radojčić (2009) concept is identical to the narrow body concept presented in Sect. 10.3.1.

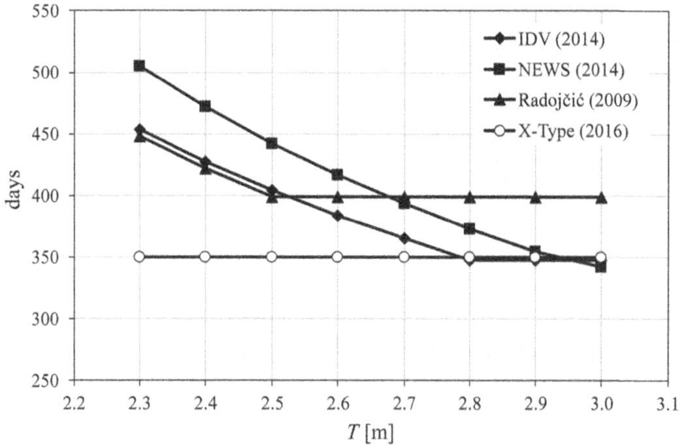

Fig. 10.34 Number of days necessary to transport the reference cargo by each of the examined vessels, depending on potential draught. *Data from* Bačkalov et al. (2016)

When navigating in shallow water conditions, the values change with the variation of draught; i.e. cargo carrying capacities of NEWS, IDV and Radojčić concepts decrease considerably, whereas as expected, the X-Type's performance stays unaffected (see Fig. 10.34). Accordingly, for the draught of 2.3 m, IDV and Radojčić vessels need 3.5 and NEWS vessel 5 months more to transport the same annual reference cargo as the X-Type vessel. Moreover, in the case of restricted navigation, the mass of containers on these three deeper draught vessels also decreases.

Since the speed of the vessel influences its performance, and due to the potential squat effect, the same vessels are analyzed in that regard as well. It was evaluated that the wide body, low draught vessel has an advantage, as in shallow waters it could potentially navigate at higher speeds than the other designs.

Energy Efficiency of X-Type Vessel in Limited Water Depth

Energy efficiency analysis is given in Bačkalov et al. (2016), but it should be noted that it is a bit unusual and probably questionable. Namely, in the basic *EEDI* equation, the mass of deadweight (m_{DWT}) figures in the denominator, while here the number of TEU containers is taken instead (see Chapt. 6, particularly Sect. 6.2.3 where Modified *EEDI**, which accounts for the waterway depth, is discussed). Thus, $EEDI_{TEU}$ is given in g CO_2/TEU · km, which is different from the common *EEDI* given in g CO_2/t · km. Note that this may impact some conclusions regarding the energy efficiency of X-Type vessel, i.e. number of transported containers is favoured not pertinent to their mass. According to the assumptions given in Bačkalov et al. (2016), $EEDI_{TEU}$ indices are calculated for the water depth of 3.5 m and are shown in Fig. 10.35. The diagram shows that the X-Type concept is as energy efficient as deeper draught vessel IDV, and more efficient than NEWS at all speeds. Only the Radojčić design has better efficiency.

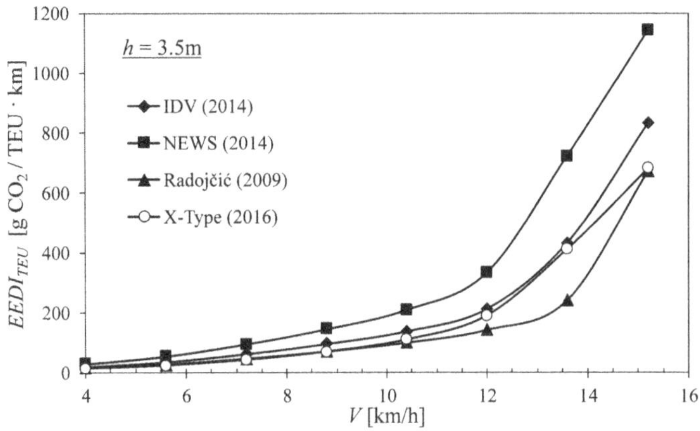

Fig. 10.35 Energy efficiency of the vessels. *Data from* Bačkalov et al. (2016)

Stability and Safety of Wide Body X-Type Vessel

Intact stability Rules of the Directive 2006/87/EC (EC 2006) limit the static angle of heel of a ship simultaneously exposed to constant beam wind and heeling moment due to turning. For a fully laden vessel carrying 4 container layers and unsecured containers (the usual case) the minimal metacentric height is $GM_{min} \approx 2.6$ m. However, a more advanced probabilistic approach based on stochastic analysis was applied here (see Sect. 8.3).

Container ships are vulnerable to gusting beam wind (not constant as assumed in the deterministic rules)—i.e. its speed fluctuates relative to mean value. Therefore, ship dynamics is simulated by coupled nonlinear equations of sway and roll. After evaluations that are beyond the book's scope, the probability of a critical event was estimated. Similarly, the probability of sliding of non-fixed containers in the uppermost layer was evaluated and presented graphically (evaluation procedure is given in Hofman and Bačkalov 2010; Bačkalov 2015).

From the presented diagrams (in the case of very strong winds) the following was concluded:

- *4 container layers* (metacentric height above 2.5 m)—roll amplitudes are practically negligible, thus the probability of flooding of the cargo hold is insignificant and sliding of containers would occur predominantly due to wind gusts. The empty containers in the upper layer would not have to be lashed even for wind up to 24 m/s.
- *5 container layers* (the metacentric height may fall below 2 m, the lateral area exposed to wind is increased by some 25%)—a bit heavier rolling may occur, still insignificant to cause cargo hold flooding, but may increase the probability of container sliding. The empty containers in the upper layer would have to be lashed If the wind speed exceeds 12 m/s.

Structural Strength of Wide Body Vessel

The vessel's main structural particularity is significantly increased *L*/*H* ratio (41.52) compared to "standard" inland vessels. The DNV-GL Rules (GL 2011), according to which the structure may be calculated if one wants to avoid direct calculations for determination of hull structure scantlings, limits $L/H < 35$. Thus, X-Type vessel is regarded as "unusual" and "unconventional" and its strength has to be considered "on a case by case basis" by more sophisticated (and more complicated) methods.

Increased *L*/*H* means that the vessel is too lengthy for its height and could be subjected to large bending moments and stresses in the structure. Apart from the length issue, the vessel is also unusually shallow. Length affects the increase of loads, whereas height lowers the structural resistance. Due to extremely low height, most remote structural elements of the cross section (bottom, deck and hatch coaming) are still very close to the neutral axis, giving a very small contribution to hull cross section modulus. Thus, additional strengthening of those structures, compared to the other Danube vessels, is necessary.

Rules still require the use of their deterministic formulae, but expect additional direct calculations of longitudinal strength, including buckling. These calculations are usually labelled as strength checks and used in order to increase or, sometimes, even reduce the Rule based scantlings.

The scantling determination procedure includes employing the prescribed Rule formulae for the net thickness of the plating and the hull modulus of the structural members, on which the corrosion margin is added to obtain final gross dimensions. Net scantlings are used for further strength assessments (checks) and gross scantlings are to be built into the vessel. Corrosion addition is usually in the range of 1–3 mm, depending on the position of the elements and their function (part of the ballast, oil, cargo space, etc.). X-Type vessel is chosen to be longitudinally framed in order to increase its hull modulus as much as possible and to lower the hull weight. Web frames spacing $S = 1.53$ m is governed by the ADN rules and TEU container positioning. Additional strengthening is provided in way of container corners.

Longitudinal strength calculation is performed for several load cases, as required by the rules; respective bending moment distributions are shown in Fig. 10.36. However, since these directly calculated bending moments are lower than the rule bending moments derived according to DNV-GL Rules' formulae, the later ones are used for further assessment of the stresses and buckling. Longitudinal strength and buckling analyses are coupled since the bending moments and shear forces obtained from longitudinal strength calculation produce normal and shear stresses, as the main input for buckling assessments. Buckling is especially critical for the bottom structure (in hogging condition) and for the deck and hatch coaming structure (in case of sagging).

Buckling assessments are performed on plates and stiffeners. Plate buckling is assessed for the unstiffened plate between the stiffeners and girders. Lateral and torsional buckling of stiffeners is analysed. Lateral buckling comprises bending and compressive stresses. Torsional buckling takes into account the torsional properties of the stiffeners. These assessments led to a more strengthened deck and hatch coaming

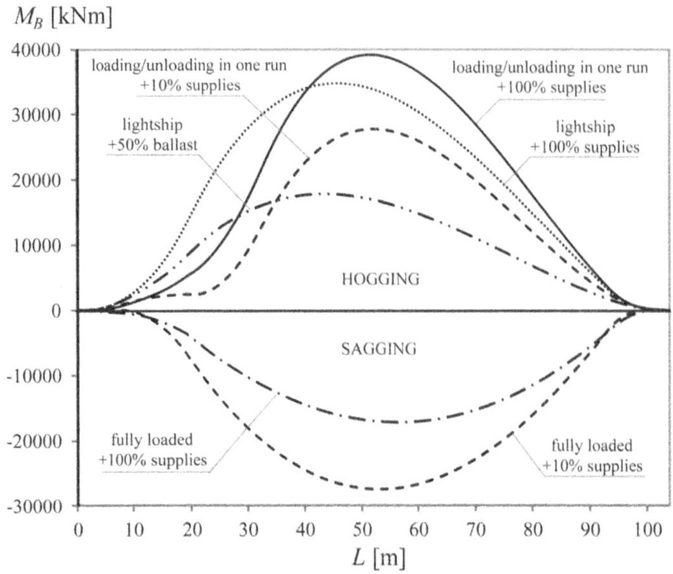

Fig. 10.36 X-Type vessel bending moments calculated according to direct longitudinal strength calculation. *Source* Bačkalov et al. (2016)

structure. Final (gross) cross section scantlings of the midship cross section are given in Fig. 10.37.

The calculated inner bottom gross thickness is 7 mm. However, most of the Danube vessels, in practise, have increased inner bottom thickness by at least 2 or 3 mm, then required by the classification society. This is due to the shipowners' aim to extend the life of inner bottom structure, which is very prone to wear and tear from

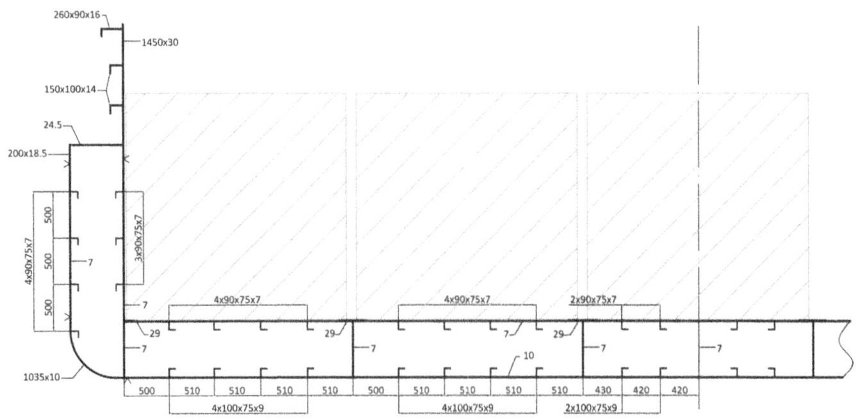

Fig. 10.37 Gross scantlings of the midbody cross section. *Source* Bačkalov et al. (2016)

frequent loading and unloading operations when using dry cargo grabs. Therefore, the inner bottom thickness was varied from 7 mm to 10 mm, which corresponds to the lightweight increase or deadweight decrease, of approximately 10 t per millimetre of inner bottom thickness increase. 9 mm was adopted, resulting in the lightweight and deadweight mass of 671 t and 2248 t, respectively.

Final Notes on Structural Strength of Wide Body Vessel

Wide body X-Type vessel presents the concept of shallow draught vessel with a high L/H ratio, making the design vulnerable from the viewpoint of longitudinal strength and buckling issues. Such structure is expected to be additionally reinforced due to increased stresses compared to designs having a lower (standard) L/H ratio. The stress peaks are in deck and hatch coaming structure and, for some load cases, in bottom plating. Therefore, for the vessels not satisfying the $L/H < 35$ condition, Rules generally require additional, direct calculations to prove their structural strength. This can affect the hull weight and thus present a disadvantage, i.e. for the same displacement, the higher hull weight reduces the deadweight. However, in the case of X-Type vessel the direct calculations showed that, although partially additionally strengthened, the vessel's lightweight mass (and hence, the deadweight) is still compatible with "standard" designs.

10.3.3 Barge Train for Transport of Bulk Cargo[8]

The main advantage of a push convoy, or barge transport, compared to selfpropelled ship transport, is that *cost-effective navigation with a reduced draught of partly loaded barges* may be utilized. Usually, it is the draught of a pushboat that poses the main problem, as it cannot be reduced below a certain level (the transom and propellers must have designed minimal draught, otherwise they cannot work properly, see Sect. 8.2.2). Conventional Danube pushboats with the power of around 2000 kW usually have a draught of more than 1.7–1.8 m, meaning that this *draught is actually a limiting factor*. Due to that, pulling technology prolonged its life on the Danube. Towing vessels have lower draught (usually below 1.5 m) and are therefore even today used on some Danube sectors during the dry seasons. Consequently, a *low draught pushboat with a power of around 2000 kW* would be more than advantageous on the Danube.

If navigation with a reduced draught is required, then, to substitute for reduced carrying capacity, the number of barges in a convoy might be increased; power needed for pushing this convoy would not increase proportionally (see Sect. 4.2); this is the main advantage of pushboat technology. Suggested power of around 2000 kW would be sufficient for sailing along the whole Danube at usual push train speeds with up to six fully loaded Danube barges (carrying 1500–1600 t each). The tonnage capacity of a typical Danube barge (77 × 11 × 2.8 m) at reduced draughts are given in Table 10.3.

[8]The material presented in this section originates from Radojčić (2009).

Table 10.3 Tonnage capacity at reduced of a danube barge 77 × 11 × 2.8 m

T	[m]	~ 0.5	1.0	1.5	2.0	2.5
Tonnage	[t]	-	300-400	700-800	1100-1200	1500-1600

Proposed Features of a Low Draught Pushboat Concept

- *Draught (T)—1.4 m maximum.* A larger draught would certainly be desirable from a hydrodynamic point of view, but if there is a need to push a convoy at extremely low waters (see Sect. 2.4), then 1.4 m is probably the maximum allowable draught. It still allows the installation of a propeller in a nozzle with a diameter of 1.5 m which could accept the power of up to 700 kW. This makes a three-propeller installation feasible. Furthermore, a standard Danube barge with draught of 1.5 m will carry a bit less than 800 t, which is approximately half of the carrying capacity of a fully loaded barge at 2.5 m. It is a discussible subject, but sailing at a lower draught than 1.5 m would probably not be cost-effective. From this point of view, the pushboat's draught of 1.4 m is also justified. Obviously, the choice of draught is the most important technical compromise in pushboat design.
- *Triple-screw propulsion, (skewed) propellers in nozzles with a diameter (D) of 1.5 m,* should be located in a relatively shallow tunnel. A somewhat larger propeller diameter would be allowable (and desirable), but taking into account the limited breadth of 11 m and high-speed diesel engines, it is believed that 1.5 m would be just sufficient. With an engine power of 700 kW, propeller loading would be 375 kW/m^2, which is high, but still acceptable. Special attention should be paid to the design of tunnels, propellers and nozzles with the aim to increase thrust and reduce vibrations (model experiments and CFD analysis are recommended).
- *Breadth (B)—11 m,* which is the same as of a standard Danube barge. A somewhat larger breadth would not be harmful, as the pushboat is usually pushing a much wider barge convoy. Even if only one barge is pushed, a pushboat somewhat wider than a barge would not be so disadvantageous. Barge packing, however, is easier if both the pushboat and a barge have the same width. Nevertheless, if draught is limited, then either the length or width (or both) should substitute the needed buoyancy. Consequently, it appears that the best choice is to fix the breadth to 11 m.
- *Length (L)—of around 30 m,* under the condition that there is enough space for all necessary machinery and crew. A somewhat longer vessel (if *B* and *T* are fixed) would be acceptable. With $L = 30$ m, the overall length of a convoy of two barges and a pushboat would be $2 \times 77 + 30 = 184$ m, which is still acceptable for passing through Danube locks (see Tables 2.2 and 4.1b).
- *Height (H)—2.5 m* is considered to be minimal for fitting engines and other necessary machine-room equipment below the deck.
- *Weight* (dry) is estimated to be 270 t taking into account lightweight engines and other equipment and machinery. A larger value might compromise the draught and therefore the project itself. A fully loaded pushboat with fuel and other provisions should weigh around 350 t (at a level draught of 1.4 m). Weight saving should

be considered wherever possible (SPS technology might be employed for the superstructure).

- *Ship form,* particularly the tunnels, is of utmost importance as relatively large power needs to be installed within an extremely shallow draught hull (see Figs. 9.18 and 10.38). The transom and propellers should always have a draught of around 1.4 m, while weight variations (due to fuel consumption) should change the bow draught only. Model experiments and CFD analysis are recommended.
- *Propulsion plant—Low emission Stage V diesel engines of 3 × 700 kW* with exhaust gas after-treatment systems (see Sect. 9.2.3) and relatively high power to weight ratio should be considered. Transmission of power should be via a *conventional horizontal shaftline* and a gearbox (i.e. diesel-direct). Main engines and gensets should be flexibly mounted to the foundations to reduce noise and vibrations. With an installed power of around 2000 kW, sailing along the whole Danube with a push train of six fully loaded barges (at $T = 2.5$ m, carrying around 1500–1600 t of cargo each) at usual convoy speeds is possible during most of the navigable season. Expected fuel consumption would be around 10 t/day.
- *Shore-to-ship-power supply* (of electricity while in port, often called *cold ironing*) should be considered with the aim to reduce onboard diesel emissions.
- *Steering—three fish-tail rudders* located behind propellers (without flanking rudders, see Sect. 9.4.2) and a *gondola type bow thruster,* with an electric motor of around 300 kW, should be considered.
- *Provisions—for max 7 days,* meaning that 70 t (around 85 m^3) of fuel should be provided. Nevertheless, although on the Danube it was accustomed to carry relatively large quantities of fuel (often for a roundtrip), much smaller quantities and refuelling on the way can be considered nowadays.
- *Crew members—8,* according to statutory regulations. Accommodation premises should be in one-tier superstructure on the deck (comprising 4 single and 2 double cabins, although this depends on the shipowner's needs/request). Living premises should be fully air-conditioned. Resiliently mounted superstructure (on pneumatic shock absorbers) for reduced vibrations, noise and increased comfort should also be considered (however that might require a somewhat different cabin arrangement than on hereby given GA plan).
- *Wheelhouse—with the possibility to be raised* to increase visibility to at least 250 m ahead of the ship, as requested by statutory regulations. When lowered, maximal height above the water level should be below 6.3 m, allowing sailing below all Danube bridges (except one on the Upper Danube at HWL, see Table 2.2).
- *Electronics* and computerization should be of the latest technology, providing one-man watch operation of the vessel with the engine and ship-system monitoring and recording, voyage optimization, etc.; see Sect. 9.6.

Everything else should be as usual on a pushboat of this size, intended for navigation along the Danube River. Nevertheless, modern lightweight equipment and materials should be considered wherever possible, as larger weight (displacement, hence draught) than predicted can easily compromise every pushboat.

General Arrangement Plan of a Pushboat Concept

The General Arrangement (GA) plan of a pushboat concept is shown in Fig. 10.38. The pushboat's main particulars are given in Table 10.4.

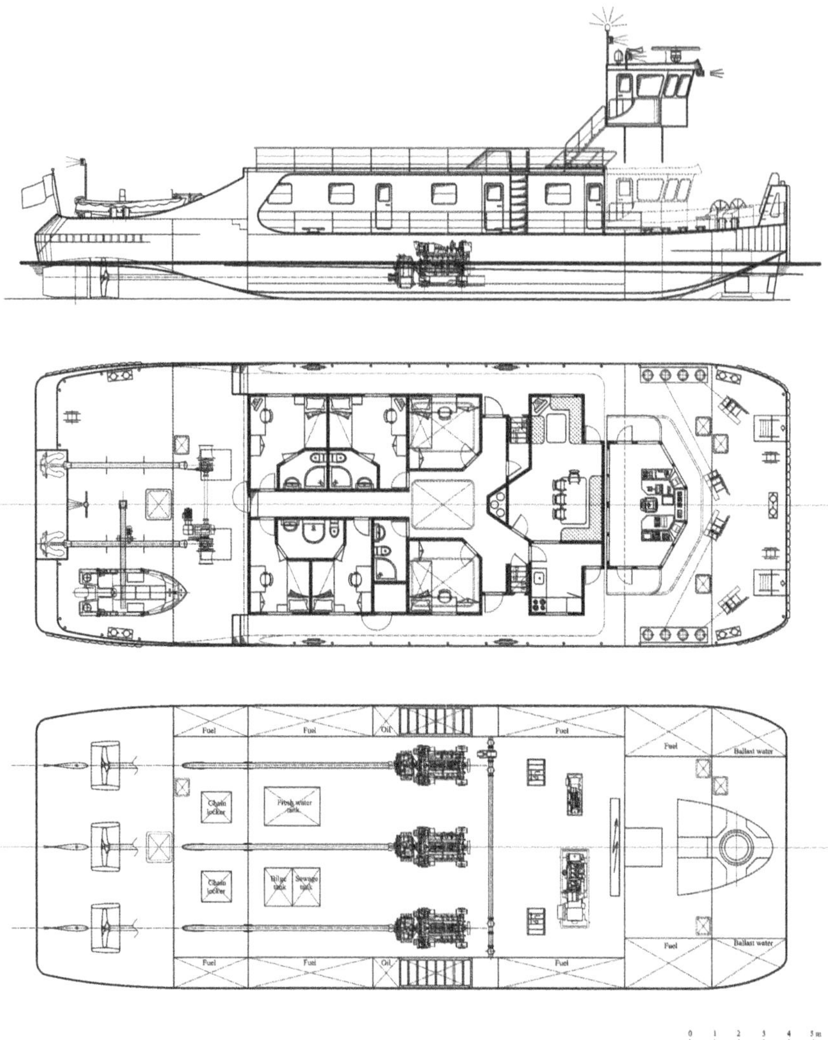

Fig. 10.38 General arrangement plan of a pushboat concept

Table 10.4 Main particulars of proposed pushboat concept

L_{OA}	[m]	30.0
B_{OA}	[m]	11.0
H	[m]	2.5
T	[m]	1.4
Height above basis line	[m]	6.0
P_B	[kW]	3 x 700
Bow thruster	[kW]	250 ÷ 300
Crew	[-]	8

Advantages of the Concept Compared to Conventional Pushboats

- The main advantage of the proposed pushboat is its *low draught of only 1.4 m* (compared to a draught of above 1.7 m of similar conventional pushboats). This enables navigation with partly loaded barges on the whole Danube even at LNWL.
- The latest generation of efficient and green engines, presently STAGE V with exhaust gas aftertreatment system.
- A gondola-type *bow thruster of 250–300 kW enables enhanced manoeuvring capabilities*, eliminating the necessity for conventional flanking rudders and improving stopping abilities. Due to the absence of flanking rudders, *unobstructed water inflow to the nozzles* can be achieved (hence higher efficiency), which is very important particularly for highly loaded propellers (due to limited propeller diameter, which is a result of draught limitation).

Note that the benefits of the application of the latest technological achievements that increase efficiency, safety, cleanliness and comfort, are not a result of the proposed pushboat concept, but rather of a modern era. Namely, almost all Danube pushboats were built 30 or so years ago and therefore were equipped according to the standards belonging to the previous technological generation, so a newly built pushboat of any design or concept will be advantageous compared to the old (conventional) ones.

10.4 Concluding Remarks

Researched concepts of innovative vessels for IWW (Sect. 10.1) mostly include attempts to reduce transport costs and environmental impacts of vessel navigation. Hence, they are related to novel technologies and generally consider propulsion innovations (LNG, electric), a decrease of resistance (air lubrication) and various shallow draught novelties. These concepts are not operational yet, but some of them are only a step from the ones that came to life.

Some already built advanced vessels are also reviewed here (Sect. 10.2) as their innovative features may be used on the Danube waterway. In most cases, advanced solutions refer to LNG and electrical propulsion, whether they are applied together or independently. Fully electric vessels still lack a satisfactory range of operation, so they are mostly used within the restricted navigational areas. For some recent demonstrators, hydrogen was used as power storage (for small pilot applications), coupled with diesel or electric systems.

Besides the introduction of new power sources, improvements are made through optimization of hull form and/or application of some unusual solutions, e.g. air lubrication or adjustable retractable tunnel for decreasing the resistance and better propulsion. The key goal of all mentioned designs is to reduce operational costs and make vessels with significantly smaller environmental impact. It was often demonstrated that success is achievable but at rather high investment costs.

Concerning both the environmental impacts and cost-effectiveness, the decisions made in the early design phases are the most influential (compared to various retrofitting options that are treated in Chap. 11). Thus three new design concepts suited particularly for the Danube waterway are subject of Sect. 10.3. These are:

- *Selfpropelled, narrow-body* general cargo and container vessel
- *Selfpropelled, wide-body* general cargo and container vessel
- *Pushboat* (for bulk cargo transport by conventional barge convoys).

It is important to emphasize that the *present-day technology and current Danube river infrastructure* are assumed for all three designs, which was not the case for concepts treated in Sects. 10.1 and 10.2. Moreover, the focus of Sect. 10.3 is on the naval architecture issues and not on the innovative prime movers, fuels and other innovative equipment which are researched and developed by other engineering disciplines and industries.

The accent on the *narrow-body* selfpropelled vessel is to demonstrate how the new design should be accommodated to various intermodal loading units, fairway restrictions, available technological innovations etc. The *wide-body* X-type vessel is a kind of shallow draught Danube alternative to a large cargo vessel (GMS, class Va[9]) which is regarded as a backbone ship for container transport on the Rhine. Both are low draught vessels (compared to those on the Rhine) designed for container transport. Note that although containerization is not yet fully established in the Danube corridor, it would be thoughtless to ignore them nowadays as the vessel life is long (50 or so years).

The *pushboat* is a state-of-the-art concept, although with nothing particularly innovative, except that the last available generation of engines should be considered. The emphasis is on a relatively low draught which may be accomplished with the low-draught convoys (i.e. usage of partly loaded barges), when the pushboat draught becomes critical. Namely, adding a few more barges to a convoy and thus significantly reducing its draught (assuming the cargo is not increased), does not result in a great

[9]Class Va max. width is 11.4 m (see Tables 4.1a and 4.3), so from that viewpoint X-type ship belongs to class Vb.

increase of needed power. Note that it is not the same to repower an old pushboat with relatively large draught,[10] and have a new one designed to have a low draught, although the engines might be of the latest generation in both cases.

References

Argonon (2020) Argonon, sustainable ship. http://www.wpci.nl/lng/sites/default/files/Deen%20Shipping%2C%20Argonon%2C%20%E2%80%98Sustainable%20Ship%E2%80%99.pdf. Accessed 01 Feb 2021

AUTOSHIP (2019) AUTOSHIP—autonomous shipping initiative for european waters. www.autoship-project.eu. Accessed 01 Nov 2020

Bačkalov I (2015) Impact of contemporary ship stability regulations on safety of shallow-draught inland container vessels. Saf Sci 72

Bačkalov I, Kalajdžić M, Momčilović N, Rudaković S (2016) A study of an unconventional container vessel concept for the Danube. In: PRADS 2016 conference, Copenhagen 2016

Bačkalov I, Kalajdžić M, Momčilović N, Simić A (2014) E-type selfpropelled vessel: a novel concept for the Danube. In: EIWN 2014 conference, Budapest, 2014

Barge EV (2020) This EV barge concept takes battery-swapping to a whole new level. https://insideevs.com/news/427049/ev-barge-concept-battery-swapping-zes. Accessed 20 Dec 2020

Bilen B, Bilen-Katić B (1997) Modified pusher tug design for the Danube river. Int Shipbuilding Prog 44(438)

Bilen B, Žerjal M (1998) A new concept of pushboat design. Practical Design of Ships and Mobile Units, Elsevier Science B.V

Concordia Damen (2020) Concordia Damen to deliver 40 inland waterway barge for JP Morgan. Accessed 24 Feb 2021

DAMEN (2015) LNG-powered Ecoliner delivers dramatic fuel savings. https://www.damen.com. Accessed 01 Nov 2020

DAMEN (2020) Towards zero-emission shipping. https://www.damengreen.com/-/media/Damen-Green-Solutions/Download/DACS_Brochure_2020.pdf. Accessed 01 Nov 2020

e4ships (2021) ELEKTRA. https://www.e4ships.de/deutsch/projekte-binnenschifffahrt/elektra. Accessed 08 Feb 2021

EC (2006) Directive of the European parliament and of the council laying down technical requirements for Inland waterway vessels (2006/87/EC). Official J European Union

ECCONET (2012) Deliverable 2.1.1: IWT fleet and operation. Report for European Commission, Directorate-General for Energy and Transport, Brussels

E-Ferry (2020) Curious to know how it is possible to operate a car- and passenger ferry across a distance of 22 nautical miles on 100% electricity? https://www.conf.eferry.eu. Accessed 08 Dec 2020

EST-Floattech (2021) World's first hydrogen tug in Germany with ESS from EST-Floattech. https://www.est-floattech.com/worlds-first-hydrogen-tug-in-kiel-with-ess-from-est-floattech. Accessed 08 Feb 2021

FLAGSHIPS (2021) World's first hydrogen cargo vessel set for Paris debut | Flagships. Accessed 8 Apr 2021

[10]Repowering an old pushboat with the new green and *light engines*, does not inevitably mean that the draught can be significantly reduced, i.e. the pushboat's propeller tunnels require approximately the same submergence as they were initially designed for (see Sect. 8.2.2). Consequently, the pushboat's draught cannot be reduced by decreasing its lightweight.

Foeth EJ, Eggers R, Quadvlieg FHHA (2010) The efficacy of air-bubble lubrication for decreasing friction resistance. In: International conference on ship drag reduction, SMOOTH-SHIPS, Istanbul 2010

GL (2011) GL rules for classification and construction of inland navigation vessels. Germanisher Lloyd

Greenstream (2021) Green Stream. https://www.vethpropulsion.com/references/inland-marine/tankers/greenstream. Accessed 8 Feb 2021

Godjevac M, Drijver M (2014) Power configuration of an inland pusher. In: ICMT 7th–9th July, Glasgow 2014

Gorbachev YN (1993) Decrease of ship's fuel consumption by means of air layer under its bottom. In: 14th Duisburg colloquium, duisburg, May 1993

GRENDEL (2020) Innovative & greening inland vessel concepts of NAVROM & SDG (WP4.1). http://www.interreg-danube.eu/uploads/media/approved_project_public/0001/42/5455f26b2c36f91626fc58a833100d76bb9d28f2.pdf. Accessed 01 Feb 2021

Guesnet T, Delius A, Jastrzepski T (2004) Efficient freight transport on shallow inland waterways—results of the INBAT R&D project. In: 9th Symposium PRADS, Luebeck-Travemuende, 2004

HARTENERGY (2011) World's first LNG/diesel dual-fuelled inland waterway tanker ship launches. https://www.hartenergy.com/news/worlds-first-lngdiesel-dual-fuelled-inland-waterway-tanker-ship-launches-54085. Accessed 8 Feb 2021

Hofman M, Bačkalov I (2010) Risk based analysis of inland vessel stability. In: Proceedings of the 11th international ship stability workshop, Wageningen

Hofman M, Maksić I, Bačkalov I (2006) Some disturbing aspects of Inland vessel stability rules. J Ship Technol, New Delhi 2(2)

IDV (2014) Innovative danube vessel—main project results. Danube region strategy

IMO (2013) MEPC 1/Circ.815 2013: Guidance on treatment of innovative energy efficiency technologies for calculation and verification of the attained EEDI

IMMA (2019) IMMA's PASCAT concept. http://imaa.associates/website_000001.htm. Accessed 8 Feb 2021

INBISHIP (1998) New Opportunities for Inland Waterway Transport. BRITE EURAM III Project

INE (2019) EMELI. Flyer within the Inland Navigation Europe (INE) website. www.inlandnavigation.eu/media/92260/Flyer-Emeli.pdf. Accessed 8 Feb 2021

Inovation Origins (2020) Electric push boat helps inland shipping towards emission-free future. https://innovationorigins.com/electric-push-boat-helps-inland-navigation-towards-emission-free-future/. Accessed 30 Dec 2020

Internationales Verkehrswesen (2018) Hydroville vessel: CMB received port of antwerp sustainability award. https://www.internationales-verkehrswesen.de/hydroville-vessel-cmb/. Accessed 20 Dec 2020

Müller E (2003) Innovative types of inland ships and their use on the river Rhine, its tributaries and the adjacent canals. SPIN, Version 1, VBD Duisburg

NEW-LOGISTICS (2007) Futura Carrier. Presentation. www.marine-marchande.net/Petits_Reportages/Gwenaelle/Ciudad/Kaja.PDF. Accessed 13 Feb 2021

NEWS (2014) Concept and specification of a Novel Inland container vessel. In: Development of a next generation european inland waterway ship and logistics system, Report, Vienna, 2014

NOVIMAR (2017) https://novimar.eu. Accessed 20 Dec 2020

Pavlov GA, Yun L, Bliault A, He SL (2020) Air lubricated and air cavity ships. Development, design, and application. Springer. ISBN 978-1-0716-0423-6

Pospiech P (2013) Full speed ahead with gas. https://www.marinelink.com/news/speed-ahead-ful1359396. Accessed 01 Feb 2021

Provoost S (2018) First-hand experience with LNG as fuel for inland vessels: MS Ecoliner. Workshop on modernisation of danube vessels. https://indanube.eu/2018/04/19/workshop-modernisation-danube-vessels/. Accessed 01 Feb 2021

Radojčić D (2009) Environmentally friendly inland waterway ship design for the Danube river. World Wide Fund for Nature International, Danube-Carpathian Programme (WWF-DCP)

Renner V (2007) Der Futura Carrier—Schiffstyp der Zukunft? Binnenschifffahrt—ZfB—Nr. 4, Apr 2007

Semper Fi (2012) Innovative propulsion concept and hull design. https://cdn-static.veth.net/media/709-Semper_Fi_Innovative_Propulsion_concept_and_hull_design.pdf. Accessed 20 Dec 2020

Ship & Boat Int. (1995) Austrian river icebreaker with azipod propulsion. https://www.rina.org.uk/sbi.html. Accessed 20 Dec 2020

Ship & Boat Int. (2020) Pushing in a new direction. RINA Dec. 2020, https://mcusercontent.com/e5b1642e15270ded3e42150d1/files/3033df6e-5b1a-45f1-9e5a-069ee7e10bc7/SDG_Artcile_final.pdf. Accessed 01 Feb 2021

Ship Technology (2020) LNG greenstream tanker. https://www.ship-technology.com/projects/lng-greenstream-tanker/. Accessed 01 Feb 2021

SWZ Maritime (2019) Low-emission Sendo liner becomes KNVTS Ship of the year 2019. https://www.swzmaritime.nl/news/2019/11/04/champion-in-low-emission-sailing-sendo-liner-becomes-knvts-ship-of-the-year-2019/. Accessed 20 Dec 2020

Thill CH, Toxopeus SL, Walree van F (2005) Project Energy-saving air-Lubricated Ships (PELS). In: ISSDR 2nd international symposium on seawater drag reduction, Busan

UNECE (2015) European agreement concerning the international carriage of dangerous goods by inland waterways. United Nations Economic Commission for Europe, Geneva

Zenczak W, Michalski R, Jastrzebski T (2003) Conceptions of power plant of innovative push-boat on shallow waters. EIWN 2003 conference, June 2003

ZES (2020) Wärtsilä, Engie, ING bank develop emissions-free barge business with interchangeable energy containers. https://cleantechnica.com/2020/06/07/wartsila-engie-ing-bank-develop-emissions-free-barge-business-with-interchangeable-energy-containers. Accessed 20 Dec 2020

Zibell HG, Müller E (1996) Binnenschiffe fur extrem flaches Wasser. Ergebnisse des VEBIS-Projektes, Handb. D. Werften, vol XXIII

Chapter 11
Conversion and Retrofitting Options

11.1 Introduction

Various aspects that influence the design of contemporary Danube vessels are discussed in Part I. Shallow water hydrodynamics and design guidelines are given in Part II. They both converge towards Chaps. 9 and 10 that treat the innovative greening technologies and various design concepts of IW vessels. The present chapter focuses on options for improvement of the existing IW vessels through modernization, i.e. conversions and retrofitting using innovative technologies that are nowadays available (mainly issues treated in Chap. 9).

Furthermore, the majority of the Danube vessels were built between 1960 and 1990 (see Sect. 1.1.3). The replacement rate of the fleet is low and the recent tendency is to purchase pre-owned selfpropelled ships from the Rhine Corridor, rather than to build new ones. Taking into account that the lifetime of ships can easily extend 50 years, it sounds reasonable to consider various improvement measures. This primarily concerns the improvement of environmental and economic performances of the existing fleet through conversion and retrofitting measures that include:

- replacement of outdated engines with new ones,
- replacement of propulsors with better ones,
- increase of ship cargo-carrying capacity, and
- application of various state-of-the-art technologies.

The optimal solution often involves a combination of the above measures. Note that in some cases, conversion and/or retrofitting is required in order to fulfil current environmental, safety and other regulations. As in most cases, the candidates for the retrofitting and improvements are relatively older vessels, probably only the medium-term investment should be considered, with the expected return period of up to 5 years, rarely more. Concerning greening the return period is usually longer.

Efficient modernization of the IW fleet is recognized as an ambitious task and thus is considered within the EU projects such as MoVeIt!, PROMINENT and GRENDEL for instance. Some of their findings are given in the following.

D. Radojčić et al., *Design of Contemporary Inland Waterway Vessels*,
https://doi.org/10.1007/978-3-030-77325-0_11

11.2 Retrofitting: New Engines and Propulsors

Engine Replacement

The lifespan of the main engines is shorter (approximately 20 years) than of the IW vessels (50 or so years, see Sect. 1.1.3). Hence, the over-aged engines have to be replaced with new, more efficient and cleaner and in every respect superior engines. This can extend the vessel's service life for 10 or so years and can significantly reduce engine emissions. Fortunately, in most cases, the refitting of the main engine is relatively simple, compared to other more complex ship improvements (conversions).

In order to replace only the existing engine (without additional drivetrain and/or vessel modifications), two essential conditions have to be met (see Giuglea et al. 2020):

(a) Power of the new engine should be similar to that of the current one, i.e. acceptable difference from the initial power is approximately between −2 and +10%
(b) Speed of the new engine should be similar to that of the existing one, within a margin of ±2% from the initial RPM.

This kind of retrofitting is common and is usually the most convenient for the shipowner as it involves the least modifications and investment costs.

In most cases, contemporary *high-speed diesel engines* should be considered for retrofitting as they are more efficient, cleaner, cheaper (for maintenance too) and lighter than conventional engines. However, the gearbox should also be considered for replacement, due to contemporary engines' higher rotational speed.[1] The new gearbox should ensure that the initial propeller speed remains the same, as otherwise the propeller, probably the shaftline too, should also be replaced.

This kind of retrofitting usually involves modifications of the engine foundations, and modifications of various systems required for a proper engine operation, i.e. fuel, cooling, lubrication, exhaust gas, ventilation and/or electrical systems.

Repowering of existing IW vessels with over-aged engines is a feasible solution in order to comply with the new emission requirements imposed by CCNR Stage V regulations. These regulations (see Sect. 9.2) are effective from 2019 (2020 for the large engines) and apply to propulsion (IWP) and auxiliary (IWA) engines.

Right-Sizing

Project PROMINENT (PROMINENT, 2018) focuses on *right-sizing*, amongst others. Namely, it was concluded that many IW vessels have much more propulsion power than it is needed for most of their activities (in this respect see Fig. 9.20), which increases emissions and reduces the aftertreatment potentials of exhausted gasses. The reasons for this are the requirements for manoeuvring and stopping actions (see Sect. 7.7). In particular, during the monitoring process of the IW selfpropelled

---

[1] They are usually derived from the general application engines applied for the road vehicles or generating set engines, thus often have 1500 or 1800 rpm.

vessels, it was found that the power load of the propulsion engines was on average between 30 and 35% and that seldom exceeded 60 to 70% of MCR. Consequently, within the project PROMINENT a new definition of a right-sized vessel drivetrain was introduced: "A right-sized vessel driveline has the *minimum power installed to fulfil its operational duties* while complying with the regulations for manoeuvrability". Therefore, it might be expected that right-sizing will impact the repowering too and that in many cases smaller and cleaner engines will be considered to replace the initially installed ones. Right-sizing or downsizing may also bring considerable cost benefits.

Replacement of Propulsors

It is not uncommon to replace old propellers with new ones, particularly when a new engine is also installed. That may be an opportunity to consider the replacement of a naked propeller with the ducted one (see Sect. 9.4.3), as well as to install a better and more efficient rudder (e.g. replace a single rudder with two rudders). That may require reconstruction of a tunnel or installation of a Flex Tunnel or some kind of ESD (see Sect. 9.4.8), which is essentially a kind of ship stern conversion. Moreover, some producers (e.g. Nautican) have integrated propulsion unit that consists of pre-swirl stator, nozzle and triple rudders (see Figs. 9.48 and 9.28) which can replace the existing propulsor and a rudder. Rarely the rudder propeller will replace direct line propulsion or vice versa, but that can happen too, see for instance discussion about Ecoliner, Sect. 10.2.1.

11.3 Conversions: Lengthening and Hull Improvements

Although the hull conversions are generally rather costly, and thus a return of investments uncertain, the lengthening of a ship hull may be considered as a measure aimed at increasing the competitiveness of some vessel types. Besides increasing the ship's cargo capacity, that leads to an increase in energy efficiency. The lengthening may also provide a draught reduction and thus enhance the operating time during low water periods (when the fright rates are considerably higher, see Fig. 2.4). The lengthening usually means the insertion of a parallel mid-body section. However, this may bring some new problems, discussed in the next section for a particular ship. Note that lengthening is hardly a novelty since the same practice is applied for a long time both in the seagoing and inland shipping industries.

Besides lengthening, conversions of stern and bow hull sections are not unusual. Some of the reasons for stern conversions are mentioned in the previous section, while the bow conversion may be necessary, for instance, for the accommodation of a gondola with a side thruster (see Sect. 11.5).

11.4 Numerical Example of Retrofitting and Conversion

The list of retrofitting and conversion possibilities is long. The improvements have to be assessed on a *case-by-case basis*, thus an example of a particular ship is examined in the following.

Within the MoVeIT! (2014) project (Modernization of Vessels for Inland waterway freight Transport), in WP 6 "New Scales & Services", a systematic analysis of lengthening of two typical IW ships belonging to CEMT II and III/IV classes was done (a shorter version of this study is given in Bačkalov et al. 2014). The lengthening analysis of a larger vessel MV Hendrik, built in 1975, is presented here. Her main particulars are given in Table 11.1.

In order to estimate the benefits of the lengthening (extension of the parallel middle body, with the bow and stern parts of the vessel remaining unchanged), various aspects of lengthening were considered: the strength of the lengthened hull structure, powering requirements and manoeuvring capabilities of the converted and retrofitted vessel. Systematic analyses were conducted with *a lengthening step set to 6.0 m*, which approximately corresponds to the length of one TEU. The maximum considered length was 94.99 m because 95.0 m leads to a higher CEMT vessel class with requirements for additional crew. Particulars of considered cases are shown in Table 11.2.

Table 11.1 Main ship particulars of the parent vessel MV Hendrik

Length over all	[m]	69.98
Breadth	[m]	8.60
Draught	[m]	2.95
Depth	[m]	3.00
Displacement	[t]	1360
Propeller diameter	[m]	1.5

Table 11.2 Particulars of considered lengthening cases

Dimensions		Initially built	Lengthening 1	Lengthening 2	Lengthening 3	Lengthening 4
L_{OA}	[m]	69.98	76	82	88	94.99
B	[m]	8.60	8.60	8.60	8.60	8.60
T	[m]	2.95	2.95	2.95	2.94	2.93
Δ	[t]	1360	1508	1664	1812	1988
m_{DWT}	[t]	1114	1241	1372	1502	1653

Ship Structure

In general, the main challenge concerning the ship lengthening, from the structural point of view, is to determine the required structure scantlings of the lengthened hull, according to the current rules of classification societies. This is followed by the calculation of the cargo-carrying capacity increase and estimation of conversion costs. Hull lengthening, however, is not a straightforward procedure as it requires a certain level of optimization, depending on the existing ship structure and the requirements of the rules.

The determination of hull structure scantlings of MV Hendrik was done according to GL (2011). Detailed calculations and analysis are given in Wilcke et al. (2014).

Summary of results related to the thickness of the plating,[2] for considered lengthening steps (note, $L_{OA} = 90$ m is added to existing four 6 m lengthening steps), is shown in Fig. 11.1. The analysis indicated that the lengthening for more than 18 m could be technically challenging and hence possibly not feasible. Beyond this length, several structural elements would have to be significantly upgraded. The replacement of bottom steel plates would drastically increase conversion costs. Additional masses of structural components and additional cargo-carrying capacity, for each lengthening step, are shown in Fig. 11.2.

To evaluate the considered lengthening steps and to identify the most beneficial ones, the ratio of additional cargo-carrying capacity to additional lightship weight is taken as an indicating value. Accordingly, the length of $L_{OA} = 82$ m exhibits the lowest ratio of 5.7 (due to additional masses of bulkheads and stiffeners) while all other considered lengthening steps refer to a ratio above 6. Moreover, the relative increase of cargo-carrying capacity, for each lengthening step per lengthening meter, up to $L_{OA} = 88$ m is around 21.5 t/m, but it drops for the next two considered lengths. Finally, relative costs of hull lengthening (including material and labour costs, hatch covers, docking costs per lengthening meter etc.) also indicated that *88 m is the upper practicable limit* of lengthening the vessel. Therefore, 88 m is taken as the

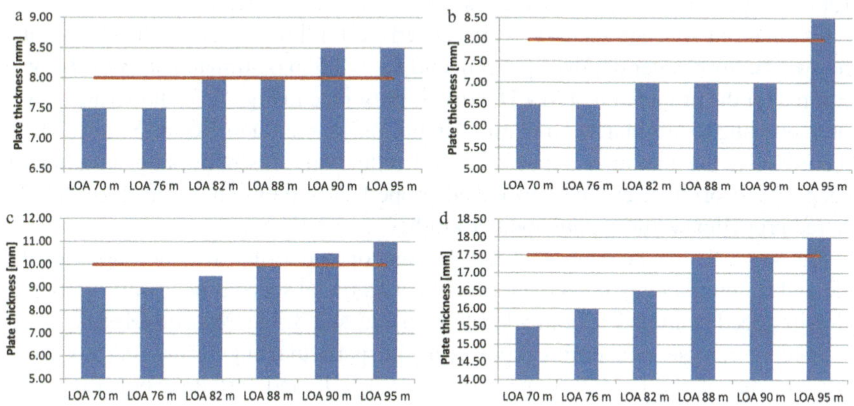

Fig. 11.1 Thickness of **a** bottom plate, **b** side plating, **c** hatch coaming and **d** sheer strake, for different lengthening steps as required by GL (2011); Horizontal lines correspond to as-built condition. *Source* Wilcke et al. (2014)

[2]Note that, according to a good shipbuilding practice and owner's choice, as-built thicknesses are greater than required by the rules.

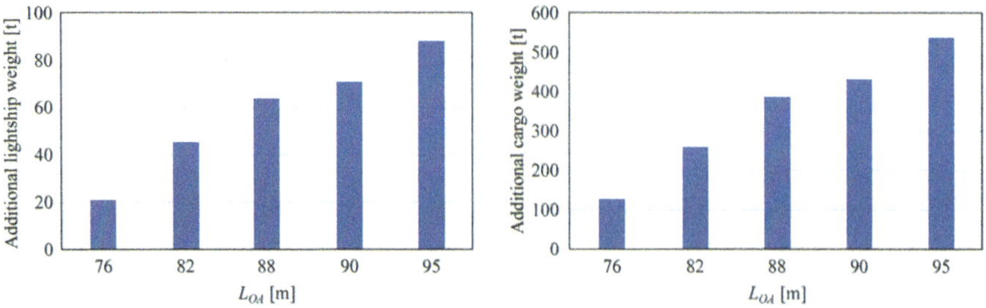

Fig. 11.2 Additional lightship and cargo weight for different lengthening steps. *Source* Wilcke et al. (2014)

new maximal length for further analysis. For more details and further explanations see Wilcke et al. (2014).

Repowering

The increase of the length of parallel midbody leads to an increase in ship resistance and a decrease of propulsive efficiency (due to higher propeller loading). Both are additionally pronounced in shallow water, thus special attention was paid to shallow water effects.

Powering analysis was done in *two phases*. In the *first*, the same drivetrain (engine/gearbox/propeller) was considered for all three lengthening steps, since the power needed for achieving a certain speed was not significantly influenced by the lengthening. In the *second* phase, an existing naked propeller was replaced by the ducted propeller which was optimized for new loading conditions. Thus, analysis of propulsive performances for each considered lengthening step had to be done.

Calculations for each scenario were carried out for three water depths: deep water and two shallow-water depths $h_w = 5$ m and $h_w = 3.5$ m. Estimated delivered power required for the propulsion of MV Hendrik is presented in Fig. 11.3 (shown are curves corresponding to original and maximal lengths only), see Radojčić and Simić (2014). Accordingly, the maximal expected speed reduction caused by ship lengthening (keeping the original engine and naked propeller) is between 1 and 2 km/h. The ducted propeller would regain about 1 km/h.

The squat had to be estimated too. It was concluded that it is not significantly affected by lengthening, but a reduction of navigational speed is necessary (maximal allowed speed is around 13 and 8 km/h for h = 5 and 3.5 m, respectively).

Specific analyses concerning repowering, i.e. particular engine and gearbox selection, are out of the book's scope. For more details and further explanations see Radojčić and Simić (2014).

Manoeuvrability

Generally speaking, the increase of ship length, thus its lateral area too, with the unchanged rudder tends to decrease ratio A_R/LT. Hence, the ship rudder area A_R and turning ability may become insufficient. Moreover, the displacement is increased as

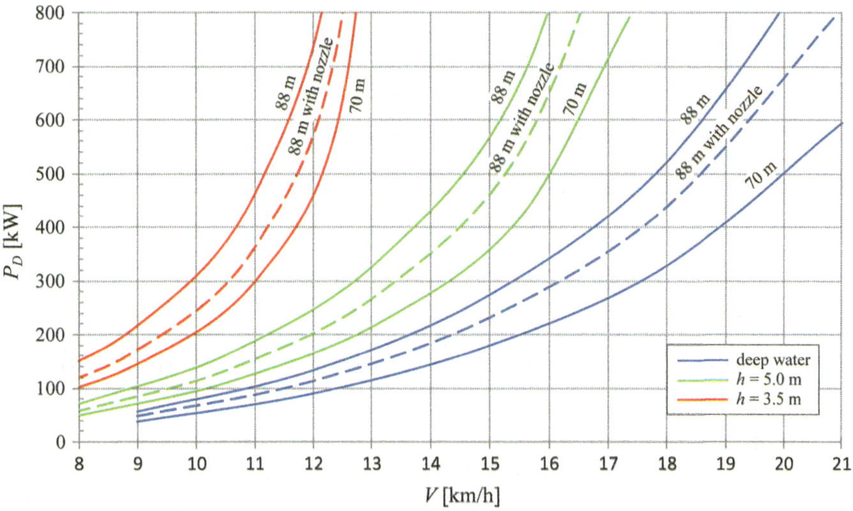

Fig. 11.3 Estimated delivered power for different retrofit options. *Source* Radojčić and Simić (2014)

the ship length is increased. Consequently, it may be expected that the lengthening, combined with shallow water effects, will worsen ship manoeuvring performances.

Therefore, the assessments of manoeuvring performances of MV Hendrik were conducted for the initially built vessel and the lengthening cases 2 and 4 (see Table 11.2). Simulations of zig-zag manoeuvre, combined turning circle/pull out, evasive manoeuvre and crash stop tests, were conducted with the MARIN's in-house software. This was done for two speeds (10 and 13 km/h) and three water depths (deep water, $h = 5$ and 3.5 m). The complete results obtained through these simulations are presented in Tonelli (2014).

The manoeuvrability standards for seagoing ships are defined by IMO Resolution MSC.137(76), see IMO (2002). However, it has limited applicability for IW vessels. On the other hand, the manoeuvrability criteria laid down by the Rhine Commission are descriptive (i.e. qualitative rather than quantitative), except for the evasive manoeuvres (see CCNR 2011). Accordingly, the evasive manoeuvre simulations were conducted for Hendrik; results for all considered lengthening steps and for $V =$ 13 km/h and $\delta_R = 45°$ are shown in Fig. 11.4. Thick horizontal lines (depending on waterway depth to draught ratio) refer to CCNR (2011) requirements. Follows that CCNR (2011) limits are breached for $L_{OA} > 84$ m, i.e. lengthening steps 3 and 4.

Nevertheless, the overall conclusion was that the hull extension did not significantly affect the manoeuvring features (including the stopping ability), except in extremely shallow water ($h = 3.5$ m). Thus, to improve the turning ability in shallow water, a decrease of the approach speed was proposed. Rudder replacement or

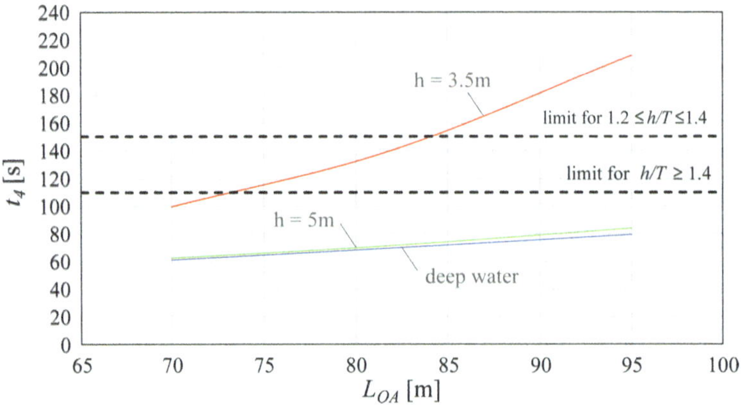

Fig. 11.4 Evasive manoeuvre performance for $V = 13$ km/h and $\delta_R = 45°$ for a range of lengthening steps compared with CCNR (2011) limits. *Data from* Tonelli (2014)

increase of rudder area was suggested as well. For a more detailed analysis and further explanations see Tonelli (2014).

Economic Assessment and Environmental Impact of Conversion and Retrofitting Options

To make lengthening a feasible option, the benefits need to outweigh the investment costs. However, various presumptions needed for the economic assessment are out of the book's scope. Moreover, economic analysis depends on the particular operational scenario. Thus, *under the specific assumptions* outlined in Gille and de Swart (2014), it was concluded that the payback time for MV Hendrik, for both retrofitting options (naked and ducted propeller), would be 2–6 years for, respectively, 88 and 76 m lengths. The environmental impacts[3] of the conversion and retrofitted options should be considered as well. These are expressed as yearly emissions of ship engine (carbon dioxide CO_2, nitrogen oxide NO_X, particulate matter PM, hydrocarbon HC, carbon monoxide CO, and sulphur dioxide SO_2) evaluated for a naked and ducted propeller for each lengthening step. According to Gille and de Swart (2014) lengthening of 18 m increased emissions by some 30% for a speed of 10 km/h; however, the emissions would increase only 3% if the lengthened ship was retrofitted with the ducted propeller. Similar results were obtained for a speed of 14 km/h.

[3]The ship-owners usually do not care much for the environmental impacts (emissions), except conforming with current engine emission limits.

11.5 Application of State-of-the-Art Technologies

Aside from mentioned retrofitting and conversion options, various innovative solutions, based on contemporary technologies, may be applied to improve the environmental footprint, energy efficiency and economic performances of existing IW vessels. According to various projects (e.g. CREATING, ECONET, MoveIt!, PROMINENT, GRENDEL, etc.—see Sect. 9.6), the list of conversion and retrofitting options to modernize existing vessels is endless. Some of them, thoroughly analyzed within the project MoveIt! are summarized in Hekkenberg and Thill (2014).

Note that the two simplest, cheapest and the most beneficial energy consumption and emission reduction technologies will not be discussed here as they have already found their place elsewhere in the book. Namely, various engine add-ons that suppress emissions (e.g. particulate filter, selective catalyst reduction etc.) are discussed in Sect. 9.2.3; they usually go hand-in-hand with engine retrofitting (see Sect. 11.2). The same holds good for the ducted vs. naked propeller, different kinds of nozzles etc. (see Sects. 9.4.3 and 11.4).

As already mentioned, there is no single retrofit solution that would be effective for all vessel types. Retrofitting measures have to be done on a case-by-case basis. Nonetheless, some promising retrofitting options, not mentioned before, are given in the following.

Energy Saving Devices (ESD)

Some kind of ESD (see Sect. 9.4.8) should probably be the first refit option that should be considered, as ESD may be effective, simple and relatively cheap to install. In that respect, the already mentioned integrated propulsion unit that consists of a pre-swirl stator, nozzle and rudders (see Figs. 9.48 and 9.28) looks promising. See for instance a refit program discussed in Mulder and Williams (2006), which treats the replacement of nozzles and rudders on the Canadian shallow draft tugs. Another type of relatively simple ESD—a pre-swirl fin, is discussed in Peters (1994); reported efficiency increase of a very shallow draught pushboats was around 15%.

Adjustable Retractable Tunnel

The shallow draught performance may be significantly improved with the adjustable retractable tunnel, see Sect. 9.4.8. Its installation on an existing 110 m ship is discussed in Gille and Heumen (2014) for instance. It was estimated that the transport costs per ton could be lowered by some 7%, while a payback time in the particular case would be 3 years.

Trapezoid Wedge

Trapezoid wedge is an artificial pre- or aft-body modification placed between the vessels (pushed or coupled convoys) forming a stump-end connection, see Fig. 11.5. It reduces convoy resistance by around 10% by minimizing the vortices in the coupling

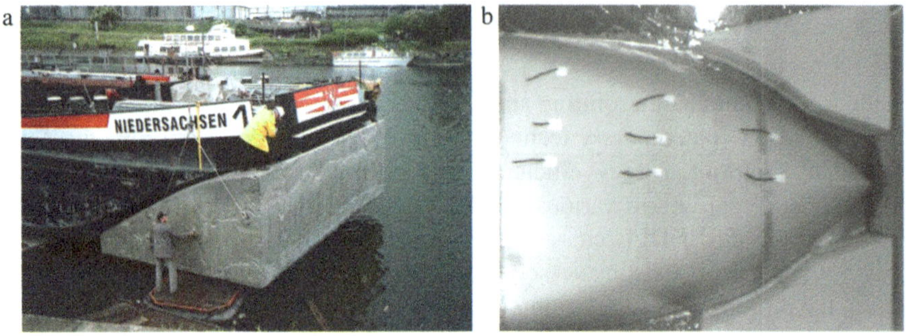

Fig. 11.5 A foam filling match the stump transom of a barge **a** For wedged bow (full-scale demonstration) and **b** conventional shaped bow (model-scale). *Source* Zöllner (2009) and Hekkenberg and Thill (2014)

area between the hulls, i.e. provides a smoother transition between them. Different fillings may be used, as depicted in, Fig. 11.5.

Gondola Type Bow Thruster

The manoeuvrability of IW vessels may be enhanced with the gondola type bow thruster or pump-jet. It is an add-on device that can be fitted on the pushboat's underwater part. Bow thruster can replace conventional flanking rudders (see Sects. 8.2.2 and 9.4.2); due to better and unobstructed water inflow to ducted propellers, it indirectly increases the pushboat's propulsive efficiency.

Information Technologies (IT)

IT are used in IWW navigation for better utilization of the fairway, planning of the voyage, optimization of the time schedule and speed, and therefore improved safety and efficiency of the navigation. It leads to a reduction of fuel consumption and environmental friendliness of the IWW transport. Some state-of-the-art services of the kind (RIS, COVADEM etc.) are presented in Sect. 9.6. Similar is the Advising Tempomaat (ATM) system based on the onboard computer program which provides guidance for an economically optimized operation of the vessel.

The high end of IT application in IW navigation are Autonomous Ships—vessels that are partly or fully remotely controlled and, thus can sail with reduced or no crew at all. That can significantly reduce shipping costs. Development of such concepts, which are still not the state-of-the-art of IW shipbuilding, is reviewed in Sect. 9.6.3.

11.6 Concluding Remarks

The list of retrofitting and conversion possibilities is long and miscellaneous. Besides those explained in the above sections, other kinds of measures might be undertaken also. The analyses demonstrated that some improvements are relatively simple and cheap (e.g. engine replacement) or, at least within certain boundaries, economically feasible (e.g. lengthening of vessels). However, one should bear in mind that some essential features of existing vessels still cannot be changed, nor their characteristics noticeably improved, whichever reasonable technical measures would be applied. The most important recommendation is that intended improvements have to be assessed on a case-by-case basis.

References

Bačkalov I, Radojčić D, Molter L, Wilcke T, Meij van K, Simić A, Gille J (2014) Extending the life of a ship by extending her length: technical and economic assessment of lengthening of inland vessels. In: EIWN 2014 conference, Budapest, 2014

CCNR (2011) Administrative instructions to the Inspection Commissions. https://www.ccr-zkr.org/. Accessed 01 Apr 2021

Gille J, de Swart L (2014) Deliverable 6.4: Synthesis and impact assessment. modernization of vessels for inland waterway freight transport (MoVeIT!)

Gille J, Heumen van E (2014) Deliverable 6.5: Adjustment to climate change, modernization of vessels for inland waterway freight transport. (MoVeIT!)

Giuglea V, Chirica I, Angheluta C, Danaila I (2020) Output 4.1: Innovative & greening inland vessel concepts of NAVROM & SDG. GRENDEL—Green and Efficient Danube Fleet

GL (2011) Rules for Classification and Construction of Inland Navigation Vessels. Germanisher Lloyd

Hekkenberg R, Thill C (2014) Retrofit solutions for inland ships—MoVeIT! Approach. EIWN 2014 conference, Budapest 2014

IMO (2002) Resolution MSC.137(76). Annex 6. Standards for ship manoeuvrability. The maritime safety committee

Mulder C, Williams P (2006) Improving the performance of shallow draft tugs in Northern Canada. SNAME Trans. 114

Peters HE (1994) Propulsion improvements of inland going ships with ducted propellers. WEGEMT 21st graduate school, Gerhard Mercator University, Duisburg

Radojčić D, Simić A (2014) Deliverable 6.3: powering. Modernization of vessels for inland waterway freight transport (MoVeIT!)

Tonelli R (2014) Deliverable 6.2: maneuverability. Modernization of vessels for inland waterway freight transport (MoVeIT!)

Wilcke T, Chirica I, Giuglea V (2014) Deliverable 6.1: Ship structure. Modernization of vessels for inland waterway freight transport (MoVeIT!)

Zöllner J (2009) Strömungstechnische Möglichkeiten zur Reduzierung des Kraftstoffverbrauchsund der CO_2-Emissionen von Binnenschiffen. ZKR Kongress Rheinschifffahrt und Klimawandel

The manufacturer's authorised representative in the EU is Springer
Nature Customer Service Centre GmbH, Europaplatz 3, 69115 Heidelberg,
Germany. If you have any concerns regarding our products, please
contact ProductSafety@springernature.com

Printed and bound by CPI Group (UK) Ltd, Croydon, CR0 4YY
28/09/2026
02224089-0014